Agricultural and Rural Development: Current Trends and Future Prospects

Agricultural and Rural Development: Current Trends and Future Prospects

Edited by Jamie Kelley

Syrawood
PUBLISHING HOUSE
New York

Published by Syrawood Publishing House,
750 Third Avenue, 9th Floor,
New York, NY 10017, USA
www.syrawoodpublishinghouse.com

Agricultural and Rural Development: Current Trends and Future Prospects
Edited by Jamie Kelley

International Standard Book Number: 978-1-68286-858-4 (Hardback)

Cataloging-in-Publication Data

Agricultural and rural development : current trends and future prospects / edited by Jamie Kelley.
     p. cm.
Includes bibliographical references and index.
ISBN 978-1-68286-858-4
1. Agriculture. 2. Rural development. 3. Agricultural development projects. 4. Agriculture--Forecasting.
5. Rural development--Forecasting. I. Kelley, Jamie.
S521 .A37 2020
630--dc23

TABLE OF CONTENTS

PREFACE

Rural development is a concept centred on the improvement in the quality of life and the economic health of communities living in rural areas. Traditionally, the focus of rural development has been solely on the exploitation of natural resources such as forestry and agriculture. In modern times, with increased urbanization and improved global production networks, the focus has shifted to tourism, recreation and niche manufacturing. However for much of the developing world today, agricultural development remains the most potent weapon to end extreme poverty and boost rural development. It is crucial for economic growth, accounting for nearly 1/3rd of the global gross domestic product. This book studies, analyzes and upholds the pillars of agricultural development and its utmost significance in modern times. It elucidates the concepts and innovative models around prospective developments with respect to rural development. This book, with its detailed analyses and data, will prove immensely beneficial to professionals and students involved in this area at various levels.

The researches compiled throughout the book are authentic and of high quality, combining several disciplines and from very diverse regions from around the world. Drawing on the contributions of many researchers from diverse countries, the book's objective is to provide the readers with the latest achievements in the area of research. This book will surely be a source of knowledge to all interested and researching the field.

In the end, I would like to express my deep sense of gratitude to all the authors for meeting the set deadlines in completing and submitting their research chapters. I would also like to thank the publisher for the support offered to us throughout the course of the book. Finally, I extend my sincere thanks to my family for being a constant source of inspiration and encouragement.

Editor

Nutrient utilisation and blood chemistry of Red Sokoto bucks fed on diets with different inclusion levels of raw and soaked roselle (*Hibiscus sabdariffa* L.) seeds

Taofik Adam Ibrahim [a,*], Salisu Bakura Abdu [a], Muhammad Rabiu Hassan [a], Suleiman Makama Yashim [a], Hanwa Yusuf Adamu [a], Owolabi S. Lamidi [b]

[a]*Department of Animal Science, Ahmadu Bello University, Zaria, Nigeria*
[b]*National Animal Production Research Institute, Shika, Zaria, Nigeria*

Abstract

This study evaluated nutrient utilisation and blood chemistry of Red Sokoto bucks fed a 10 and 20 % inclusion level of raw, water- and lime-soaked *Hibiscus sabdariffa* L. seeds in rice bran based diets. 21 Red Sokoto bucks aged 8–10 months and weighing 9–13 kg were randomly allotted into six treatments with three bucks each, while a seventh dietary treatment with zero inclusion of seeds served as control in a 2×3 factorial arrangement using a complete randomised design. The results indicated that increase in dietary inclusion levels of soaked *H. sabdariffa* seeds increased ($P < 0.05$) the nutrient utilisation of bucks as compared to the control, while a decrease was observed with increasing dietary inclusion levels of raw seeds. Dietary inclusion of both raw and water-soaked *H. sabdariffa* seeds increased ($P < 0.05$) the packed cell volume. Soaking also influenced the white blood cell value which increased with increasing inclusion levels of *H. sabdariffa* seeds. However, values of haemoglobin and red blood cells were only affected by 20 % inclusion of raw and water-soaked *H. sabdariffa* seeds ($P < 0.05$) compared to control. Inclusion of *H. sabdariffa* seeds furthermore reduced serum protein, albumin, globulin, glucose and urea levels compared to control. It is therefore concluded that *H. sabdariffa* seeds support haematopoiesis in Red Sokoto bucks. While both inclusion levels of water-soaked and 10 % raw *H. sabdariffa* seeds improved nutrient utilisation compared to control and 20 % inclusion of raw seeds, the 20 % inclusion of water-soaked *H. sabdariffa* seeds recorded the best nitrogen utilisation efficiency.

Keywords: Hibiscus sabdariffa seeds, soaking, Red Sokoto bucks, nutrient utilisation, blood chemistry, goats

1 Introduction

Nigeria hosts an estimated 19.5 million cattle, 41.3 million sheep and 72.5 million goats (National Agricultural Sample Survey, 2011). From this estimate, goats represent about 54.4 % of total ruminant livestock. The indigenous goat breeds in order of importance are Red Sokoto (50 %), West African Dwarf (45 %) and Sahel (5 %) (Ajala *et al.*, 2008). Goats contribute about 24 % of meat supply in Nigeria (Oni, 2002).

Goats, like other herbivores in the tropics and sub-tropics, experience marked seasonal fluctuations in feed supply which results in a seasonal pattern of wet season live weight gains and dry season live weight losses until animals reach marketable weight (Poppi & McLennan, 1995). This is due to the scarcity of good quality feed during the dry season. Feed intake is one of the important factors that influence animals' lifetime productivity, health and carcass characteristics (Bawa *et al.*, 2003). The increased demand and high cost of conventional animal feed ingredients like soybean or ground nut cake makes it necessary to search for alternative indigenous feed resources which are readily available and cheaper than the conventional feed ingredients (Sodeinde

* Corresponding author
Taofik Adam Ibrahim (taibrahim@abu.edu.ng)

et al., 2007). The search for alternative feed resources has over the past decades rekindled research interest in the use of tropical browses, herbs and medicinal plants as nutrient sources for ruminants (Okoli *et al.*, 2002).

H. sabdariffa L. or roselle is one of these alternative feed resources that is cultivated on a wide range of tropical soil sand performs satisfactorily well on relatively infertile soils (Adanlawo & Ajibade, 2006). It is popularly called "Yakuwa" in Hausa and belongs to the family of Malvaceae; it is a popular vegetable in Indonesia, India, West Africa and many other tropical countries (Tindall, 1986; Babatunde, 2003). The vegetable is widely grown in the North-Eastern and middle belt regions of Nigeria (Akanya *et al.*, 1997).

H. sabdariffa is cultivated for its pleasant red colour calyx, used for making a local drink (*sobo*) and wine (Al-Wandawi *et al.*, 1984). Although abundant seeds are produced, they are highly underutilised. The seeds have been reported to have a high content of oil and protein. Al-Wandawi *et al.* (1984) reported that seeds contain 25.20 % crude protein (CP), while Abdu *et al.* (2008) reported a value of 23.46 % CP. However, the utilisation of *H. sabdariffa* seeds as an alternative feed source for ruminant livestock may be limited due to the presence of some antinutritional factors such as tannins, phytic acid, and trypsin inhibitor (Abu-Tarboush & Basher, 1996; Abdu *et al.*, 2008), as well as gossypol (Abu-Tarboush & Basher, 1996). Nutrient utilisation by ruminants is limited if the antinutrient concentration in the feed offered is above a certain threshold, despite the detoxifying activities of rumen microbes on many anti-nutrients (Kama, 2005; Isah *et al.*, 2011). Hence, there is a need for feed processing in order to improve the quality and utilisation of such plant materials by ruminants. Different processing methods have been carried out on roselle seeds with quite a number of successes recorded (Abdu *et al.*, 2008; Hambidge *et al.*, 2005; Ibrahim & Yashim, 2014) while the economic efficiency of feeding water-soaked *H. sabdariffa* seeds to Red Sokoto bucks has been reported (Ibrahim *et al.*, 2016).

According to Afolabi *et al.* (2011), changes in haematological parameters are often used to determine the physiological status of the body and to determine stresses due to environmental, nutritional and/or pathological factors. Haematological components are valuable in monitoring feed toxicity especially with feed constituents that affect blood formation (Oyawoye & Ogunkunle, 2004). It was reported that roselle calyces have a haematopoietic effect in farm animals (Mahadevan *et al.*, 2009; Olusola, 2011).

The objective of this study was to evaluate the growth performance and selected blood parameters of Red Sokoto bucks fed different inclusion levels of raw and soaked

H. sabdariffa seeds in rice bran based diets, so as to establish baseline information on the nutrient utilisation of roselle seeds in bucks.

2   Materials and methods

2.1   Study site

The study was conducted at the Small Ruminant Unit of the Department of Animal Science, Teaching and Research Farm, Ahmadu Bello University, Zaria. Zaria is located in the northern Guinea Savanna Zone of Nigeria, on latitude 11°14′44″ N and longitude 7°38′65″ E, at an altitude of 610 m a.s.l. The climate is relatively dry with annual rainfall of 700–1400 mm occurring between the months of April and September (Ovimaps, 2014).

2.2   Sourcing and processing of Hibiscus sabdariffa seeds

The seeds of the red variety of *H. sabdariffa* were purchased on a public market in Yobe state during the harvest period (October / November); they were cleaned to remove impurities before processing. A first batch of 20 kg of *H. sabdariffa* seeds was soaked in 50-litre plastic containers containing a 6 % lime solution (0.06 g CaO per g of water) for 24 h. The seeds were removed, washed and sundried for 72 h. A second batch of 20 kg of *H. sabdariffa* seeds were completely submerged and soaked in 60 litres clean water and stored in a 100-litres plastic container. The water was drained every 24 h and again 60 litres of clean water were added. On the fourth day the seeds were removed, washed and sundried for 72 h. The dried seeds of both water and lime treatments were stored in airtight polythene containers until required for diet formulation.

A third batch of 20 kg cleaned but untreated *H. sabdariffa* seeds was also stored in an airtight polythene container for diet formulation.

2.3   Experimental setup

The raw and treated *H. sabdariffa* seeds were each included at a level of 10 and 20 % into a diet consisting of cotton seed cake, maize bran, rice bran, bone meal and common salt to obtain iso-nitrogenous and iso-caloric diets (Table 1). The diets were fed as a total mixed ration for a period of 8 weeks for the haematological study and an additional 7 days for the digestibility study, in a 2 × 3 factorial arrangement using a complete randomised design with three different diet types (i.e. raw, water- and lime-soaked roselle) and two inclusion levels (10 and 20 %).

The experimental animals were purchased from a public market in Anchau, Kaduna State. Twenty one growing Red

Sokoto bucks aged 8–10 months and weighing 9–13 kg were randomly allotted to 7 groups with 3 animals per group. The animals were treated against endo- and ecto-parasites using Albendazole® and acaricides respectively, according to the manufacturers' recommendation.

The animals were housed in individual metabolic crates for total faecal and urine collection. Prior to the experiment they were allowed 14 days to adjust to the metabolic crates and to the experimental diets before the commencement of the digestibility trial. The experimental diets were offered at 3 % of their live weight in a single meal at 8:00 a.m. daily. During the 7-day digestibility study, the daily faecal output was quantified (amount of fresh matter); 10 g of the homogenized fresh faeces were sub-sampled daily and oven-dried at 60 °C for 48 h for dry matter determination. After that, the daily samples of each animal were pooled for laboratory analysis. The daily urine output was collected into a plastic container placed under the metabolic crates and containing 100 ml of 0.1 NH_2SO_4 to prevent nitrogen loss by volatilisation. The collected urine was strained through a layer of glass wool to remove detached hair fragments and/or other solid contaminants. A 10 % aliquot of the total daily urine output of each animal was stored in a refrigerator at 4 °C for nitrogen determination (Osuji et al., 1993). All the aliquots per animal were pooled after the experimental period for nitrogen assay.

2.4 Sample analysis

The proximate analysis of samples of the experimental diets, faeces, raw, water-soaked and lime-soaked H. sabdariffa seeds and nitrogen determination in urine was conducted according to standard methods (AOAC, 2005). The residual dry matter of the samples was determined by oven-drying at 60 °C for 48 h. Nitrogen was determined by the micro Kjeldahl method with Tecator Product apparatus (KjeltecTM2100), while crude protein was calculated by multiplying $N \times 6.25$. The Soxhlet extraction procedure was used for determination of crude fat (ether extract) using electromantle ME. The ash was measured by combustion of the dried material in a muffle furnace at 600 °C for 8 h. Crude fibre, sequential neutral detergent fibre (NDF) and acid detergent fibre (ADF) were determined using Tecator Line (FT 122 FibertecTM) according to the method described by Van Soest & Wine (1967). The concentration of phytic acid was determined according to Wheeler & Ferrel (1971). A standard curve of ferric nitrate was plotted. Phytate phosphorus was calculated from the standard curve assuming a 4:6 Fe to P molar ratio. The concentration of total tannins present in raw and soaked roselle seeds was determined colorimetrically as described in AOAC (2005), whereby tannic acid was used as a reference standard. The

total oxalates concentration was determined by calcium oxalate precipitation (titrimetric method) of Oke (1966); the method involved titration of acidic aqueous extracts of the sample with a standard solution of potassium permanganate. The analysis of minerals (calcium, phosphorus, potassium, iron and magnesium) was carried out using atomic absorption spectrophotometry in wet feed samples digested by concentrated nitric and pelrchloric acids, using PG instrument AA500 model. The content of metabolisable energy (ME) in each diet was determined using the equation of Pauzenga (1985).

$$ME \; (Kcal / kg \; DM) = 37 \times \%CP + 81.8 \times \%EE + 35 \times \%NFE$$

The digestibility coefficients (d) of nutrients were calculated as follows:

$$d = \frac{\text{Nutrient intake (g/d)} - \text{Faecal nutrient excretion (g/d)}}{\text{Nutrient intake (g/d)}}$$

The animals' nitrogen balance was calculated as follows:

$$N\text{-balance (g/d)} = N\text{-intake (g/d)} - \text{Faecal N-excretion (g/d)} - \text{Urine N-excretion (g/d)}$$

2.5 Haematological analysis

Blood samples of 20 ml volume were taken from the jugular vein of each animal using a hypodermic needle after eight weeks since the start of the feeding trial (and before the commencement of the digestibility trial) into two types of test tubes; 10 ml into sterile plain test tubes for serum preparation and the remnant 10 ml into Ethylene Diamine Tetra Acetic acid (EDTA) anticoagulant bottles containing 0.5 ml EDTA for determination of haemoglobin (Hb), packed cell volume (PCV), red blood cells (RBCs), white blood cells (WBCs), lymphocytes and neutrophils according to standard methods (Coles, 1986). The blood in EDTA anticoagulant bottles was immediately inserted in ice containers. RBCs count was done in a haemocytometer chamber with Natt and Herdrics diluents to obtain a 1:200 blood dilution. The number of leucocytes was estimated as total $WBC \times 200$. PCV was determined as micro haematocrit with 75×16 mm capillary tubes filled with blood and centrifuged at 3000 rpm for 5 minutes. The differential count of leucocytes was obtained from blood stained with Wrights dye and the neutrophil and lymphocyte cells were counted with a laboratory counter while the Hb concentration was also calculated. The serum biochemical parameters determined were blood urea nitrogen, glucose, total serum protein, globulin, albumin, creatinine and cholesterol. The blood in the plain test tubes was immediately centrifuged for serum preparation so as to ensure optimum glucose de-

Table 1: *Ingredients and calculated nutritional values of the experimental diets.*

Ingredients (% of DM)	Control	Raw		WSS		LSS	
		10 %	20 %	10 %	20 %	10 %	20 %
H. sabdariffa seeds	0	10.0	20.0	10.0	20.0	10.0	20.0
Cotton seed cake	39.0	29.0	19.0	24.0	9.0	25.0	13.0
Maize bran	19.5	19.5	19.5	24.5	29.5	23.5	25.5
Rice bran	40.0	40.0	40.0	40.0	40.0	40.0	40.0
Bone meal	1.0	1.0	1.0	1.0	1.0	1.0	1.0
Common salt	0.5	0.5	0.5	0.5	0.5	0.5	0.5
Total	100	100	100	100	100	100	100
Calculated composition (% of DM)							
Crude protein	14.10	14.10	14.10	14.10	14.10	14.04	14.26
Crude fibre	21.35	23.10	24.85	23.15	24.95	23.14	24.91
ME (MJ / kg DM)	8.21	8.09	7.98	8.13	8.05	8.12	8.01
Cost (Naira / kg DM)	34.95	33.90	32.90	32.80	30.70	34.20	34.10

WSS: Water-soaked seeds, LSS: Lime-soaked seeds; DM: Dry matter.

termination. Blood glucose was measured by the glucose oxidase method. It involved the production of a coloured compound from the activities of glucose oxidase enzyme on Beta-D-glucose and the colorimetric measurement of the coloured compound (Miller, 1959). Total serum protein and serum globulin were determined using the burette method as described by Doumas (1975); urea nitrogen was analysed by the di-methyl monoxide method as described by Varley & Bell (1980). Creatinine was determined by the Jaffe reaction method (Jaffe, 1886). Albumin was measured using dye–binding technique with bromocresol green as described by Doumas & Biggs (1972), while serum cholesterol was assessed according to the reaction described by Bernhard (1918).

2.6   Statistical analysis

All data collected during the experiment were subjected to statistical analysis using the general linear models (GLM) procedure of SAS version 9.13 (SAS, 2002) according to a completely randomized model in 2×3 factorial arrangement as:

$$Y_{ijk} = \mu + P_i + G_j + (PG)_{ij} + e_{ijk}$$

where Y_{ijk} = dependent variable, μ = over all mean, P_i = effect of ith seed processing method, G_j = effect of jth inclusion level, $(PG)_{ij}$ = interaction effects of ith processing method and jth inclusion level, and e_{ijk} = random error. Significance was declared at $P < 0.05$. Significantly different means were compared using Duncan multiple range test (Duncan, 1955). As normal distribution of the individual variables was not tested, the results of statistical analysis are only indicative.

3   Results

3.1   Chemical composition of experimental diets

Table 2 illustrates the proximate components of raw and soaked *H. sabdariffa* seeds. The crude protein concentrations of water-soaked and lime-soaked seeds as well as their NFE and crude ash concentrations were numerically higher than of the raw seeds, while a decrease was observed in the concentrations of crude fibre, fibre fractions and the ether extract of water-soaked and lime-soaked seeds.

The concentrations of anti-nutrients and minerals in raw and soaked *H. sabdariffa* seeds is shown in Table 3. Soaking

Table 2: *Chemical composition of raw and soaked H. sabdariffa seeds.*

Proximate constituents (% of DM)	Raw	WSS	LSS
Dry matter[†]	94.67	94.09	92.90
Crude protein	25.18	27.88	30.98
Crude fibre	27.26	26.22	23.97
Ether extract	15.18	9.22	8.97
Ash	9.18	12.00	10.66
NFE	21.40	29.12	23.42
ADF	35.98	21.32	20.33
NDF	62.89	48.81	45.40
Hemicellulose	11.59	9.01	8.41
Processing cost (Naira / kg)	0.00	7.00	15.00

[†] in % of air dry matter, WSS = Water-soaked seeds, LSS = Lime-soaked seeds; DM = Dry matter; NFE = Nitrogen free extract, ADF = Acid detergent fibre, NDF = Neutral detergent fibre.

Table 3: *Concentrations of anti-nutrients and minerals in raw and soaked H. sabdariffa seeds.*

Component (% of DM)	Raw	WSS	LSS
Phytate	0.17	0.15	0.16
Total tannins	2.40	1.17	1.58
Oxalate	1.46	0.87	1.04
Calcium	1.10	1.73	2.57
Phosphorus	4.30	5.56	7.71
Magnesium	0.56	0.41	0.44
Iron	0.36	1.80	2.16
Potassium	11.98	8.30	3.83

WSS = Water-soaked seeds, LSS = Lime-soaked seeds.

reduced the concentration of anti-nutrients, magnesium and potassium but increased the calcium, phosphorus and iron concentrations of *H. sabdariffa* seeds.

Table 4 presents the chemical composition of the experimental diets. For all dietary treatments, the analysed crude protein concentration of the diets was about 2 % lower than the calculated concentrations, but the analyses verified the iso-nitrogenous and iso-energetic nature of the formulated diets.

3.2 Nutrient digestibility and nitrogen balance

Table 5 shows the interaction between processing methods and inclusion levels of *H. sabdariffa* seeds in rice bran based diets on nutrient digestibility in Red Sokoto bucks. The diet containing 10 % of lime-soaked *H. sabdariffa* seeds had a significantly higher ($P < 0.05$) dry matter digestibility than all other processing methods and inclusion levels. The crude protein digestibility for the 20 % dietary inclusion levels of water-soaked (82.4 %) and 10 % inclusion level of lime-soaked (79.9 %) *H. sabdariffa* seeds were not significantly different, but were significantly higher ($P < 0.05$) than in all other treatments. The crude fibre digestibility at 20 % inclusion of lime-soaked seeds was significantly higher ($P < 0.05$) than in all other treatments. The digestibility of the ether extract was similar ($P > 0.05$) for the 10 % inclusions level of water-soaked and lime-soaked seeds but significantly higher ($P < 0.05$) than of all other treatments and the control, while the digestibility of soluble carbohydrates (NFE-fraction) was higher ($P < 0.05$) for the 20 % inclusion of water-soaked seeds. The ADF, NDF and hemicelluloses digestibility for both dietary inclusion levels of water-soaked and lime-soaked seeds was higher ($P < 0.05$) compared to the control and the diets containing raw seeds.

Table 6 summarizes the interaction between processing method and inclusion level of raw, water-soaked and lime-

soaked *H. sabdariffa* seeds on the nitrogen balance of Red Sokoto bucks. For the 10 % inclusion level of raw seeds, higher ($P < 0.05$) values for nitrogen intake, urinary nitrogen excretion, total nitrogen excretion and absorbed nitrogen were found. Across all treatments and the control, the highest ($P < 0.05$) values for retained nitrogen, both absolute and in percent of ingested nitrogen, was determined at 20 % inclusion of water-soaked seeds.

3.3 Haematology and serum chemistry

Table 7 shows the impact of processing method and inclusion level of *H. sabdariffa* seeds on the haematology and serum chemistry of Red Sokoto bucks. The value for PCV increased with increasing inclusion of the raw and water-soaked seeds ($P < 0.05$), while there was only a numerical increase with increasing inclusion level of lime-soaked seeds. The WBC values also increased ($P < 0.05$) with increasing inclusion level of both raw and soaked seeds, whereby the highest value was determined at 20 % inclusion of raw seeds ($P < 0.05$). Total serum protein ranged from 4.8 to 7.6 g / 100 ml, with lowest values ($P < 0.05$) obtained for the 20 % inclusion level of raw seeds. The concentration of albumin, blood glucose and blood urea of the control was higher ($P < 0.05$) than of all inclusion levels of raw and soaked seeds, whereas the globulin value was similar to the values measured in bucks receiving water-soaked seeds. At 20 % inclusion of raw seeds, the cholesterol was lower ($P < 0.05$) than with a 20 % inclusion of water-soaked seeds. Creatinine ranged from 47.0 to 82.7 mmol / l and increased ($P < 0.05$) with increasing inclusion of both water-soaked and lime-soaked seeds but not with raw seeds ($P > 0.05$). Blood urea was unaffected by soaking method and inclusion level.

4 Discussion

4.1 Nutrient digestibility

The inclusion of water-soaked and lime-soaked *H. sabdariffa* seeds effectively increased nutrient digestibility compared to the inclusion of raw *H. sabdariffa* seeds and the control diet only. However, while the 10 % inclusion of raw *H. sabdariffa* seeds competed favourably with the control diets, the nutrient digestibility, but not the dry matter digestibility declined at a 20 % inclusion. Increasing dietary inclusion level of water-soak *H. sabdariffa* seeds had no effects on dry matter digestibility, while a reducing effect was recorded for lime-soaked *H. sabdariffa* seeds. The crude protein digestibility was positively influenced by 10 and 20 % inclusion of water-soaked *H. sabdariffa* seeds but negatively affected by the inclusion of raw and lime-soaked *H. sabdariffa* seeds. This may be attributed to an improved

Table 4: *Chemical composition of the experimental diets.*

Constituent (% of DM)	Control	Raw 10 %	Raw 20 %	WSS 10 %	WSS 20 %	LSS 10 %	LSS 20 %
Dry matter	89.60	90.30	89.93	89.53	90.23	90.20	89.80
Crude protein	12.60	12.38	12.50	12.88	12.31	12.98	12.64
Crude fibre	22.70	23.80	22.90	24.80	20.90	18.70	25.50
Ash	8.90	8.70	9.00	7.90	8.00	9.80	7.80
Ether extract	4.00	4.20	4.40	4.10	4.20	4.40	4.60
NFE	51.90	50.80	51.20	50.40	54.60	54.20	49.40
ADF	27.40	29.50	27.80	28.00	29.60	28.80	30.00
NDF	46.60	48.20	49.00	44.80	50.00	49.30	49.00
Hemicellulose	10.60	11.10	13.10	12.80	11.90	10.00	10.20
ME (MJ / kg DM)	10.90	10.80	10.90	10.80	11.30	11.40	10.80

WSS: Water-soaked seeds, LSS: Lime-soaked seeds; NFE: Nitrogen free extract,
ADF: Acid detergent fibre, NDF: Neutral detergent fibre.

Table 5: *Effects of soaking method and inclusion level of H. sabdariffa seeds on nutrient digestibility in Red Sokoto bucks.*

Digestibility (%)	Control	Raw 10 %	Raw 20 %	WSS 10 %	WSS 20 %	LSS 10 %	LSS 20 %	SEM
Dry matter	58.12^c	59.4^c	58.2^c	68.2^b	70.0^b	74.4^a	68.6^b	1.38
Crude protein	70.8^c	71.7^{bc}	65.8^d	75.7^b	82.4^a	79.9^a	74.2^b	1.33
Crude fibre	54.1^e	57.9^d	56.3^{de}	68.1^b	64.4^c	66.0^{bc}	72.3^a	1.48
Ether extract	41.6^e	44.8^e	48.9^d	59.9^a	54.6^c	58.6^{ab}	55.4^{bc}	1.75
NFE	69.3^d	73.5^{bc}	62.1^e	75.8^{ab}	77.3^a	74.1^{bc}	72.5^c	1.18
ADF	64.4^c	60.7^d	52.2^e	69.0^{ab}	70.0^a	66.6^{bc}	68.0^{ab}	1.50
NDF	59.4^d	63.7^{cd}	61.0^d	67.2^{ab}	68.2^a	65.3^{ab}	64.8^{bc}	1.49
Hemicellulose	65.2^b	59.7^c	54.0^d	71.1^a	72.2^a	63.9^b	69.6^a	1.45

$a-e$ Means with different superscript within rows are significantly different ($P < 0.05$), SEM: Standard Error of
Mean, WSS: Water-soaked seeds, LSS: Lime-soaked seeds; NFE: Nitrogen free extract, ADF: Acid detergent
fibre, NDF: Neutral detergent fibre.

Table 6: *Effects of soaking method and inclusion level of H. sabdariffa seeds on the nitrogen balance of Red Sokoto bucks.*

Parameter (g / day)	Control	Raw 10 %	Raw 20 %	WSS 10 %	WSS 20 %	LSS 10 %	LSS 20 %	SEM
Nitrogen intake	31.9^f	40.9^a	19.0^g	38.7^b	37.3^d	34.8^e	38.4^c	0.12
Urinary nitrogen	1.9^c	5.6^a	3.5^b	1.3^e	1.3^e	1.1^f	1.5^d	0.04
Faecal nitrogen	9.5^a	5.4^c	4.2^e	7.0^b	4.7^d	7.0^b	9.5^a	0.07
Total nitrogen excretion	11.4^b	11.5^a	7.6^f	8.3^d	6.0^g	8.1^e	11.0^c	0.07
Nitrogen absorbed	22.4^f	35.5^a	14.8^g	31.6^c	32.6^b	27.8^e	28.9^d	0.13
Nitrogen retained	20.5^f	29.9^c	11.4^g	30.3^b	31.3^a	26.7^e	27.3^d	0.13
Nitrogen retained (% of N-intake)	64.9^d	73.2^c	59.9^e	78.4^b	83.9^a	76.6^b	70.1^c	1.68

$a-g$ Means with different superscript within rows are significantly different ($P < 0.05$), SEM: Standard Error of Mean,
WSS: Water-soaked seeds, LSS: Lime-soaked seeds.

Table 7: *Effects of soaking method and inclusion level of raw and soaked H. sabdariffa seeds in rice bran based diets on haematology and serum chemistry of Red Sokoto bucks.*

Parameter	Control	Raw		WSS		LSS		SEM
		10 %	20 %	10 %	20 %	10 %	20 %	
Packed cell volume (%)	32.0 [c]	35.3 [b]	39.3 [a]	36.0 [b]	40.3 [a]	33.7 [bc]	35.7 [b]	1.40
Haemoglobin (g / 100 ml)	10.6 [bc]	11.8 [b]	13.7 [a]	13.1 [ab]	13.9 [a]	11.0 [bc]	10.0 [c]	0.76
Red blood cells (g / 100 ml)	5.3 [b]	6.0 [ab]	7.0 [a]	7.8 [a]	7.9 [a]	5.8 [b]	5.6 [b]	0.50
White blood cells ($\times 10^3$ / mm^3)	5.7 [d]	10.1 [b]	12.0 [a]	7.0 [c]	9.3 [b]	7.4 [c]	8.7 [b]	0.62
Neutrophils (%)	17.7 [c]	23.7 [a]	24.3 [a]	15.3 [d]	21.7 [b]	13.7 [d]	17.7 [c]	0.86
Lymphocytes (%)	78.0 [a]	53.0 [d]	68.7 [c]	71.7 [b]	71.0 [bc]	78.0 [a]	66.0 [c]	1.42
Total protein (g / 100 ml)	7.4 [a]	5.7 [c]	4.8 [d]	6.2 [bc]	7.6 [a]	6.6 [b]	6.6 [b]	0.38
Albumin (% of TP)	48.3 [a]	41.0 [b]	42.0 [b]	35.7 [d]	38.0 [c]	41.3 [b]	38.3 [c]	1.09
Globulin (% of TP)	39.0 [a]	35.2 [b]	36.0 [b]	38.0 [ab]	38.9 [a]	35.9 [b]	36.3 [b]	1.07
Glucose (mmol / l)	30.3 [a]	27.0 [b]	25.0 [bc]	23.3 [cd]	23.7 [c]	21.3 [d]	24.0 [c]	1.05
Cholesterol (mmol / l)	4.8 [ab]	4.6 [ab]	4.0 [b]	4.6 [ab]	4.9 [a]	4.5 [ab]	4.5 [ab]	0.42
Creatinine (mmol / l)	56.3 [c]	47.0 [d]	49.0 [d]	57.3 [c]	62.7 [b]	54.7 [c]	72.7 [a]	1.45
Blood urea (mmol / l)	4.5 [a]	3.5 [b]	2.7 [b]	2.9 [b]	2.9 [b]	3.2 [b]	3.2 [b]	0.39

[a-e] Means with different superscript within rows are significantly different ($P < 0.05$), SEM: Standard Error of Mean, WSS: Water-soaked seeds, LSS: Lime-soaked seeds.

diet quality – in physical forms and chemical terms due to reduced antinutriens – with the water-soaked *H. sabdariffa* seeds. Also, the economic efficiency of feeding water-soaked *H. sabdariffa* seeds has been reported (Ibrahim *et al.*, 2016). No effect of inclusion level and seed treatment was observed with regard to the digestibility of soluble carbohydrates (NFE fraction), whereas inclusion of raw seeds had a reducing effect. While there was negative effect associated with the digestibility coefficient of ADF and hemicelluloses with increasing inclusion of raw *H. sabdariffa* seeds, no effect was noted for both inclusion levels of the water-soaked seeds and an almost similar trend was observed for lime-soaked seeds. Elaigwu (2008) reported a declined trend of the digestibility of dry matter, crude protein and crude fibre in weaner rabbits with increasing dietary inclusion of raw *H. sabdariffa* seeds from 10 to 30 % as a replacement for groundnut cake.

4.2 Nitrogen balance

The 20 % inclusion level of raw *H. sabdariffa* seeds had a significantly ($P < 0.05$) negative effect on the animals' nitrogen balance, whereas the 10 % inclusion of raw seeds improved nitrogen utilisation as compared to the control. Both inclusion levels of lime-soaked seeds improved nitrogen utilisation. However, nitrogen retained as a percentage of intake, which is a major indicator for protein nutrition (Owen & Zinn, 1988) tended to be lower with 20 % inclusion level of raw *H. sabdariffa* seeds. The authors reported that high protein retention in animals is related to high biological value of the protein (good amino acids pro-

file with less antinutrients), readily digestible and absorbable. Both dietary inclusion levels of water-soaked *H. sabdariffa* seeds reduced nitrogen intake, faecal nitrogen and total nitrogen output but favoured the absorption and retention of nitrogen. Especially the 20 % inclusion of water-soaked seeds yielded the highest nitrogen utilisation efficiency. These results illustrate that both inclusion levels of the two soaking methods as well as the 10 % inclusion of raw *H. sabdariffa* seeds improved nitrogen utilisation when compared to the control diet. This may be due to complementary/synergistic effects of the combination of the two protein sources: *H. sabdariffa* seeds and cotton seed cake, which resulted in better digestibility (Gibson *et al.*, 1998; Dewey, 2003). The poor nitrogen utilisation at 20 % inclusion of raw seeds might be attributed to low nitrogen intake and low digestibility of crude protein of the raw seeds. Also, Yankugh *et al.* (1986) observed a decrease in nitrogen intake and nitrogen retention in pigs when fed high levels of brewer's dried grain, a feed with a high tannin concentration, as a replacement for maize.

4.3 Haematology and serum chemistry

The dietary inclusion of both the raw and soaked *H. sabdariffa* seeds increased pack cell volume, haemoglobin, red blood cells, white blood cells and neutrophils both significantly and numerically, except for haemoglobin and red blood cells, which were unaffected by lime-soaked seed inclusions. However, these parameters are all within the range reported by Research Animal Resources (2009). The increase in PCV with dietary inclusion of *H. sabdariffa* seeds may be due to the increase in the number of red blood cells, which was similar to the values reported by Isaac *et al.* (2013). These observations suggested that a chemical component in raw and water-soaked *H. sabdariffa* seeds might support haemopoiesis, since the RBC value depends on those of haemoglobin and PCV. The findings are also in agreement with the work of Mahadevan *et al.* (2009) who discovered the presence of iron minerals and vitamins in roselle calyxes, and the study of Olusola (2011) who reported an antioxidative potency of roselle calyxes in rabbits, resulting in a gradual increase in PCV, Hb and RBC as the dietary roselle concentration increased. The increase in PCV with increasing dietary inclusion of treated *H. sabdariffa* seeds will probably improve nutrient absorption and transportation and thus result in an increased primary and secondary polycythemia as earlier reported by Isaac *et al.* (2013). The higher values of white blood cells and neutrophils for the 20 % inclusion of raw *H. sabdariffa* seeds as compared to the other treatments, suggest the presence of antinutritional factors to which the animals reacted by increasing phagocytising blood cells and antibodies (Soetan *et al.*, 2013). Dietary nutrients are an important factor affecting haematological parameters in farm animals (Yeong, 1999; Iheukwumere & Herbert, 2002).The increase in blood creatinine levels with inclusion of treated *H. sabdariffa* seeds suggests an increase in the animals' muscular activities which may be due to increase breakdown of creatine in muscle or decrease excretion in the urine (Barcelos *et al.*, 2016).

5   Conclusions

(1) Water-soaked and lime-soaked *H. sabdariffa* seeds can replace cotton seed cake in rice offal based diets up to 20 % for better nutrient utilisation while the raw seeds can be included up to 10 % in such a diet.

(2) The 20 % inclusion of water-soaked *H. sabdariffa* seeds as a replacement for cotton seed cake recorded the best nitrogen utilisation efficiency.

(3) Inclusion levels of *H. sabdariffa* seeds in rice offal based diets as a replacement for cotton seed cakes was noted to support haematopoiesis in Red Sokoto bucks while reduction in serum protein, albumin, globulin, glucose and urea levels was observed.

Acknowledgements

Sincere appreciation goes to the technical staff in the biochemical laboratory and all the staff in the Departmental farm for their technical support during the course of the research.

References

Abdu, S. B., Adegoke, T. F., Abdulrashid, M., Duru, S., Yashim, S. M. & Jokthan, G. E. (2008). Effect of cooking duration on chemical composition of Roselle (*Hibiscus sabdariffa*) seed. *In:* Bawa, G. S., Akpa, G. N., Jokthan, G. E., Kabir, M. & Abdu, S. B. (eds.), *13th Proc. of the Animal Science Association of Nigerian (ASAN), at A.B.U. Zaria, 15th September 2008.* pp. 521–523, ASAN, Zaria, Nigeria.

Abu-Tarboush, H. M. & Basher Ahmed, S. A. (1996). Studies on Karkade (*Hibiscus sabdariffa*) protease inhibitors, phytate, in vitro protein digestibility and gossypol content. *Journal of Food Chemistry*, 56, 15–19.

Adanlawo, I. G. & Ajibade, V. A. (2006). Nutritive values of two varieties of *Hibiscus sabdariffa* seeds calyxes soaked in wood ash. *Pakistan Journal of Nutrition*, 5 (6), 555–557.

Afolabi, K. D., Akinsoyinu, A. O., Olajide, R. & Akinleye, S. B. (2011). Haematological parameters of the Nigerian local grower chickens fed varying dietary levels of palm kernel cake. *Poljoprivreda*, 17 (1), 74–78.

Ajala, M. K., Lamidi, O. S. & Otaru, S. M. (2008). Peri-Urban small ruminant production in Northern Guinea Savanna, Nigeria. *Asian Journal of Animal and Veterinary Advances*, 3 (3), 138–146.

Akanya, H. O., Oyeleke, S. B., Jigm, A. A. & Lawal, F. F. (1997). Analysis of sorrel drink Zoborodo. *Nigeria Journal of Biochemistry*, 12, 77–81.

Al-Wandawi, H., Aisackly, K. & Abdulrahman, M. (1984). *Hibiscus sabdariffa* seeds: A new protein source. *Journal of Agriculture and Food Chemistry*, 510–512.

AOAC (2005). *Official methods of Analysis.* (15th ed.). Association of Official Analytical Chemistry (AOAC), Washington, D. C.

Babatunde, F. E. (2003). Intercrop productivity of *Hibiscus sabdariffa* in Nigeria. *African Crop Science Journal*, 11, 43–47.

Barcelos, R. P., Stefanello, S. T., Mauriz, J. L., Gonzalez-Gallego, J. & Soares, F. A. (2016). Creatine and the Liver: Metabolism and Possible Interactions. *Mini Reviews in Medicinal Chemistry*, 16 (1), 12–18.

Bawa, G. S., Tegbe, T. S. B., Ogundipe, S. O., Dafwang, I. I. & Abu, E. A. (2003). The effect of duration of cooking lablab seeds on the level of some anti-nutritional factors. *In:* Ogundipe, S. O., Dafwang, I. I., Abu, E. A. & Bawa, G. S. (eds.), *28th Proceedings of the Annual Conference of the Nigeria Society for Animal Production (NSAP), held in Ibadan, 16–20th March 2003.* pp. 213–215, NSAP.

Bernhard, A. (1918). The determination of Cholesterol in blood serum. *Journal of Biological Chemistry*, 35, 15–18.

Coles, E. H. (1986). *Veterinary clinical pathology.* (4th ed.). W.B. Saunders Company, Philadelphia.

Dewey, K. G. & Brown, K. H. (2003). Update on technical issues concerning complementary feeding of young children in developing countries and implications for intervention programs. *Food and Nutrition Bulletin*, 24, 5–28.

Doumas, B. T. (1975). Standards for total Serum Protein assay. *Clinical Chemistry*, 21, 11–59.

Doumas, B. T. & Biggs, H. G. (1972). Determination of serum albumin. *In:* Cooper, G. R. (ed.), *Standard Methods of Clinical Chemistry. Vol. 7.* pp. 175–188, Academic Press, New York.

Duncan, D. B. (1955). Multiple range and Multiple F-tests. *Biometrics*, 11, 1–42.

Elaigwu, S. (2008). Effect of graded levels of Hibiscus (*Hibiscus sabdariffa*) seeds as replacement for groundnut cake on performance and nutrient utilization in weaner rabbits. Undergraduate Thesis: Department of Animal Science, Ahmadu Bello University, 30 pp.

Gibson, R. S., Ferguson, E. L. & Lehrfeld, J. (1998). Complementary foods for infant feeding in developing countries: their nutrient adequacy and improvement. *European Journal of Clinical Nutrition*, 52, 764–770.

Hambidge, K. M., Krebs, N. F., Westcott, J. L., Sian, L., Miller, L. V., Peterson, K. L. & Raboy, V. (2005). Absorption of calcium from tortilla meals prepared from low-phytate maize. *The American Journal of Clinical Nutrition*, 82 (1), 84–87. doi:10.1093/ajcn/82.1.84.

Ibrahim, T. A., Abdu, S. B., Hassan, M. R., Yashim, S. M. & Adamu, H. Y. (2016). Growth Performance of Red Sokoto Bucks Fed Inclusion levels of Raw and Soaked Roselle (*Hibiscus sabdariffa* L.) Seeds in Rice Offal Based Diets. *Nigeria Journal of Animal Science*, 18 (2), 434–443.

Ibrahim, T. A. & Yashim, S. M. (2014). Growth response, nutrient digestibility and haematological parameters of Red Sokoto bucks fed lime treated maize cob supplemented with concentrate diet. *Nigerian Journal of Animal Science*, 16 (2), 264–271.

Iheukwumere, F. C. & Herbert, U. (2002). Physiological responses of broiler chickens to quantitative water restrictions: haematology and serum biochemistry. *Journal of Poultry Science*, 2, 117–119.

Isaac, L. J., Abah, G., Akpan, B. & Ekaette, I. U. (2013). Haematological properties of different breeds and sexes of rabbits. *In:* 18th Proceedings of the Annual Conference of Animal Science Association of Nigeria (ASAN), at University of Uyo, 6th September 2013. pp. 24–27, ASAN.

Isah, O. A., Aderinboye, R. Y. & Enogieru, V. A. (2011). Effect of Anti-nutritional factors on Rumen Bacteria of West African Dwarf Goats Fed Tropical Browse Species and Crop By-products. *Journal of Agricultural Science and Environment*, 11 (1), 50–58.

Jaffe, M. (1886). Über den Niederschlag, welchen Pikrinsäre in normalem Harn erzeugt und über eine neue Reaction des Kreatinins. *Zeitschrift für physiologische Chemie*, 10 (5), 391–400.

Kama, D. N. (2005). Rumen microbial ecosystem. Special section, Microbial diversity. *Current Science*, 89 (1), 124–139.

Mahadevan, N., Shivali & Kamboj, P. (2009). *Hibiscus sabdariffa* Linn.–An overview. *Natural Product Radiance*, 8 (1), 77–83.

Miller, G. L. (1959). Use of dinitrosalicylic acid reagent for determination of reducing sugar. *Analytical Chemistry*, 31 (3), 426–428.

National Agricultural Sample Survey (2011). Collaborative Survey on National Agriculture Sample Survey (NASS), 2010/2011. National Bureau Of Statistics/ Federal Ministry of Agriculture and Rural Development. Available at: http://nigeria.countrystat.org/documents/detail/en/c/454834/

Oke, L. O. (1966). Chemical Composition of some Nigeria leafy vegetables. *Journal of the American Dietetic Association*, 53, 130–132.

Okoli, I. C., Ebere, C. S., Uchegbu, M. C., Uddah, C. A. & Ibeaewuchi, I. I. (2002). Survey of the diversity of plant utilized for small ruminant feeding in south eastern Nigeria. *Agriculture Ecosystem and Environment*, 45 (6), 25–29.

Olusola, A. O. (2011). Evaluation of the Anti-oxidant Effects of Hibiscus Sabdariffa Calyx Extracts on 2, 4-Dinitrophenylhydrazine-Induced Oxidative Damage in Rabbits. *Webmedcentral Biochemistry*, 2 (10), WMC002283.

Oni, O. O. (2002). Breeds and genetic improvement of small ruminants. Small ruminant production training workshop. National Animal Production Research Institute, Ahmadu Bello University, Shika-Zaria, Nigeria. pp. 1–7.

Osuji, P. U., Nsahlai, I. V. & Khalili, H. (1993). Feed evaluation. ILCA manual 5: ILCA, Addis Ababa, Ethiopia. Pp. 40

Ovimaps (2014). Ovi location map. Ovi earth imagery date; July 15th, 2014.

Owen, F. N. & Zinn, R. (1988). Protein metabolism in ruminant animals. *In:* Church, D. C. (ed.), *The Ruminant Animal Digestive Physiology and Nutrition*. pp. 227–249, Waveland Press Inc., Prospects Hights, IL, USA.

Oyawoye, B. M. & Ogunkunle, H. N. (2004). *Biochemical and haematological reference values in normal experimental animals*. Masson, New York. pp. 212–218

Pauzenga, U. (1985). Feeding parent stock. *Zootecnica International*, 22–24.

Poppi, D. P. & McLennan, S. R. (1995). Protein and Energy Utilization by Ruminants at Pasture. *Journal of Animal Science*, 73, 278–290.

Research Animal Resources (2009). Reference values for laboratory animals: Normal haematological values. RAR Websites, Research Animal Resources (RAR), University of Minnesota. Available at: http://www.ahc.umn.edu/rar/refvalues.html

SAS Institute (2002). SAS/STAT User's guide. 6.03 Edition. SAS Institute Inc., Cary, NC, USA.

Sodeinde, F. G., Asaolu, V., Oladipo, M. A., Akinlade, J. A., Ige, A. O., Amao, S. R. & Alalade, J. A. (2007). Mineral and antinutritional contents of some forage legumes consumed by small ruminants in the derived savanna of Nigeria. *Research Journal of Agronomy*, 1 (1), 30–32.

Soetan, K. O., Akinrinde, A. S. & Ajibade, T. O. (2013). Preliminary studies on the hematological parameters of cockerels fed raw and processed guinea corn (*Sorghum bicolor*). *In:* Proceedings of the 38th Annual Conference of Nigerian Society for Animal Production. pp. 49–52, Nigerian Society of Animal Production (NSAP).

Tindall, H. D. (1986). *Vegetable in the Tropics*. Macmillan Education Ltd., Houndmills, Basingstoke. pp. 256

Van Soest, P. J. & Wine, R. H. (1967). Use of Detergents in the Analysis of Fibrous Feeds. IV. Determination of Plant Cell-Wall Constituents. *Journal of the Association of Official Analytical Chemists*, 50, 50–55.

Varley, H. G. & Bell, M. (1980). *Determination of serum urea using biochemistry*. (5th ed.). William Heineman Medical Books, Ltd., London.

Wheeler, E. I. & Ferrel, R. E. (1971). A method for phytic acid determination in wheat and wheat fractions. *Cereal Chemistry*, 48, 312–320.

Yankugh, I. O. I. (1986). *The feeding value of brewers dried grains in the diet of growing and finishing pigs*. Master's thesis, Department of Animal Science, Ahmadu Bello University, Zaria, Nigeria. Unpublished, 93 pp.

Yeong, S. W. (1999). Effect of dietary protein on growth performance of village chicken. *In:* Proceedings of the National IRPA (Intensification of Research in Priority Areas) Seminar (Agricultural Sector). pp. 519–520.

2

Assessment of varietal diversity and production systems of cowpea (*Vigna unguiculata* (L.) Walp.) in Southwest Nigeria

Jelili Olaide Saka *, Opeyemi Adeola Agbeleye ,
Olukemi Titilola Ayoola , Bosede Olukemi Lawal ,
Johnson Adedayo Adetumbi , Qudrah Olaitan Oloyede-Kamiyo

Institute of Agricultural Research and Training (IAR&T), Obafemi Awolowo University, Moor Plantation, P.M.B. 5029, Ibadan, Nigeria

Abstract

Cowpea (*Vigna unguiculata* (L.) Walp.) is the most important source of plant protein consumed in Nigeria with major supplies coming from the northern part of the country. However, reduction in supplies due to insurgency resulted in sharp increases in price, especially in Southwest Nigeria where cowpea is relished in different delicacies. Sustainable production increase in suitable Southwest agro-ecologies depends on suitability of cultivated varieties and production practices of the farmers. A study was conducted to identify cowpea varieties cultivated by farmers, the varietal attributes, farmers' preferences, and production constraints. Data were generated through a farm survey of 120 farmers selected by multi-stage sampling technique in Ondo and Oyo States of Southwest Nigeria. Cowpea production was male dominated, with 20.8 % of cultivated area allocated to its production, averaging 0.96 ha per farm household, fragmented over circa three locations. Local varieties were cultivated by 51.6 % of the farmers with seeds sourced mainly from local markets (62.7 %). Cowpea was mainly cultivated as intercrop (55.1 %) notably with cassava. Herbicides and insecticides were prominently used by the farmers while fertiliser was hardly used for cowpea production (12.8 %). Notable attributes cherished by farmers included brown or white coat colour, smooth texture and medium sized grains, erect or creeping growth pattern, and long pod length. Average yield of cowpea on farmers' field was 530 kg ha^{-1} while inadequate access to quality seeds, incidence of field insect pests, and rodents (storage pest) were identified as the most severe production constraints by the farmers. Community-based seed production systems should be introduced for improved access to quality seed.

Keywords: cowpea attributes, farmers' preferences, production constraints

1 Introduction

Cowpea (*Vigna unguiculata* (L.) Walp.) is a leguminous crop grown mainly in the savannah regions of the tropics and subtropics. Cowpea is a very important crop in the semiarid farming systems of the West African countries. The crop is unique in that it provides food, cash and fodder (Kormawa *et al.*, 2002). It is an indigenous African grain legume rated as one of the most economically important crops and a veritable source of plant protein. It is therefore considered cru-

cial for reduction of malnutrition among children and resource poor rural households (Phillips & McWatters, 1991; Langyintuo *et al.*, 2003; Kristjanson *et al.*, 2005; Owolabi *et al.*, 2012).

Available data (FAO, 2017) indicate that West Africa subregion produced about 81 % (4,525,891 metric tonnes) of the global production of cowpea (5,589,216 metric tonnes) in 2014. Nigeria's production of 2,137,900 metric tonnes for the same period accounted for 38.3 % and 47.2 % of global and West African production, respectively. Consequently, Nigeria has remained the largest producer of the commodity globally despite the fall in production of about 58.5 % between 2012 and 2016 which was largely attributed to in-

* Corresponding author – saka_sakang@yahoo.com

surgency in the North-Eastern part of the country which dislodged many farmers from their farms. Nigeria is also the largest consumer of cowpea globally, arising from the economic importance of the crop among households. However, due to the inability of the country to match its population growth rate with commensurate increase in production, persistent gap has been reported between the country's supply and demand for cowpea. Langyintuo *et al.* (2003) gave Nigeria's cowpea supply deficit as 469,000 metric tonnes in 1999 while Sanni *et al.* (2014) reported a supply deficit of 518,400 metric tonnes in 2007. Similarly, per capita cowpea consumption of 23 kg per annum (Coulibaly & Lowenberg-DeBoer, 2002) for a 2014 population of 181.4 million people (NBS, 2018) implies a cowpea demand estimate of about 4.2 million metric tonnes for Nigeria. Consequently, cowpea production of 2.1 million metric tonnes in 2014 (FAO, 2017) suggests a supply deficit of 2.1 million metric tonnes for the same period. Nigeria has thus remained a net importer of the commodity alongside countries such as Ghana, Togo, Cote d'Ivoire and Mauritania while Niger, Burkina Faso, Benin, Mali, Cameroon, Chad and Senegal are net exporters according to Langyintuo *et al.* (2003).

The increasing population of Nigeria without a commensurate increase in cowpea production is likely to further widen the demand-supply gap, thus resulting in persistent increase in price of this commodity among other staple crops. A price increase of 44.2 % between May 2016 and May 2017 (NBS, 2017) is indicative of the persistent pressure on the supply of the commodity. Therefore, reversing this trend would require expanded local production of the commodity to satisfy the growing demand.

In Nigeria, Petu-Ibikunle *et al.* (2008, in Aluko *et al.*, 2016) attributed the bulk of the production of the crop to the semi-arid zones of northern Nigeria despite the increasing economic importance of the commodity in the southern States. However, the current security challenges in the Northeastern part of Nigeria has been identified by Aluko *et al.* (2016) as a major factor behind the reduction in cowpea supply to the southern part of the country. The attendant threat on food security has thus become an impetus to the need to increase production in other suitable agro-ecologies. Southwest Nigeria, comprising of six states (Ekiti, Lagos, Ogun, Ondo, Osun and Oyo) is endowed with some spread of savannah agro-ecologies suitable for cowpea production and also blessed with vast genetic pool of local varieties of cowpea cultivated by farmers under different production systems. Research efforts have consistently produced cowpea varieties that are of different maturity periods and adaptability to diverse agro-ecologies of Nigeria (Brader,

2002; Ewansiha & Tofa, 2016). Many of the released varieties are however not readily adopted by the local farmers who tend to stick to their known varieties and landraces. It therefore becomes imperative to re-appraise the production systems of cowpea in this region with the aim of developing participatory breeding strategies targeted at improving the adoption of developed varieties or developing more adaptable and acceptable varieties for the region.

This study therefore assessed the diversities in the cowpea production system in the Southwest region by highlighting the relative importance, attributes and potentials of cowpea varieties cultivated by farmers; identifying cowpea traits preferred by the farmers and the implication for breeding. The study also explored the challenges experienced by farmers in the adoption of improved varieties and cowpea production, generally.

2   Materials and methods

2.1   Study area and data collection

Southwestern Nigeria represents a geographical land area of 79,665 km^2 spreading between latitudes 6° N and 4° S and longitudes 4° W and 6° E. The region has a largely agrarian population with a climate classified as typically equatorial with distinct dry and wet seasons and a main growing season lasting up to 9 months. Average annual rainfall is 1480 mm with a mean monthly temperature range of 18–24 °C during the rainy season and 30–35 °C during the dry season (FMA & NR, 1997). The region also has four distinct sub-ecologies, comprising of swamp mangrove forest, moist and dry lowland forest, woodland forest and Savannah mosaic. The soil has low to medium productivity potential. The farming system is dominated by arable crops such as cassava, maize, yam, cowpea, sorghum, millet and soybean while notable tree crops are cocoa, kola tree, oil palm, citrus, cashew and mango among others.

Data for this study were generated from a survey of 120 farmers selected from two prominent cowpea producing states (Ondo and Oyo States) in Southwest Nigeria. Oyo and Ondo states have 33 and 18 Local Government Areas (LGAs), respectively. These LGAs are grouped into Agricultural Development Programme (ADP) administrative zones (ADP zones) based on agro-ecological attributes, with the states having 4 ADP zones each. Saki and Ogbomosho ADP zones in Oyo State and Owo ADP zone in Ondo State were selected purposively as main cowpea producing areas in each of the states at the second stage of the sampling process. Subsequently, LGAs were selected in each of the states based on probability proportionate to the number of LGAs in each of the state. Four and six LGAs were selected randomly in Ondo and Oyo States, respectively. Farmers were thereafter randomly selected from the list of cowpea farmers in each of the selected LGAs based on probability proportionate to size of farming households in each of the LGAs. Southwest REFILS (2006) estimated the number of farming households in Oyo State as 415,030 while Ondo State has 250,000 farming households. Consequently, 75 cowpea farmers were selected from Oyo State while 45 farmers were selected from Ondo State.

Data were collected with the aid of structured interview schedule on socio-economic characteristics of cowpea farmers, sources of land holding, cowpea-based cropping system and land allocation to cowpea production. Data were also collected on varieties cultivated by farmers, sources of cowpea seed and production practices, farm size, cowpea output (based on farmers' estimates) and constraints experienced in cowpea production. Farmers were subsequently asked to identify and rank attributes observed on cowpea varieties on the basis of their preferences. Constraints were also identified and ranked on the basis of prevalence and severity.

2.2   Data analysis

Data analyses were carried out using descriptive statistics such as frequency, percentages and mean. Identified constraints were ranked on the basis of prevalence and severity scores. Mean values were also compared across states.

3   Results

3.1   Socio-economic characteristics of cowpea farmers

The distribution of cowpea farmers was male dominated, with an average age of 48.8 years (Table 1). None of the farmers was below 20 years of age while about a quarter of the population (23.7 %) were above the productive age of 60 years. Most of the farmers (65.2 %) have been cultivating cowpea for more than five years while the average years of experience of the farmers in cowpea production was 13.8 years. Majority of the farmers are educated, with average of 7.9 years in formal schools but technical training in cowpea production was predominantly lacking (Table 1).

Cowpea farmlands were mostly owned by the farmers (59.9 %) with such land acquired through inheritance (56.3 %) and outright purchases (3.6 %). However, the relative importance of these modes of land acquisition varied between the two states, with inheritance being more common in Oyo State (78.1 %) while land for cowpea production were mostly rented (51.3 %) in Ondo State (Table 2). Cowpea production enjoyed substantial allocation of land resource; with 20.8 % of the cultivated land area allocated

Table 1: *Socio-economic characteristics of cowpea farmers.*

Characteristics	Ondo (%) n = 45	Oyo (%) n = 73	Total (%) n = 118	T-Stat
Sex				
Male	82.2	72.6	76.3	
Female	17.9	27.4	23.7	
Age				
21–40	60.0	20.6	35.6	
41–60	11.1	58.9	40.6	
Above 60 years	28.9	20.6	23.7	
Average	45.3 (16.6)	51.1 (11.2)	48.9 (13.7)	5.25**
Cowpea experience (years)				
Not more than 5 years	33.3	35.6	34.8	
6–10	17.8	24.7	22.0	
11–20	22.2	20.6	21.2	
21–30	26.7	12.3	17.8	
Above 30 years		6.9	4.2	
Average	13.8 (12.0)	13.9 (13.3)	13.8 (12.6)	0.00
Formal education				
Had no formal education	42.2	35.3	37.3	
Had formal education	57.8	65.8	62.7	
Average years of education	6.4 (6.5)	7.3 (8.7)	6.9 (7.9)	0.37
Training in cowpea production				
Had no training	80.0	84.9	83.1	
Had training	20.0	15.3	16.9	

Source: Field Survey, 2015.
Values in parentheses are standard deviations
** Significant at $P < 0.05$

to this enterprise and this was similar across the two states (Table 3). Average cowpea field was 0.96 ha, with larger cowpea fields (1.10 ha) found in Oyo State than in Ondo State (0.73 ha). Cowpea fields were spread averagely over three locations thereby showing the high level of land fragmentation characterised by an average plot size of 0.33 ha.

Table 2: *Method of land acquisition.*

Method of land acquisition	Ondo (%) n = 45	Oyo (%) n = 73	Total (%) n = 118
Inheritance	15.4	78.1	56.3
Purchased	10.3		3.6
Rented	51.3	15.0	27.7
Pledged	23.0	6.9	12.5

Source: Field Survey, 2015.

Cowpea was cultivated either as sole crop or intercropped especially with cassava. However, there was no significant difference in the average plot size of cowpea cultivated as sole- (1.15 ha) or as inter-crop (0.81 ha).

Table 3: *Land holding and allocation to cowpea production.*

Attributes	Ondo	Oyo	Pooled	F. stat
Number of farm plot locations	2.80 (1.25)	2.97 (2.61)	2.91 (2.19)	0.17
Average size of land holding (ha)	5.32 (4.18)	5.53 (5.00)	5.45 (4.69)	0.05
Average size of land area cultivated (ha)	3.73 (3.45)	5.14 (5.72)	4.61 (5.01)	2.24
Average cultivated area under cowpea (ha)	0.73 (0.95)	1.10 (0.92)	0.96 (0.95)	4.41**
Area allocation to cowpea (%)	19.6	21.4	20.8	
Average size of cowpea plot (ha)	0.26	0.37	0.33	

Source: Field Survey, 2015.
Values in parentheses are standard deviations
** Significant at $P < 0.05$

3.2 Cowpea varieties, types and sources of seed cultivated by farmers

Cultivation of local cowpea varieties (51.7 %) dominated the cowpea cropping system in the two states (Table 4). Cowpea seeds were commonly sourced from local markets (62.7 %) while 29.7 % sourced their seed mainly from the public extension agencies (ADPs) in each of the states (Table 4). Fifteen cowpea varieties were cultivated by the farmers; four of these varieties (Ife Brown, Ife Bimpe, TVX 3236 and Ife-98-14) were known as improved varieties with Ife brown being the only prominent improved variety (Table 5).

Table 4: *Type of cowpea seeds cultivated and sources.*

Type of Seed	Ondo (%) n = 45	Oyo (%) n = 73	Pooled (%) n = 118
Local	40.0	58.9	51.7
Improved	60.0	41.1	48.3
Sources of Seed			
ADP[†]	33.3	27.4	29.7
Research Institute		4.1	2.5
Fellow farmers	2.2	4.1	3.4
Local market	64.4	61.6	62.7
Previous harvest		2.7	1.7

Source: Field Survey, 2015.
[†] ADP: Agricultural Development Programme

3.3 Production systems and input use in cowpea production

Cowpea was cultivated either as sole crop or intercropped especially with cassava. However, there was diversity in cowpea cropping systems between the two states. Intercropping was more common in Ondo State (64.4 %) whereas both sole (49.3 %) and intercropping (49.3 %) were equally prominent among cowpea farmers in Oyo State. Where intercropping was the choice of the farmer, cowpea was intercropped with cassava (22 %) or as cowpea/cassava/maize

Table 5: *Distribution of farmers by cowpea varieties cultivated.*

Varieties	Ondo (%)	Oyo (%)	Pooled (%)
Improved varieties			
Ife 98-14		1.4	0.9
Ife Bimpe		2.7	1.7
TVX 3236	8.9	12.3	11.0
Ife Brown	37.8	36.9	37.3
Landraces/ local varieties			
Igibira White	42.2	28.7	33.9
Oloyin	31.1	9.6	17.8
Saadu	44.4	20.6	29.7
Sokoto White	24.4	5.5	12.7
Tede	22.2	17.8	19.5
Abewehe	4.4	15.1	11.0
Wewe	15.6		5.9
Abalaye	8.9	6.9	7.6
Kawoleri	4.4		1.7
Big Brown (Drum)	2.2	8.2	5.9
Gbomogungi		1.4	0.9

Source: Field Survey, 2015.

(9.3 %). Most farmers plant cowpea on ridges at the recommended rate of two seeds per hole (Table 6). Cassava is a very prominent crop in the farming system of Southwest Nigeria. Its prominence in most cropping systems arises from its strategic position in the food security of households in the zone. Cassava is also considered less susceptible to crop failure but largely sensitive to market dynamics. Generally, intercropping is widely adopted by farmers as a cropping strategy against risk associated with irregular rainfall and market dynamics.

Cowpea planting was done between July and August in consonance with recommended practices as reported by Dugje *et al.* (2009). However, some of the farmers (31.4 %) planted as early as April at the onset of the rains especially for some local varieties. Weeding was predominantly done

Table 6: *Management practices in cowpea production.*

	Ondo (%)	Oyo (%)	Pooled (%)
Planting period			
March	17.8	6.9	11.0
April	33.3	30.1	31.4
May	4.4	12.3	9.3
June	13.3	1.4	5.9
July	37.8	30.1	33.1
August	60.0	35.6	44.9
Planting method			
On ridges	51.1	73.9	64.9
Flat	24.4	20.3	21.9
Ridges & flat	24.4	5.8	13.2
No. of seeds/hole			
Two	91.9	85.0	87.6
Three	8.1	15.0	12.4
No. of weeding			
One	33.3	43.5	39.5
Two	64.4	49.3	55.3
Three		5.8	3.5
Method of weeding			
Manual	15.6	20.3	18.4
Chemical	64.4	73.4	71.1
Manual & chemical	20.0	4.4	10.5

Source: Field Survey, 2015.

Table 7: *Attributes preference in cowpea.*

Attributes	Ondo (%)	Oyo (%)	Total (%)
Seed texture			
Smooth	77.8	61.6	67.8
Coat colour			
Brown	48.9	58.3	54.7
White	51.1	23.6	34.2
Eye colour			
Black	46.7	50.7	49.1
Seed size			
Small	31.0	24.7	27.0
Medium	50.0	28.8	36.5
Days to maturity			
Medium	53.3	63.5	59.3
Growth habit			
Erect	55.6	6.9	25.4
Creeping	40.0	20.6	28.0
Pod length			
Long	83.3	26.0	47.0

Only values indicating high preference by farmers are presented across varietal attributes.
Source: Field Survey, 2015.

with the use of herbicides by 71.1 % of the farmers while supplementary weeding was done twice by 55.3 % of the farmers before harvesting. However, 3.5 % of the farmers weeded thrice. Use of herbicides and insecticides for cowpea production was very prominent among the farmers. However, fertiliser application was sparingly practiced for cowpea production, although more farmers in Ondo State were favourably disposed to its application with organic manure being the most prominent fertiliser used followed by NPK and Urea. It is known that cowpea does not require much fertiliser as it fixes nitrogen unless the soil is markedly depleted of nutrients.

3.4 Farmer's preference rating for cowpea traits

Assessment of the varietal attributes preferred by farmers showed that brown or white seed coat colour, smooth texture and medium sized grains were more preferred by cowpea farmers. Also, erect and creeping growth pattern were more preferred to semi- erect type (Table 7). Ife brown was preferred by majority of the farmers for its good taste, early maturity, attractive colour and market price while Abewehe and Igbira black were noted for good taste and high yield. Sokoto white was the only variety preferred by farmers for

its swelling and short cooking time attributes during processing while Oloyin was preferred by most farmers that cultivated it for its good taste (Table 8).

3.5 Cowpea yields on farmers' fields

Mean yield of cowpea varieties by farmers' assessment was estimated as 529 kg ha^{-1}. Yield across varieties ranged from 688 kg ha^{-1} for Igbira black to 417 kg ha^{-1} for Wewe (Table 9). There was no significant difference in cowpea yield cultivated by farmers as sole crop (532 kg ha^{-1}) or as intercrop (526 kg ha^{-1}) while overall yield of improved (520 kg ha^{-1}) and local varieties (535 kg ha^{-1}) was also comparable. Also, there were no significant differences between the individual yields of the varieties but data obtained from farmers on Igbira black, Oloyin, Saadu, and Abewehe gave an indication of high yield potential of these local varieties. This however requires further investigation as estimates given by the farmers were based on memory recall and use of local measures.

3.6 Constraints in cowpea production

The most prevalent limiting constraints experienced by the farmers in cowpea production were poor access to quality cowpea seeds (87.2 %), incidence of insect pests (86.2 %), poor germination (83.6 %), and irregular rainfall pattern (82.7 %). Other constraints included low yield

Table 8: *Specific attributes cherished by farmers in cowpea varieties cultivated.*

Variety	Good Taste (%)	High Yield (%)	Swelling ability (%)	Early maturing (%)	Shorter cooking time (%)	Erect Growth pattern (%)	Attractive colour (%)	Market Premium (%)
Ife Brown	68.9	17.7		66.7	28.9	24.4	42.2	68.9
Tede Local	36.4							
Saadu	40.0	35.0						
Igbira White	100.0	20.0			10.0			
TVX 3236	42.9							
Abewehe	40.0	60.0	20.0	20.0				
Oloyin	100.0							
Abalaye	25.0		25.0					
Sokoto White	25.0		100.0		100.0			
Igbira black	50.0	50.0			50.0			

Source: Field Survey, 2015.

(81.0 %), incidence of diseases (80.2 %), storage losses (80.2 %) and shattering on the field (80.2 %). In addition, the farmers ranked poor access to quality seed, incidence of diseases, rodents (storage pest), and high cost of pesticides as those constraints having the most severe effect on their productivity.

4   Discussion

Farmers in the study area allocated about 21 % of their cultivated land area to cowpea production thereby showing that this commodity is highly valued in the farming system. However, the average land area of 0.96 ha cultivated with cowpea per household was lower than the cowpea area of 2 ha reported for Kano state by Sanni *et al.* (2014).

Cowpea production is well adapted to both sole and inter-cropping systems, and mostly found with cassava when cultivated as intercrop.

High yielding and disease resistant cowpea varieties are constantly developed to ensure sustained increase in productivity of cowpea. Dugje *et al.* (2009) listed eight improved cowpea varieties with different agronomic attributes as recommended for the Nigerian farming systems. However, only two of these varieties (Ife Brown and TVX 3236) featured prominently among the varieties cultivated by farmers in this study. The remaining cultivars in farmers' hands were local varieties. Cowpea seeds were largely sourced from local markets hence, the quality of such seeds cannot be ascertained. Similar occurrence in Benin Republic was attributed to near non-existence of dedicated cowpea seed suppliers across farming communities (Agyekum *et al.*, 2016).

Table 9: *Cowpea yield across cropping systems and varieties.*

Item	Yield (kg ha^{-1})	F-Statistics
Cropping System		
Sole	532 (269.3)	
Intercrop	525 (397.4)	0.0
Class of varieties		
Improved	520 (263.2)	
Local	535 (400.4)	0.1
Improved		
TVX-3236	525 (335.4)	
Ife Brown	520 (258.4)	
Local		
Igbira Black	688 (88.4)	
Oloyin	675 (813.2)	
Saadu	635 (611.8)	
Abewehe	583 (381.9)	
Tede Local	458 (239.4)	
Sokoto White	456 (62.5)	
Abalaye	433 (275.4)	
Igbira White	424 (225.0)	
Wewe	417 (64.6)	
Average	529 (346.1)	
F-Statistics	0.5	

Source: Field Survey, 2015.
Values in parentheses are standard deviations.

Production efficiency is a necessary precursor for increased productivity and sustainable growth in agricultural production. This is in turn enhanced by timely conduct of farm operations and the use of modern inputs. Planting of cowpea was done between July and August in consonance with recommended practices. The best period of planting cowpea is August, a time when the rains would have ceased but soil moisture is still enough for the crop to be established before the resumption of rains in September (Dugje et al., 2009). However, planting at earlier periods especially between May and June as done by few of the farmers, exposes cowpea to heavy rains and seed may rot causing loss of harvest and income. Incidence of pests and diseases is also high during this period (Omongo et al., 1997; Karungi et al., 2000) and much labour is needed to weed and spray pesticides.

Timely use of pesticides enhances productivity of cowpea by creation of environments which disallow yield depressing bio-factors from thriving (Dugje et al., 2009; Awunyo-Vitor et al., 2013; Awotide et al., 2015). For instance, weeds constitute serious problems in cowpea production through competition for air and soil nutrients in addition to harbouring pests thereby reducing both grain yield and quality. The prominent use of herbicides and insecticides among the farmers in this study was a reflection of the relative economic importance attached to these constraints in cowpea production. The insecticides are largely targeted at insects such as aphids, leaf hoppers, flower trips, Ootheca mutabilis and other cowpea beetles.

Major attributes used in describing cowpea varieties include the seed coat colour, texture, days to maturity, growth pattern, pod length, size of grains and yield among others. These attributes strongly determine the adoption pattern of crop varieties among farmers.

The erect and creeping traits are two widely contrasting extremes in cowpea growth pattern. The almost equal preference of the farmers to erect or creeping growth may perhaps be connected with the convenience of weed control and harvesting. The fast growth and spreading habit of traditional cowpea varieties supress weeds, and soil nitrogen is increased which improves cereal growth (Gómez, 2004).

The erect types were likely selected mainly due to the ease of harvest, and increased plant population because they are usually planted at closer spacing. The creeping varieties on the other hand were considered by the farmers as having the potential to smother weeds thereby reducing the frequency and cost of weeding. Kamara et al. (2010) attributed the continuous use of a local cowpea variety Kanannado Brown by farmers in North East Nigeria to its suitability for relay intercropping as well as its creeping and weeds smothering abilities. Also, the varieties with long pod length are widely cherished by the farmers. The long pod is possibly indicative of a variety's ability to bear more seeds and consequently a higher yield potential.

Kasali et al. (2018) identified weevil tolerance, taste (sweetness), time to cook, swelling ability and coat colour as the five most preferred varietal attributes by consumers of cowpea in Osun State, Nigeria. Similar studies in Nigeria have also documented considerable evidences of the positive influence of varietal attributes on the adoption or dissemination of improved cowpea varieties by farmers in Nigeria. Mbavai et al. (2015) identified such attributes as including high yield, resistance to drought and early maturity. In a study on factors influencing farmer-to-farmer transfer of an improved cowpea variety, Kormawa et al. (2004) identified threshing quality of cowpea varieties as a significant factor in explaining willingness of farmers to transfer improved cowpea varieties to other farmers.

The higher market price attracted by brown seeded Ife brown cowpea could be considered as an influencing factor in the farmers' preference for brown coat seed colour. Kamara et al. (2010) also reported that brown seeds fetch higher market prices. Considering the significant influence of basic attributes of improved varieties on adoption (Omonona et al., 2005; Fashola et al., 2007, Saka and Lawal, 2009), it is important to target such farmer-cherished attributes in the selection and breeding programme for cowpea in the region. It is therefore pertinent that breeding objectives for the development of cowpea varieties suitable for cultivation in the Southwest agro-ecologies continues to target improved varieties that can combine early or medium maturity period, brown coat colour, smooth texture, medium sized grains, long pod length with either erect growth pattern efficient for high yield, harvesting convenience or creeping growth pattern for weed control and cost efficiency. Such varieties should also compare favourably in good taste, swelling ability and short cooking period, as well as tolerant to insect pests.

The average cowpea yield in the study area was greater than the 2014 yields given by FAO (2017) for West Africa, Burkina Faso or Mali. Although there was no significant difference in yields across varieties, the yields of some local varieties such as Igbira black and Saadu compared favourably with yields from improved varieties thereby pointing to a potential requiring further investigation.

This study has identified poor access to quality cowpea seeds, incidence of insect pests, poor germination and irregular rainfall pattern as the most limiting constraints in cowpea production in the Southwest agro-ecology of Nigeria. These set of constraints are of agronomic importance to the performance of the crop on the field and consequently held contributory to low yield identified by the farmers as the fifth most prevalent constraint in cowpea production. Mohammed & Mohammed (2014) also reported incidence of pests and disease as the most serious constraint faced by cowpea farmers even in Kano State which is in the drier savannah agro-ecology. In the southern agro-ecology of Nigeria, incidence of insect pests and diseases has been identified (Sangoyomi & Alabi 2016; Ezeaku *et al.*, 2017) as major limiting constraint in cowpea production. Insect pests have been known to cause maximum damage to cowpea from seedling stage to grain storage. In the midst of high cost of chemicals however, farmers are confronted with limiting capabilities to adopt recommended practices and optimum performance of cowpea is ordinarily threatened. This inadvertently makes development of varieties that are tolerant to insect pests and diseases more desirable.

5　Conclusion

Cowpea production in Southwest Nigeria is characterized by diversities in wide cultivation of landraces of differing preferred attributes among the farmers. Therefore, breeding objectives should target developing improved varieties that meet farmers' needs. Use of a participatory varietal selection breeding system would enable incorporation of cowpea attributes cherished by farmers into the cowpea breeding objectives. These efforts will entrench wide ownership of the developed technologies, wider adoption among cowpea farmers and improved cowpea cultivation in this region. In addition, the low level of adoption of improved varieties observed in the region was mainly due to inadequate access to quality seed. Consequently, introduction of community-based seed production system or out-grower schemes by the National Agricultural Research and Extension system is recommended for farmers to have adequate access to quality seeds for enhanced productivity.

References

Agyekum, M., Donovan, C. & Lupi, F. (2016). Novel IPM Intervention for West Africa: Smallholder Farmers' Preferences for Biological versus Synthetic Control Strategies for Cowpea Pests. Selected Paper prepared for presentation at the 2016 Agricultural & Applied Economics Association Annual Meeting, Boston, Massachusetts, July 31–August 2. 26 pp., Available at: http://ageconsearch.umn.edu/bitstream/235993/2/Selected%20paper_2016%20AAEA%20Meeting_Agyekum%20et%20al.pdf

Aluko, O. J., Osikabor, B., Adejumo, A. A. & Sumade, S. (2016). Perceived Effect of Boko-Haram Insurgency on Means of Accessing Cowpea from North-East Nigeria to Bodija Market, Ibadan, Oyo State, Nigeria. *Open Access Library Journal*, 3, e2723. doi:10.4236/oalib.1102723.

Awotide, D. O., Ikudaisi, O. J., Ajala, S. O. & Kaltungo, J. H. (2015). Input Use and Profitability of Arable Crops Production in Nigeria. *Journal of Sustainable Development*, 8 (3), 139–146.

Awunyo-Vitor, D., Bakang, J. & Smith, C. (2013). Estimation of Farm Level Technical Efficiency of Small-Scale Cowpea Production in Ghana. *American-Eurasian Journal of Agricultural & Environmental Sciences (AEJAES)*, 13 (8), 1080–1087. doi:10.5829/idosi.aejaes.2013.13.08.11013.

Brader, L. (2002). Forward. *In:* Fatokun, C. A., Tarawali, S. A., Singh, B. B., Kormawa, P. M. & Tamò, M. (eds.), *Challenges and opportunities for enhancing sustainable cowpea production. Proceedings of the World Cowpea Conference III held at the International Institute of Tropical Agriculture (IITA), Ibadan, Nigeria, 4–8 September 2000*. p. vi, IITA, Ibadan, Nigeria.

Coulibaly, O. & Lowenberg-DeBoer, J. (2002). The economics of cowpea in West Africa. *In:* Fatokun, C. A., Tarawali, S. A., Singh, B. B., Kormawa, P. M. & Tamò, M. (eds.), *Challenges and opportunities for enhancing sustainable cowpea production. Proceedings of the World Cowpea Conference III held at the International Institute of Tropical Agriculture (IITA), Ibadan, Nigeria, 4–8 September 2000*. pp. 351–366, IITA, Ibadan, Nigeria.

Dugje, I. Y., Omoigui, L. O., Ekeleme, F., Kamara, A. Y. & Ajeigbe, H. (2009). *Farmers' Guide to Cowpea Production in West Africa*. International Institute of Tropical Agriculture (IITA), Ibadan, Nigeria.

Ewansiha, S. U. & Tofa, A. I. (2016). Yield Response of Cowpea Varieties to Sowing Dates in a Sudan Savannah Agroecology of Nigeria. *Bayero Journal of Pure and Applied Sciences*, 9 (1), 62–67. doi: 10.4314/bajopas.v9i1.10.

Ezeaku, I. E., Mbah, B. N. & Baiyeri, K. P. (2017). Response of cowpea (*Vigna unguiculata* (L.) Walp genotypes to sowing dates and insecticide spraying in south eastern Nigeria. *The Journal of Animal & Plant Sciences*, 27 (1), 239–245.

FAO (2017). FAOSTAT Online Statistical Services: Crop production data. Food and Agriculture Organization of the United Nation (FAO), Rome.

Fashola, O. O., Oladele, O. I., Alabi, M. O., Tologbonse, D. & Wakatsuki, T. (2007). Socio-economic factors influencing the adoption of sawah rice production technology in Nigeria. *Journal of Food, Agriculture and Environment*, 5 (1), 239–242.

FMA&NR (1997). Nigeria: National Agricultural Research Strategy Plan: 1996–2010. Department of Agricultural Sciences, Federal Ministry of Agriculture and Natural Resources, Abuja , Nigeria.

Gómez, C. (2004). *Cowpea: post-harvest operations*. Food and Agriculture Organization of the United Nations (FAO), Rome, Italy. Available at: http://www.fao.org/3/a-au994e.pdf

Kamara, A. Y., Ellis-Jones, J., Ekeleme, F., Omoigui, L., Amaza, P., Chikoye, D. & Dugje, I. Y. (2010). A participatory evaluation of improved cowpea cultivars in the Guinea and Sudan savannah zones of north east Nigeria. *Archives of Agronomy and Soil Science*, 56, 355–370.

Karungi, J., Adipala, E., Kyamanywa, S., Ogenga-Latigo, M. W., Oyobo, N. & Jackai, L. E. N. (2000). Pest management in cowpea. Part 2. Integrating planting time, plant density and insecticide application for management of cowpea field insect pests in eastern Uganda. *Crop Protection*, 19, 237–245.

Kasali, R., Oyewale, A. Y. & Yesufu, O. A. (2018). Analysis of consumer's WTP for cowpea varieties in Osun State, Nigeria: the Hedonic pricing approach. *Turkish Journal of Agriculture – Food Science and Technology*, 6 (9), 1120–1128. doi:10.24925/turjaf.v6i9.1120-1128.1832.

Kormawa, P. M., Chianu, J. N. & Manyong, V. M. (2002). Cowpea demand and supply patterns in West Africa: the case of Nigeria. *In:* Fatokun, C. A., Tarawali, S. A., Singh, B. B., Kormawa, P. M. & Tamò, M. (eds.), *Challenges and opportunities for enhancing sustainable cowpea production. Proceedings of the World Cowpea Conference III held at the International Institute of Tropical Agriculture (IITA), Ibadan, Nigeria, 4–8 September 2000.* Ch. V Cowpea postharvest and socioeconomic studies, pp. 376–386, IITA, Ibadan, Nigeria.

Kormawa, P. M., Ezedinma, C. I. & Singh, B. B. (2004). Factors influencing farmer-to-farmer transfer of an improved cowpea variety in Kano State, Nigeria. *Journal of Agriculture and Rural Development in the Tropics and Subtropics*, 105 (1), 1–13. Available at: https://jarts.info/index.php/jarts/article/view/46

Kristjanson, P., Okike, I., Tarawali, S., Singhd, B. B. & Manyong, V. M. (2005). Farmers' perceptions of benefits and factors affecting the adoption of improved dual-purpose cowpea in the dry savannahs of Nigeria. *Agricultural Economics*, 32, 195–210.

Langyintuo, A. S., Lowenberg-DeBoer, J., Faye, M., Lambert, D., Ibro, G., Moussa, B., Kergna, A., Kushwaha, S., Musa, S. & Ntoukam, G. (2003). Cowpea supply and demand in West and Central Africa. *Field Crops Research*, 82, 215–231.

Mbavai, J. J., Shitu, M. B., Abdoulaye, T., Kamara, A. Y. & Kamara, S. M. (2015). Pattern of adoption and constraints to adoption of improved cowpea varieties in the Sudan savanna zone of Northern Nigeria. *Journal of Agricultural Extension and Rural Development*, 7 (12), 322–329. doi: 10.5897/JAERD2015.0694.

Mohammed, U. S. & Mohammed, F. K. (2014). Profitability analysis of cowpea production in rural areas of Zaria local government area of Kaduna state, Nigeria. *International Journal of Development and Sustainability*, 3 (9), 1919–1926.

NBS (2017). CPI and Inflation Report August, 2017. National Bureau of Statistics (NBS), Nigeria. Available at: https://nigerianstat.gov.ng/elibrary

NBS (2018). Population of Nigeria 2016. National Bureau of Statistics (NBS), Nigeria. Available at: http://nigeria.opendataforafrica.org/crhsjdg/population-of-nigeria-2016

Omongo, C. A., Ogenga-Latigo, M. W., Kyamanywa, S. & Adipala, E. (1997). The effect of seasons and cropping systems on the occurrence of cowpea pests in Uganda. *In:* African Crop Science Conference Proceedings. Vol. 3, pp. 1111–1116.

Omonona, B. T., Oni, O. A. & Uwagboe, A. O. (2005). Adoption of improved cassava varieties and its welfare impact on rural farming households in Edo State, Nigeria. *Journal of Agriculture and Food Information*, 7 (1), 39–55.

Owolabi, A. O., Ndidi, U. S., James, B. D. & Amune, F. A. (2012). Proximate, antinutrient and mineral composition of five varieties (improved and local) of cowpea, *Vigna unguiculata*, commonly consumed in Samaru Community, Zaria-Nigeria. *Asian Journal of Food Science and Technology*, 4 (2), 70–72.

Petu-Ibikunle, A. M., Abba-Mani, F. & Od, P. E. (2008). Determination of Rate and time of Nitrogen Application on Cowpea variety in the Sudan Savannahh zone. *International Journal of Academic focus series*, 1 (1), 13–21.

Phillips, R. D. & McWatters, K. M. (1991). Contribution of cowpeas to nutrition and health. *Food Technology*, 45, 127–130.

REFILS (2006). Technology Generation and Dissemination. *In:* Oluokun, J. A., Oluwatosin, G. A. & Adesehinwa, A. O. K. (eds.), *Proceeding of the 19th OFAR/ Extension Workshop held at the Institute of Agricultural Research and Training, Moor Plantation, Ibadan 21st February, 2006.* pp. 102–186, Research, Extension Farmers Input Linkage Systems (REFILS), Nigeria.

Saka, J. O. & Lawal, B. O. (2009). Determinants of adoption and productivity of improved rice varieties in southwestern Nigeria. *African Journal of Biotechnology*, 8 (19), 4923–4932.

Sangoyomi, T. & Alabi, O. (2016). Field evaluation of cowpea varieties for adaptation to the forest/savannah transition agroecology of Osun state, Nigeria. *African Journal of Agricultural Research*, 11 (49), 4959–4963. doi:10.5897/AJAR2016.11710.

Sanni, A., Abubakar, B. Z., Yakubu, D. H., Atala, T. K. & Abubakar, L. (2014). Socio-economic Factors Influencing Adoption of Dual-purpose Cowpea Production Technologies in Bichi Local Government Area of Kano State, Nigeria. *Asian Journal of Agricultural Extension, Economics & Sociology*, 3 (4), 257–274.

3

Genetic variability among wheat (*Triticum aestivum* L.) germplasm for resistance to spot blotch disease

Batiseba Tembo [a,b,*], Julia Sibiya [a], Pangirayi Tongoona [c]

[a] *African Centre for Crop Improvement, University of KwaZulu-Natal. College of Agriculture, Engineering and Science, School of Agricultural, Earth and Environmental Sciences, Private Bag X01, Scottsville 3209, Pietermaritzburg, South Africa*
[b] *Zambia Agricultural Research Institute (ZARI), Mt. Makulu Research Station, P/B 7, Chilanga, Zambia*
[c] *West African Centre for Crop Improvement, University of Ghana, PMB 30 Legon, Ghana*

Abstract

Spot blotch caused by *Bipolaris sorokiniana* (Sacc.) Shoem. is the most devastating disease limiting wheat productivity in warm and humid environments. One hundred and fifty wheat genotypes were evaluated under field conditions in 2013 and 2014 in six different locations in Zambia. The genotypes showed different levels of resistance to spot blotch. Genotypes 19HRWSN6 (Kenya Heroe), 19HRWSN7 (Prontia Federal) and 19HRWSN15 (BRBT2/METSO) were resistant lines across environments. The genotype plus genotype by environment (GGE) biplot grouped the six environments (E) into three mega-environments (ME) with respect to spot blotch severity. ME I contained Golden Valley Agricultural Research Trust (GART) (E6) only. Mpongwe (E4), Mt. Makulu (E5 and E2) and GART (E3) formed ME II, while ME III contained only Mutanda (E1). Genotypes 16HRWYT5, SB50 and 20HRWSN33 were the most susceptible genotypes in ME I, II and III, respectively. Genotype 19HRWSN7 was the most resistant across test locations. The locations in ME III were highly correlated indicating that they provided similar information on genotypes. This suggests that one location could be chosen among the locations in ME III for screening spot blotch resistance each year if the pattern repeats across years. This could aid in reducing the cost of genotype evaluation and improve efficiency as genotypes would be handled in fewer environments.

Keywords: Bipolaris sorokiniana, disease, management, screening, host resistance

1 Introduction

Spot blotch caused by *Bipolaris sorokiniana* (Sacc.) Shoem. is the most important disease limiting wheat yields in warm and humid environments (Srivastava & Tewari, 2002; Mikhailova *et al.*, 2004; Khan & Chowdhury, 2011). It occurs worldwide especially in areas with high relative humidity (Mikhailova *et al.*, 2004; Acharya *et al.*, 2011). In Africa, the disease has been reported to occur in Kenya, Malawi, Sudan, South Africa, Zimbabwe (Acharya *et al.*, 2011), Madagascar (Rakotondramanana, 1981), and Zambia (Mukwavi *et al.*, 1990; Tembo *et al.*, 2016). The disease is most severe and damaging under temperatures of between 18 and 32 °C, high relative humidity (Duveiller & Gilchrist,

1994; Mehta, 1997). Spot blotch attacks all plant parts and can cause large yield losses. Yield losses due to spot blotch disease range from 25–43 % in South Asia, 18–22 % in India, 70–100 % in Nepal, 15 % in Bangladesh (Alam *et al.*, 1994) and 15–85 % in Zambia (Mukwavi *et al.*, 1990). Under severe infections the disease spreads to the spikes resulting in shrivelled grains with low grain weight and black points (Raemaekers, 1988; Gubiš *et al.*, 2010). Apart from these effects, spot blotch disease also reduces the grade and quality of wheat (Kumar *et al.*, 2002).

The management of spot blotch disease involving the use of fungicides is not only costly for small-scale farmers, but also difficult in its application and is not environmentally friendly (Iftikhar *et al.*, 2009; Eisa *et al.*, 2013). Use of proper crop rotation is also not feasible amongst small-scale farmers due to small farm sizes. Use of resistant cultivars is considered the most economical, cheap, sustainable and environmentally safe method of controlling the disease (Duveiller & Sharma, 2009; Iftikhar et al., 2009; Iftikhar et al., 2012), highlighting the need for the screening of wheat germplasm to identify sources of resistance for use in breeding programmes. Host immunity to *Bipolaris* have not yet been reported in wheat (Duveiller & Sharma, 2013). However, there are different reports on inheritance of resistance to spot blotch. Some research indicated oligogenic dominant resistance while others indicated polygenic resistance (Duveiller & Sharma, 2009). The objective of this study, was thus to screen wheat germplasm in different environments in Zambia to identify sources of resistance that could be used in breeding for resistance against spot blotch disease.

2   Materials and methods

2.1   Plant materials, experimental sites and experimental design

One hundred and fifty wheat genotypes from the Zambia Agriculture Research Institute and the International Maize and Wheat Improvement Centre (CIMMYT, Mexico) were evaluated in Zambia under natural field conditions in 'hot spot' sites (Sites where the disease occurs every year, naturally). The genotypes were screened over two successive years, 2013 (2012/13) and 2014 (2013/14) summer seasons, at three sites in each year. In 2013, the genotypes were evaluated at Mutanda Research Station (Environment 1 – E1), Mt. Makulu Research Station (E2) and Golden Valley Agricultural Research Trust (GART) (E3) (Table 1). In 2014, the genotypes were assessed at Mpongwe Seed-Co Research Farm (E4), Mt. Makulu Research Station (E5) and GART (E6). The genotypes were planted in the second week of November in each year, so that the anthesis coincided with warm temperatures and high humidity that favour disease development and spread. The experimental field was laid out in a 10×15 alpha lattice design with two replications. Each genotype was planted in a 2.5 m long plot of two rows, 20 cm inter row spacing with a plant to plant distance of 10 cm. One row of Loerrie I, a susceptible spreader was planted in the alleyways and borders to create enough disease pressure (Joshi et al., 2004). Standard agronomic practices were followed for good crop management. Fertiliser application involved basal fertiliser (8 % N, 24 % P_2O_5, 16 % K_2O, 0.5 % Zn, 5 % S and 0.1 % B) applied at planting at a rate of 300 kg ha^{-1} and four weeks after planting urea (46 % N) was applied as topdressing to all plots at 150 kg ha^{-1}. Neither pesticides nor fungicides were applied. Weeding was done by hand to eliminate any possible weed competition with the crop.

2.2   Disease assessment

Disease presence was evaluated based on foliar symptoms. Five random plants were tagged at the onset of infection and were checked for disease throughout the experiment. Nonetheless, plants were scored for disease severity at Zadoks' stage ZGS77 (late milking) (Eyal et al., 1987). The disease severity score was based on Saari & Prescott's (1975) scale for assessing foliar disease as cited by Eyal et al. (1987). Disease severity on leaves (Nagarajan & Kumar, 1998) of each plot was estimated by averaging the severity ratings of the tagged plants (Joshi & Chand, 2002). The severity was recorded on a 0–9 scale where 0 was scored on leaves with no symptoms while 9 on leaves having many extensive necrotic spots with pronounced chlorosis (Fetch Jr. & Steffenson, 1999). Genotypes falling in the 1–3 category were considered as resistant, 4 as moderately resistant, 5–6 as moderately susceptible and 7–9 as susceptible (Chaurasia et al., 1999).

2.3   Data analysis

A combined analysis of variance for spot blotch severity score was performed using the general linear model procedure (PROC GLM) in SAS version 9.3 (SAS Institute, 2011). The following linear statistical model for combined analysis was used (Annicchiarico, 2002):

$$Y_{ijkm} = \mu + g_i + l_j + (gl)_{ij} + y_k + r_m (l_j y_k) \\ + (gy)_{ik} + (ly)_{jk} + (gly)_{ijk} + e_{ijkr},$$

where Y_{ijkm} = observation of genotype i in location j in year k and replication m, μ = overall mean, g_i = effect of genotype i, l_j = effect of location j, y_k = effect of year k, $r_m (l_j y_k)$ effect of replication m within location j and year k, $(gy)_{ik}$ = genotype $i \times$ year k interaction, $(ly)_{jk}$ = location $j \times$ year k interaction, $(gly)_{ijk}$ = genotype $i \times$ location $l \times$ year k interaction and e_{ijkr} = residual effect.

A genotype main effect (G) plus Genotype × Environment interaction (GE) (GGE) biplot was used to visualize patterns amongst genotypes as either resistant and/or susceptible in each environment (location x year) and group of environments, to distinguish mega-environments and to explore relationships among test environments in their ranking of genotypes in relation to spot blotch (Yan & Tinker, 2006). A mega-environment refers to a group of environments that consistently share the best genotypes (Yan et al., 2007). The vertex genotype for this sector is the winning genotype for these environments. The discriminating ability of the test environment was also determined by the length of the vector. The length of the environment vector measures the discriminating ability of the test environment. Test environments with long vectors have more discriminating ability compared to those with shorter ones (Badu-Apraku

Table 1: *Mean climatic conditions for the six environments during 2012/13 and 2013/2014 season.*

Location	Environ-ment	Temperature (°C)		Mean RH[†] (%)	Rainfall (mm) Seasonal total	Number of days with rain	Latitude (South)	Longitude (East)
		Maximum	Minimum					
2012/13								
Mutanda	1	26.0	17.0	80.6	941.9	80	12°25.959′	26°12.620′
Mt.Makulu	2	27.0	17.0	76.5	868.8	57	15°32.946′	28°15.078′
GART [‡]	3	26.0	17.0	77.0	695.8	50	14°58.185′	28°06.134′
2013/14								
Mpongwe	4	28.8	20.4	87.5	1292.7	105	12°06.622′	28°09.181′
Mt. Makulu	5	28.7	17.5	79.5	931.2	67	15°32.946′	28°15.078′
GART	6	27.1	17.9	79.5	737.8	57	14°58.185′	28°06.134′

[†] RH: Relative humidity, [‡] GART: Golden Valley Agricultural Research Trust.

et al., 2013). The cosine of the angle between the vectors of two locations estimates the relationship between these (Yan & Tinker, 2006) with respect to spot blotch severity. According to Yan *et al.* (2007), an angle of < 90°(acute angle) indicates positive correlation, an angle of 90°or −90°, correlation of zero, an angle of > 90°, negative correlation, while wide obtuse angles indicates strong negative correlation. The GGE biplots were computed in Genstat version 14 computer software VSN International Ltd (Payne *et al.*, 2011). The GGE biplot analysis model equation was:

$$Y_{ij} - \mu_j = \lambda_1 \xi i_1 \eta j_1 + \lambda_2 \xi i_2 \eta j_2 + \varepsilon_{ij} \text{ (Yan, 2001)},$$

where Y_{ij} is the average yield of ith genotype in *j*th environment; μ_j is the average disease severity score across all genotypes in *j*th environment; λ_1 and λ_2 are the singular values for principal component 1 (PC1) and PC2, respectively; ξi_1 and ξi_2 are the PC1 and PC2 scores, respectively, for *i*th genotype; ηj_1 and ηj_2 are the PC1 and PC2 scores, respectively, for *j*th environment; ε_{ij} is the residual of the model associated with the *i*th genotype in *j*th environment.

3 Results

3.1 Combined analysis of variance

Highly significant differences ($P < 0.001$) were observed among genotypes (G) for their reaction to spot blotch disease (Table 2). Locations (L), years (Y), genotype (G) × location (L), genotype (G) × year (Y), L × Y, and G × L × Y were also significant ($P < 0.001$).

3.2 Reaction of the wheat genotypes to spot blotch disease across years

During 2013, the 150 genotypes screened for spot blotch disease had a mean severity score of 4.3 with the range of

Table 2: *Analysis of variance for 150 wheat genotypes for spot blotch disease severity score tested in 2013 and 2014.*

Source of variation	Degree of freedom	Mean square
Year (Y)	1	500.56 ***
Location (L)	2	303.19 ***
Y × L	2	17.13 ***
Replication (Y × L)	6	645.64
Genotype (G)	149	2.65 ***
G × Y	149	1.38 ***
G × L	298	1.69 ***
G × Y × L	298	1.54 **
Error	894	0.67
Corrected total	1799	
CV (%)	15.40	
Mean	5.32	
R^2	93.43	

***, ** indicate significance at $P < 0.001$ and $P < 0.01$, respectively

between 2.0 and 8.0. Mutanda (E1) showed a mean severity score of 3.0, Mt. Makulu (E2) of 4.5 and GART (E3) of 5.0. In 2014, the disease severity score ranged between 3.0 and 8.0 with a mean of 7.0. Mpongwe (E4) had a mean severity score of 7.3, Mt. Makulu (E5) of 7.0 and GART (E6) of 6.7. The mean disease severity score of genotypes was higher in 2014 season than in 2013 season. For example genotype number 5 (16HRWYT20) showed a mean severity score of 5 in 2013 season and 7 in 2014 season, number 12 (19HR-WSN2) had a mean score of 5 in 2013 and 7 in 2014 season, and number 25 (20HRWYT11) had scores of 4 in 2013 and 7 in 2014 (Table S1 in the Supplement). During both years, disease symptoms were first observed on the lower leaves and progressed upwards as the season advanced. The symptoms were visibly uniform on most plant parts at flowering stage.

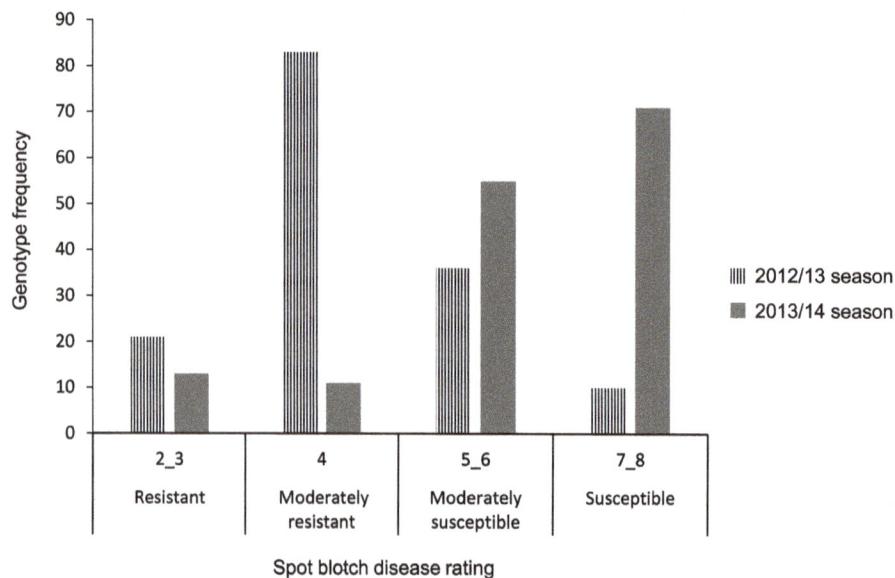

Fig. 1: *Frequency distribution for spot blotch disease severity during 2012/13 and 2013/14 seasons.*

Fig. 2: *(a) Genotype resistant to spot blotch disease. (b) Susceptible genotype.*

Based on the 0–9 scale, none of the genotypes was symptomless during both seasons. In 2012/13 season, 21 genotypes out of 150 screened were found to be resistant (R) and 83 moderately resistant (MR), 36 were moderately susceptible (MS) and 10 were susceptible (S) (Fig. 1). During 2013/14 season, 13 genotypes were found to be resistant, 11 MR, 55 MS and 71 susceptible.

The most resistant genotypes (Fig. 2a) across environments were from CIMMYT-Mexico and included 19HR-WYT6 (Kenya Heroe), 19HRWSN7 (Prontia federal) and 19HRWSN15. Some of the most susceptible genotypes (Fig. 2b) across environments were Sonalika (SB50) from CIMMYT-Mexico, and UNZAWV2, Pwele and Loerrie II from Zambia. Most of the Zambian genotypes evaluated had disease scores ranging between 5.0 and 8.0 (moderately susceptible and susceptible, respectively) across en-

vironments. No genotype from Zambia was resistant across environments.

3.3 GGE biplot analysis of environments and genotypes on spot blotch severity

The biplot (Fig. 3) explained 51.0 % (PC1 = 31.8 % and PC2 = 19.2 %) of the total genotype (G) and genotype × Environment (GE) variation. The polygon (Fig. 3) was divided by the rays into five sectors. The genotypes fell into all the sectors but the locations fell in three of them. This shows that the environments comprised of three different mega environments (I, II, and III). ME I consisted of environment 6. ME II had four environments (E) 2, 3, 4, and 5 while environment 1 appeared in mega-environment III. The vertex genotype in mega-environment I was genotype number 6 (16HRWYT5). The vertex genotypes in mega-environment

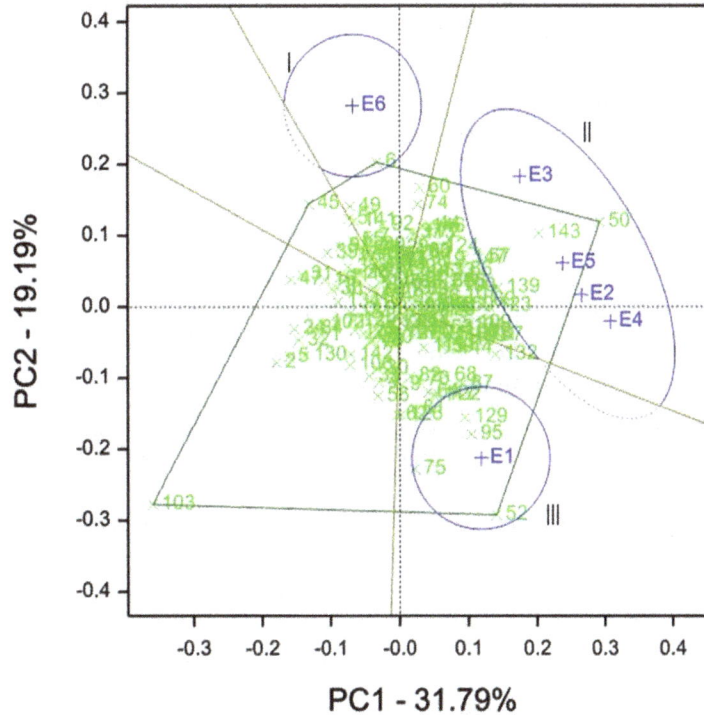

Fig. 3: *Polygon view of the total genotype (G) and genotype × environment (GE) variation (GGE) biplot based on the performance of wheat genotype with respect to spot blotch disease and also showing the mega-environments in relation to the disease. Genotypes are labelled 1 to 150. Mega-environments are labelled I, II, and III. Details for genotypes are given in Table S1. Details for environments are given in Table 1.*

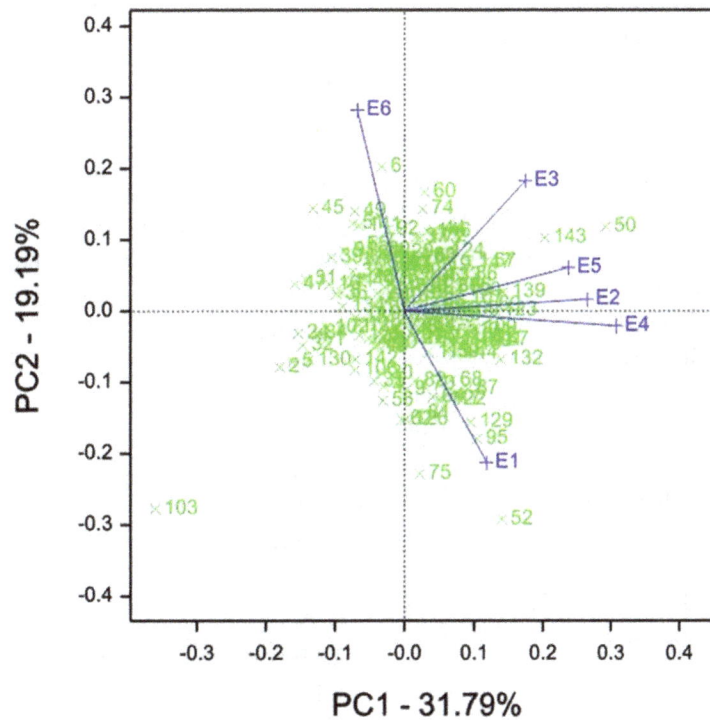

Fig. 4: *Total genotype (G) and genotype × environment (GE) variation (GGE) biplot showing relationships among test environments in discriminating genotypes in relation to spot blotch disease. Environments are labelled E1 to E6. Details for environments are given in Table 1.*

II and III were genotypes number 50 (Sonalika) and 52 (20HRWYT3), respectively. Genotype number 103 (19HR-WSN7) and 45 (20HRWYT30) were the vertex genotypes in a sector where there was no environment. However, genotype number 103 was located very far away from the test locations.

In this study, all environments except E6 had positive PC1 scores. Environment 6 had a negative PC1 but close to the origin. Environments 2, 4, and 5 had positive PC2 values close to zero. Environments 6 and 3 had large positive PC2 values while environment 1 had negative PC2 scores (Fig. 4). The angle between E2, E3, E4 and E5 was less than 90°. The largest angle (> 90°) was between E6 and E1 followed by the angle between E4 and E6. With respect to vector length from the origin of the biplot, E4 had the longest vector. This was followed by E6, E1, E3, E2, and E5.

4   Discussion

Highly significant differences observed among genotypes in their reaction to spot blotch disease indicated that genetic variability existed in the material under study which provides an opportunity for further genetic improvement. The significance of years, locations, genotype × location interaction (G × L) suggests that genotypes responded differently to locations and years. Significant genotype (G) × year (Y), G × L × Y interactions indicate that the performance of genotypes was inconstant over years (Gomez & Gomez, 1984). Therefore, screening of genotypes over locations and years is worthwhile to identify genotypes with stable resistance to spot blotch disease.

Genotypes 19HRWSN6 (Kenya Heroe), 19HRWSN7 (Prontia federal) and 19HRWSN15 were found to be resistant across seasons and sites, and therefore could be utilised in wheat breeding programme to improve resistance to spot blotch. High disease severity in 2013/14 season compared to 2012/13 could be attributed to highly conducive climatic conditions such as favourable temperatures, leaves remaining wet for quite a long period of time due to frequent rainfall and dew which favoured sporulation, multiplication and spread of the disease. The results are in line with the work done by several scientists who reported a close association between weather conditions and spot blotch disease severity (Kumar et al., 2002; Sharma & Duveiller, 2007; Duveiller et al., 2007; Acharya et al., 2011).

Genotype 6 (16HRWYT5) was the most susceptible in ME I (E6) as it is located at the vertex of the polygon. Genotype 50 (Sonalika) was the most susceptible genotype in mega-environment II (E2, E3, E4 and E5) followed by genotype 143, whereas genotype 52 was the most suscep-

tible in mega-environment III (E1). The grouping of these genotypes in separate mega-environments was very consistent with their mean performance to spot blotch disease in the aforementioned environments. Genotype 103 (19HR-WSN7) exhibited high levels of resistance to spot blotch disease across all test environments as it fell in a sector without any environment (Yan et al., 2001). Genotypes on the vertex of the polygon in each sector are either the best or worst performing as they are further from the biplot origin (Yan & Tinker, 2006). Additionally, they are the most responsive compared to those located within the polygon (Adu et al., 2013). However, those within the polygon but close to the origin, show average reaction across all environments (Yan & Falk, 2002). In this case, genotype 12, 58, 120 and 134 were some examples of genotypes that showed average reaction to spot blotch severity across all locations. Hence, therefore, GGE biplot analysis is an important tool for visualizing patterns amongst genotypes as either resistant and/or susceptible in each environment and group of environments.

In terms of environmental correlations, environments within ME II were highly correlated in their ranking of genotypes as indicated by the angle between them which was less than 90° (Yan et al., 2007). This indicates that similar information about genotypes was obtained from this mega-environment, suggesting that one location within this mega-environment could be chosen for genotype evaluation in each year if the pattern repeats across years (Yan et al., 2007). This would help to reduce on the cost of evaluating genotypes and improving efficiency of screening for resistance. The angle between environments E6 and E1, and between E6 and E4 was quite large showing that the environments were not correlated.

In terms of location versus season relationships, Mt. Makulu locations (2 and 5) in both seasons fell in one sector suggesting repeatable performance of genotypes in this location. Repeatability is very essential for assessing a test location that is representative of all test locations over years (Badu-Apraku et al., 2013). Thus, a location is considered highly representative if its genotypic rankings are repeated across years, so that genotypes selected in one year will have greater performance in forthcoming years (Yan et al., 2011). GART environments (E3 and E6) fell in different sectors both years, suggesting that there was no repeatability of genotypes in this location.

All locations except environment 6 had positive PC1 scores, an indication that they were discriminating genotypes. However, environments Mutanda (E1), Mpongwe (E4) and GART (E6) were considered highly discriminating among genotypes as shown by the length of their vectors from the biplot origin. The length of a vector of a test environment estimates the discriminating ability of genotypes (Badu-Apraku *et al.*, 2013). The longer the vector the higher the ability to discriminate genotypes and the shorter the vector the lesser the discriminating ability (Yan & Tinker, 2006). Mt. Makulu environments (E2 and E5) had short vectors indicating that they had the least discriminating ability of genotypes. Yan *et al.* (2010) indicated that environments with shorter vectors could be considered as independent test environments, treated as unique and essential test environment.

The GGE biplot showed that environments 1 (Mutanda), 3 (GART2012/13) and 6 (GART2013/14) contributed most of the genotype by environment interaction (GEI) variability in terms of genotype reaction to spot blotch disease as these were located further apart in the biplot (Joshi *et al.*, 2007). This implies that a genotype could have huge positive interaction with some environments while having large negative interactions with some other environments (Yan & Hunt, 2001). The GEI could affect the efficiency of breeding for resistance. Pinnschmidt & Hovmøller (2002) reported that GEI affects breeding for high levels of resistance due to inconsistency in the phenotypic expression of the disease. Moreover, it complicates selection of desirable genotypes (Farshadfar *et al.*, 2012).

In conclusion, genetic variation existed among genotypes to the reaction of spot blotch disease. Most of the resistant and moderately resistant genotypes were identified among CIMMYT-Mexico lines. Some of the resistant genotypes identified across locations included 19HRWSN6, 19HRWSN7 and 19HRWSN15. These resistant genotypes could be used as valuable source in breeding for resistance to the disease. The GGE biplot analysis identified genotype 19HRWSN7 as the most resistant across all test environments. Therefore, GGE biplot analysis could efficiently be used to identify genotypes resistant to spot blotch disease over locations.

Acknowledgements

We thank Alliance for Green Revolution in Africa (AGRA) for funding this research.

References

Acharya, K., Dutta, A. K. & Pradhan, P. (2011). *Bipolaris sorokiniana* (Sacc.) Shoem: The most destructive wheat fungal pathogen in the warmer areas. *Australian Journal of Crop Science*, 5, 1064–1071.

Adu, G. B., Akromah, R., Abdulai, M. S., Obeng-Antwi, K., Tengan, K. M. L. & Alidu, H. (2013). Assessment of genotype x environment interactions and grain yield performance of extra-early maize (*Zea mays* L.) hybrids. *Journal of Biology, Agriculture and Healthcare*, 3 (12), 7–15.

Alam, K. B., Shaheed, M. A., Ahmed, A. U. & Malakar, P. K. (1994). Bipolaris leaf blight (spot blotch) of wheat in Bangladesh. *In:* A., S. D. & Hettel, G. P. (eds.), *Wheat in Heat-Stressed Environments: Irrigated, Dry Areas and Rice-Wheat Farming Systems*. pp. 339–342, CIMMYT, Mexico D.F.

Annicchiarico, P. (2002). *Genotype × environment interactions – challenges and opportunities for plant breeding and cultivar recommendations.* FAO Plant Production and Protection Paper 174, Food and Agriculture Organization of the United Nations, Rome, Italy. Pp. 132

Badu-Apraku, B., Akinwale, R. O., Obeng-Antwi, K., Haruna, A., Kanton, R., Usman, I. S., Ado, G., Coulibaly, N., Yallou, G. C. & Oyekunle, M. (2013). Assessing the representativeness and repeatability of testing sites for drought-tolerant maize in West Africa. *Canadian Journal of Plant Science*, 93 (4), 699–714.

Chaurasia, S., Joshi, A., Dhari, R. & Chand, R. (1999). Resistance to foliar blight of wheat: a search. *Genetic Resources and Crop Evolution*, 46, 469–475.

Duveiller, E. & Gilchrist, L. (1994). Production constraints due to *Bipolaris sorokiniana* in wheat: current situation and future prospects. *In:* Saunders, D. A. & Hettel, G. P. (eds.), *Wheat in heat-stressed environments: irrigated, dry areas and rice-wheat farming systems*. pp. 343–352, CIMMYT, Mexico, D. F.

Duveiller, E. & Sharma, R. (2009). Genetic improvement and crop management strategies to minimize yield losses in warm non-traditional wheat growing areas due to spot blotch pathogen *Cochliobolus sativus*. *Journal of Phytopathology*, 157, 521–534.

Duveiller, E. & Sharma, R. C. (2013). Wheat Resistance to Spot Blotch or Foliar Blight. *In:* Sharma, I. (ed.), *Disease resistance in wheat*. pp. 160–189, CABI Plant Protection Series.

Duveiller, E., Sharma, R. C., Çukadar, B. & van Ginkel, M. (2007). Genetic analysis of field resistance to tan spot in spring wheat. *Field Crops Research*, 101, 62–67.

Eisa, M., Chand, R. & Joshi, A. K. (2013). Biochemical and histochemical parameters associated with slow blighting of spot blotch (*Bipolaris Sorokiniana* (Sacc.) Shoem.) in wheat (*Triticum* spp.). *Zemdirbyste-Agriculture*, 100, 191–198.

Eyal, Z., Scharen, A. L., Prescott, J. M. & van Ginkel, M. (1987). *The Septoria diseases of wheat: Concepts and methods of disease management.* CIMMYT, Mexico, D.F. (pp. 18–28)

Farshadfar, E., Mohammadi, R., Aghaee, M. & Vaisi, Z. (2012). GGE biplot analysis of genotype × environment interaction in wheat-barley disomic addition lines. *Australian Journal of Crop Science*, 6, 1074–1079.

Fetch Jr, T. G. & Steffenson, B. J. (1999). Rating scales for assessing infection responses of barley infected with *Cochliobolus sativus*. *Plant Disease*, 83, 213–217.

Gomez, K. A. & Gomez, A. A. (1984). *Statistical Procedures for Agricultural Research*. (2nd ed.). John Wiley and Sons, Inc, New York.

Gubiš, J., Masár, S. & Gubišová, M. (2010). Resistance of spring barley genotypes to *Bipolaris sorokiniana*. *Agriculture (Pol'nohospodárstvo)*, 56, 3–8.

Iftikhar, S., Asad, S. & Rattu, A. (2009). Selection of barley germplasm resistant to spot blotch. *Pakistan Journal of Botany*, 41, 309–314.

Iftikhar, S., Asad, S., Rattu, A., Munir, A. & Fayyaz, M. (2012). Screening of commercial wheat varieties to spot blotch under controlled and field conditions. *Pakistan Journal of Botany*, 44, 361–363.

Joshi, A., Kumar, S., Chand, R. & Ortiz-Ferrara, G. (2004). Inheritance of resistance to spot blotch caused by *Bipolaris sorokiniana* in spring wheat. *Plant Breeding*, 123, 213–219.

Joshi, A. K. & Chand, R. (2002). Variation and inheritance of leaf angle, and its association with spot blotch (*Bipolaris sorokiniana*) severity in wheat (*Triticum aestivum* L.). *Euphytica*, 124, 283–291.

Joshi, A. K., Ortiz-Ferrara, G., Crossa, J., Singh, G., Sharma, R. C., Chand, R. & Parsad, R. (2007). Combining superior agronomic performance and terminal heat tolerance with resistance to spot blotch (*Bipolaris sorokiniana*) of wheat in the warm humid Gangetic Plains of South Asia. *Field Crops Research*, 103, 53–61.

Khan, H. & Chowdhury, S. (2011). Identification of resistance source in wheat germplasm against spot blotch disease caused by *Bipolaris sorokiniana*. *Archives of Phytopatthology and Plant Protection*, 44, 840–844.

Kumar, J., Schäfer, P., Hückelhoven, R., Langen, G., Baltruschat, H., Stein, E., Nagarajan, S. & Kogel, K.-H. (2002). *Bipolaris sorokiniana*, a cereal pathogen of global concern: cytological and molecular approaches towards better control. *Molecular Plant Pathology*, 3, 185–195.

Mehta, Y. (1997). Constraints on the integrated management of spot blotch of wheat. *In:* Duveiller, E., Dubin, H. J., Reeves, J. & McNab, A. (eds.), *Proceedings of an International workshop, Helminthosporium blight of wheat: Spot Blotch and Tan Spot*. pp. 18–27, CIMMYT, El Batan, Mexico.

Mikhailova, L., Lianfa, S., Gogoleva, S. G. & Gultyaeva, E. I. (2004). Sources and donors of resistance to wheat spot blotch caused by *Bipolaris sorokiniana* Shoem. (*Cochliobolus sativus* Drechs. ex Dastur). *Archives of Phytopathology and Plant Protection*, 37, 161–167.

Mukwavi, M., Mooleki, S. P. & Gilliland, D. (1990). Trends, major problems and potential of wheat production in Zambia. *In:* Saunders, D. A. (ed.), *Wheat for the non-traditional warm areas*. pp. 34–43, UNDP/CIMMYT, Mexico, D. F.

Nagarajan, S. & Kumar, J. (1998). An overview of the increasing importance of research of foliar blights of wheat in India: germplasm improvement and future challenges towards a sustainable high yielding wheat production. *In:* Duveiller, E., Dubin, E., Reeves, H. J. & McNab, A. (eds.), *Proceedings of the Helminthosporium Blights of Wheat: Spot blotch and tan spot Workshop, 1997*. pp. 52–58, CIMMYT, El Batan, Mexico D.F., Mexico.

Payne, R. W., Murray, D. A., Harding, S. A., Baird, D. B. & Soutar, D. M. (2011). *GenStat for Windows (14th Edition) Introduction*. VSN International Ltd, Hemel Hempstead.

Pinnschmidt, H. O. & Hovmøller, M. S. (2002). Genotype × environment interactions in the expression of net blotch resistance in spring and winter barley varieties. *Euphytica*, 125, 227–243.

Raemaekers, R. H. (1988). *Helminthosporium sativum*: Disease complex on wheat and sources of resistance in Zambia. *In:* Wheat production constraints in tropical environments. pp. 175–185, CIMMYT, Mexico, D.F.

Rakotondramanana (1981). Problems and progress in disease resistance and breeding – Madagascar. *In:* Toogood, J. A. (ed.), *Proceedings of wheat workshop. South, Central and East Africa in conjunction with the inaugural biennial meeting of the Agricultural Association of Zambia*. pp. 36–37, Mount Makulu Research Station, Chilanga.

Saari, E. E. & Prescott, L. M. (1975). A scale for appraising the foliar intensity of wheat diseases. *Plant Disease Reporter*, 59, 377–380.

SAS Institute (2011). *The SAS sytem for Winodws, version 9.3*. SAS Insittute, Cary, North Carolina, USA.

Sharma, R. & Duveiller, E. (2007). Advancement toward new spot blotch resistant wheats in South Asia. *Crop Science*, 47, 961–968.

Srivastava, K. D. & Tewari, A. K. (2002). Fungal diseases of wheat and barley. *In:* Gupta, V. K. & Paul, Y. S. (eds.), *Diseases of field crops*. pp. 58–68, M.L. Gidwani, Indus Publishing, New Dehli.

Tembo, B., Sibiya, J., Tongoona, P. & Mukanga, M. (2016). Farmers' perceptions of rain-fed wheat production constraints, varietal preferences and their implication to rainfed wheat breeding in Zambia. *Journal of Agriculture and Crops*, 2 (12), 131–139.

Yan, W. (2001). GGE biplot: A Windows application for graphical analysis of multi-environment trial data and other types of two-way data. *Agronomy Journal*, 93, 1111–1118.

Yan, W., Cornelius, P. I., Crossa, J. & Hunt, L. A. (2001). Two types of GGE biplots for analyzing multi-environment trial data. *Crop Science*, 41, 656–663.

Yan, W. & Falk, D. E. (2002). Biplot analysis of host-by-pathogen data. *Plant Diseases*, 86, 1396–1401.

Yan, W., Fregeau-Reid, J., Pageau, D., Martin, R., Mitchell-Fetch, J., Etienne, M., Rowsell, J., Scott, P., Price, M., De Haan, B., Cummiskey, A., Lajeunesse, J., Durand, J. & Sparry, E. (2010). Identifying essential test locations for oat breeding in Eastern Canada. *Crop Science*, 50, 505–515.

Yan, W. & Hunt, L. A. (2001). Interpretation of genotype × environment interaction for winter wheat yield in Ontario. *Crop Science*, 41, 19–25.

Yan, W., Kang, M. S., Ma, B. L., Woods, S. & Cornelius, P. L. (2007). GGE biplot vs. AMMI analysis of genotype-by-environment data. *Crop Science*, 47, 643–655.

Yan, W., Pageau, D., Fregeau-Reid, J. & Durand, J. (2011). Assessing the representativeness and repeatability of test locations for genotype evaluation. *Crop Science*, 51, 1603–1610.

Yan, W. & Tinker, N. A. (2006). Biplot analysis of multi-environment trial data: Principles and applications. *Canadian Journal of Plant Science*, 86, 623–645.

4

The role of trust and networks in developing Nicaraguan farmers' agribusiness capacities

Dirk Hauke Landmann [a,b,*], Jean-Joseph Cadilhon [a,**]

[a] Policy, Trade and Value Chains Program, International Livestock Research Institute, Nairobi, Kenya
[b] Department of Agricultural Economics and Rural Development, Georg-August-Universität Göttingen, Göttingen, Germany

Abstract

The main focus of most programmes in developing countries carried out by NGOs is to develop small-scale farmers' capacities. One approach hereby is to use multi-stakeholder innovation systems, such as the 'Nicaraguan Learning Alliance' (NLA). However, tools for the evaluation of multi-stakeholder innovation systems are rare. This paper reports on the implementation of a conceptual framework to carry out an impact evaluation of multi-stakeholder innovation systems using the NLA as the object of study. The assessment focused on the business relationship constructs of trust and capacity development. Survey interviews, in-depth interviews and focus group discussions collected data from agribusiness stakeholders linked with the NLA and from a control group of stakeholders involved with other networks. The quantitative data were analysed through factor and regression analyses. Results from the quantitative analyses were triangulated with qualitative data. The analysis shows that the NLA has been successful in developing smallholder farmers' capacities as a result of trust developed through its dedicated project managers. Nonetheless, the NLA has not been more successful at developing agribusiness capacities among Nicaraguan farmers than other networks with the same goals. Results from this study point to the need for facilitating more interactions between the different networks of farmers' cooperatives and organisations with other stakeholders already active within the Nicaraguan agrifood innovation system.

Keywords: capacity development, impact evaluation, innovation systems, Latin America, value chains

1 Introduction

Traditionally, researchers and experts transferred their agricultural knowledge to their target group following a linear approach. This model has largely failed because it did not respond to the actual problems of its intended beneficiaries, it rather evaluated the knowledge of locals as inferior and did not take into account how the different stakeholders influence production one way or another (Chambers, 1994; Pretty, 1995; Klerkx *et al.*, 2012).

---

* Corresponding author – dirk.landmann@agr.uni-goettingen.de
Current affiliation: Department of Agricultural Economics and Rural Development, Faculty of Agriculture, Georg-August-Universität Göttingen, Göttingen, Germany

** Current affiliation: Head of Service, Agricultural Productions and Economics, Directorate for Territories and the Sea of Pyrénées-Atlantiques, Pau, France

This shortcoming formed the basis for 'model two' also known as 'Participatory-Research-Action' (PRA), in which more interactions between the different stakeholders occur and changes can be adopted more rapidly (Hall, 2007). This new form of capacity development could be defined generally as an approach focusing on organisational, communal and social issues. The core of this approach is to combine theory and action in the process of collaborative learning. Reflection is taking place throughout the whole process and is also prompting the next actions (Coghlan & Brydon-Miller, 2014).

The International Center for Tropical Agriculture (CIAT) has developed Learning Alliances (LA) based on this knowledge (World Bank, 2012). LAs can be assimilated to innovation platforms (IPs): a group of individuals with different backgrounds and interests, who come together to diagnose and solve problems they face (Homann-Kee Tui *et al.*, 2013). The LA concept has so far been adopted in 20 countries worldwide (Lundy & Gottret, 2005). Innovation systems as applied to agriculture rely on the interaction of different stakeholders to foster knowledge sharing and innovation (Klerkx *et al.*, 2010). The general idea is to add value and create synergistic relationships between different members and to build up a network that transcends micro, meso and macro socio-geographical levels (CRS, 2009). The Nicaraguan Learning Alliance (NLA) is an alliance of organisations formed in 2003. The alliance is training its partners on agribusiness management and access to markets to replicate this knowledge through different geographical levels along its partner network down to farmers (AdA, 2014; World Bank, 2012).

Although innovation platforms are seen as a successful tool and used in many different countries and value chains, literature on the assessment of innovation platforms is very rare. Existing literature mostly focuses on the analysis of particular cases with a specific method, thus restricting the transfer to other platforms (Nederlof *et al.*, 2011). The conceptual framework developed by Cadilhon (2013) attempts to simplify complex data within the categories of structure, conduct, and performance. The conceptual framework already embeds certain variables, factors, and other influences relevant to the development and aims of innovation platforms. This is the only conceptual framework that combines the different categories (structure, conduct, and performance) with the topics of transaction costs and marketing concepts for the purpose of analysing innovation platforms. The data of this study will help test and refine the conceptual framework for monitoring and evaluation of the impact of innovation platforms (Cadilhon, 2013, p. 2).

The focus of this study was to evaluate trust as a conduct variable and capacity development as a performance variable following the alliance's objectives. Trust is an important component in value chains and has gained more attention from scientists within the past two decades. This important component can be seen as a factor, which significantly influences individuals, organisations, partner's competence, process, characteristics and institutions, systems, calculations, economics, intentional relations, technology or services. Thus, it is described by many researchers as a complicated and multifaceted concept with no uniform definition and measurement method available up to now. However, trust has a great influence on perception and individuality, which varies with participants (Laeequddin *et al.*, 2010, pp. 53, 56).

Capacity development has also been discussed extensively in the last few decades (Watson 2010, p. 241). It is dependent on principles, dimensions, actors, levels and strategies, and each case has to be seen as a combination of different influencing factors (Neely, 2010, pp. 13–16). In an agricultural context, it often takes the form of training activities and workshops (Horton *et al.*, 2003, p. 2). Capacity development is a principal goal of the NLA to increase the replication efficiency of the knowledge being produced within the network.

However, there are no studies or data comparing NLA participants and non-participants pertaining to the cooperatives and organisations using the NLA-guides. Furthermore, there are no measures to evaluate whether the capacity of the partners is generally increasing or not. To contribute to the evaluation of the NLA's activities, this article aims to answer the following research question: How does the NLA strengthen producers' capacities through its structure and network of members and partners?

The contribution of this article to the literature on IPs as a collaborative agricultural education mechanism is the pilot-testing of a conceptual framework characterizing how IPs work in a Latin American context through mixed research methods. Although LAs and IPs are more and more common, this study aims to contribute practical tools to evaluate them using quantitative and mixed methods.

2 Literature review

2.1 Learning Alliances in the agricultural sector

The LA approach is based on the concept of 'social learning' and 'innovation systems'. Social learning is defined as an interactive process of learning-by-doing between different stakeholders for the purpose of solving problems (Bandura, 1971). LAs specifically focus on research organisations, as well as donor and development agencies. Combining these two concepts creates a process of collaborative learning, adaption, and innovation among the participants. The objectives of LAs are to develop cumulative and shared knowledge about distinct approaches, learn across different boundaries, create synergies among participants (e.g. to advance specialised knowledge), exchange information between the participants, and develop flexible mechanisms that apply to different topics (CRS, 2009). It is typical for LAs to mix traditional socio-economic research with action research. The founding principles of LAs include clear objectives, shared responsibilities, costs and benefits, outputs and inputs, differentiated learning mechanisms, and long-term trust-based relationships. Every participant will have different objectives and interests, but it is crucial that common ground can be identified. A more general objective enables participation by a wider array of members. Benefits for each stakeholder must exceed the value of their individual costs. In addition, the goals and interests of the alliance should not be in conflict with other key actions. Methods, tools, and approaches should change over time corresponding to changing situations of participants. All types of participants must be considered and respected regardless of gender, race, function, and other differentiating factors. In order to accomplish this, learning methods need to be flexible, interconnected, and viewed as long-term processes (CRS, 2009).

The main approach in the methodology of LAs is to move from a single cycle learning process to a double loop learning process. One cycle is divided into three segments. In the first step 'Reviewing our framework', problems are identified, learning topics are selected and defined, existing practices are analysed, and methods and tools are designed for adoption. The second step 'Implementing strategic actions', involves planning and implementing the approaches, methods and tools of development projects. In the third step, 'Documenting and analysing results' intervention results are systemized and evaluated before the changes in the state of development are presented to the members through workshops, training programs, platforms or other methods. After the completion of this cycle, the process starts again with the first step. This second time, the results from the first cycle are taken into account (Lundy & Gottret, 2005).

CIAT's experiences with LAs have been very positive since they were first initiated in the year 2000. Positive aspects are that stakeholders participate directly, pilot innovation occurs where help is needed, face to face information exchange occurs, and analyses throughout the entire experience help evaluate the alliance including its processes (Lundy & Gottret, 2005). However, LAs do not work for every project. One reason is member composition. Members have to be open to share information and reflect in order to enable the learning. This can be influenced by clusters or different methods of communication. Establishing an LA takes a considerable amount of time (CRS, 2009). The initiators must invest sufficient time in managing and coordinating the alliance as well as documenting, analysing, and sharing the information and results on every level. Though time commitments may be substantial,, they are crucial elements of the process. Additionally, providing funding becomes essential. It is easier to receive funding for specific projects than for projects with a wider scope. It is also vital to consider who is funding the project and to examine their motives and interests (Lundy & Gottret, 2005).

2.2 Background information on Nicaragua and the NLA

Although small in size, Nicaragua has a varied tropical landscape with fertile volcanic soil on agricultural plains, dry rangeland plateaus and hills, and humid evergreen agroforested mountains. Agriculture is an important economic sector for the country, representing 22 % of the national GDP, 32 % of national exports and 32 % of employment (FAOSTAT, 2014). However, Nicaragua is one of the poorest countries in Latin America (World Bank, 2017).

Nicaraguan farmers are organised in a dense network of cooperatives, a heritage of the former socialist Sandinista regime (Lafortezza & Consorzio, 2009). Due to the current government's connection to the previous regime, it embraces this socialist heritage and may continue to influence the structure of Nicaraguan agriculture. At present, many farmers are not well equipped to link themselves to suppliers and customers in today's market-oriented system. Agribusiness training could thus make a big difference in empowering farmers and their cooperatives to become better managers of their enterprises and livelihoods (Landmann & Cadilhon, 2016).

Having identified this training opportunity as a good long-term strategy to help rural farming communities link to markets, a partnership of ten international and local research organisations, non-government organisations, and one national-level farmers' cooperative[1] launched the NLA in 2003 (Lundy & Gottret, 2005). The objective of the NLA is farmer training on agribusiness. Much of the development of these activities and trainings in the first years were funded by aid money channelled through the international partners in the NLA. To achieve this, the NLA members first consulted each other to identify training topics and develop appropriate training methods. Based on this information, the NLA has established five training guides containing the skills and capacities farmers needed to improve[2]. These guides use methodologies designed to target Latin American farmers' cooperatives and rely on the participation of trainees in building their own understanding of the topic. The trainees first auto-evaluate the training process, before their results can be compared to training beneficiaries at different

---

[1] CATIE (Center for Tropical Agricultural Research and Education); CIAT (International Center for Tropical Agriculture); CRS (Catholic Relief Service); FUNICA (Foundation for Technological Development of Agriculture and Forestry of Nicaragua); VECO Mesoamerica (VredesEilanden Country Office Central America); GIZ (German Agency for International Cooperation); LWR (Lutheran World Relief); FENACOOP R.L. (National Federation of Agricultural Cooperatives and Agribusiness)

[2] Guide 1: Self-evaluation provided for the management of rural associative enterprises; Guide 2: Strengthening socio-organizational processes; Guide 3: Strategic orientation with a focus on value chain; Guide 4: Development of business plans; Guide 5: Strengthening of services. These are all available in Spanish from http://www.alianzasdeaprendizaje.org/metodologia (accessed 15 January 2017)

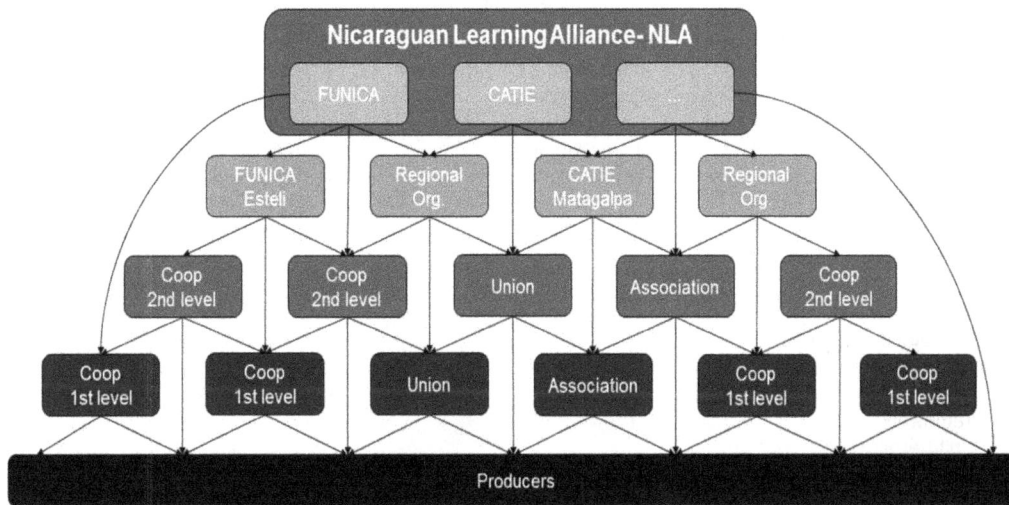

Fig. 1: *Structure of knowledge replication within the Nicaraguan Learning Alliance.*
Source: Landmann & Cadilhon (2016)

sites. Training methods and topics are adaptable to the local context of each farmer's cooperative. The process and topics of the training also promote equity across gender and social groups. Finally, the training process encourages individual and collective empowerment to engage in entrepreneurial activities. The NLA's novel idea was to use the existing network of agricultural cooperatives to snowball the training to individual farmer households. The NLA members trained regional-level cooperatives, which used the same methods to train village-level cooperatives, which in turn used them to train their individual farmer members (Fig. 1). Importantly, the NLA members assigned the training activities to one clearly identified project manager, who became the physical link between alliance members and the beneficiary cooperatives, thus creating a trusting relationship with the network of cooperatives.

Three learning cycles included 77 producer organisations and reached a total of 19,350 farming families involved in the production of various crops. Women represented 30 % of the trained farmers (AdA-Nicaragua, 2012; Landmann & Cadilhon, 2016). Although the process was at first subsidised by international partners, later on one of the NLA-members used the guides developed together with the other members independently.

3 Conceptual framework and research design

3.1 Conceptual framework to outline the analysis

Mariami *et al.* (2015) describe a conceptual framework evaluating the impact of IPs. The authors build their

framework based on three strands of literature from socio-economic theory.

First, the authors use the Structure–Conduct–Performance (SCP) model as a general outline for the study of multi-stakeholder groups such as IPs. Although it has been criticised for its use as a tool to understand the functioning of real-life markets, the authors noted the SCP model's elegant overarching logic: the structure of IPs can impact on its stakeholders' conduct, and in turn on the performance of the platform measured by reaching its objectives.

Mariami *et al.* (2015) incorporate elements from New Institutional Economics (NIE) to complement the overall SCP logic. Indeed, the NIE literature takes into account the uncertainty endemic within the food industry: technical and economic characteristics of the products due to agricultural production seasonality, weather instability, and food market cycles (Furubotn & Richter, 2010). NIE's focus on transaction costs, the organisation and development of economic activity pose as a perfect complement to the SCP model in trying to understand how IPs work to reach their objectives.

Mariami *et al.* (2015) then suggest going further into the characterisation of the way IP conduct and performance are measured by using concepts and constructs from the marketing management literature. Endorsing transaction cost economics, this strand of research studies in great detail how organisations reach more satisfactory marketing relationships by developing information sharing (Sanzo *et al.*, 2003), communication (Kumar, 1996), cooperation, coordination and joint planning (Anderson & Narus, 1984; Claro *et al.*, 2003), and trust (Kumar, 1996; Trienekens,

Structure	Conduct	Performance
IP- Structure • Membership composition and diversity • Decision making process • Committees • Source of funding • Staff availability **Individual structure** • Type of chain stakeholder • Gender • Level of education • Indicator of wealth **External environment** • Legal and regulatory framework • Cultural norms • ...	**Conduct of IP members** • Information sharing • Communication • Coordination • Joint Planning • Trust	**Value chain Performance** • Advocacy • Collective promotion • Joint quality standards • Research & development • Capacity development • Market information • Arbitration of chain conflict • Limiting transaction costs • Setting concerted marketing objectives **Other objectives set by IP** • ...

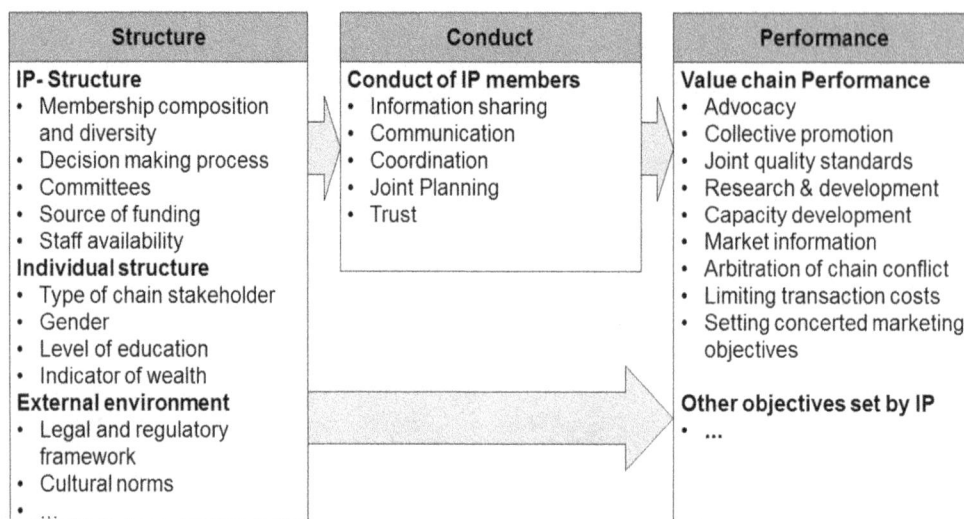

Fig. 2: *Elements of the conceptual framework to evaluate innovation platforms (IP).*
Source: Landmann & Cadilhon (2016)

2011). Though IPs are not generally a means to organise the market transactions between its members, in the case of the NLA, the alliance does help its members improve the marketing orientation of their production and planning activities. Therefore, it is still relevant to use the conceptual framework proposed by Mariami *et al.* (2015) to structure this analysis. The authors combine the three complementary theories of the SCP model, NIE, and marketing relationships management into an overarching conceptual framework to understand how IPs work and to help evaluate their impact. Figure 2 shows how the conceptual framework proposed by these authors has been adapted from the original to fit the specific context of the NLA.

Trust has already been identified as an important component of business relationships in agricultural value chains of developing countries (Trienekens, 2011). Kumar (1996) defines trust as the belief that each party in a marketing relationship is interested in the other's welfare; neither will take action without first considering its impact on the other party. Many researchers describe trust as a multifaceted concept dependent in each case on the local context. Thus, trust can be seen as a factor related to competence, process, characteristics, institutions, systems, services, and even technology (a piece of equipment). Trust can be observed in the decision and actions of participants. From a business perspective, trust is an expected outcome of a certain event or action (Laeequddin *et al.*, 2010). Applying the concept of trust as found in the business relationship literature to the context of IPs seems particularly relevant as the different stakeholders found in an IP also have to develop trust between each other to reach common objectives. At the same time, these value chain stakeholders can also be competitors or dependent on one another.

Capacity development is defined in many ways (Ubels *et al.*, 2010). The United Nations define capacity development as a 'process through which individuals [...] obtain, strengthen and maintain the capabilities to set and achieve their own development objectives over time' (UNDP, 2008, p.4). This definition includes social, political, and technical aspects. Bolger's (2000) review concluded that 'capacity development refers to approaches, strategies and methodologies used to improve performance at the individual, organisational, network/sector or broader system level'.

Farmers' capacities are mostly developed by training activities and workshops given or provided by different actors (Hall, 2007; Horton *et al.*, 2003, pp. 2–6). In this respect, capacity development describes both the process and the outcome of these activities, whereby the outcome is defined as changes in working processes and the introduction of new production methods (Hall, 2005, p. 612; Lusthaus *et al.*, 1999, pp. 15–19).

3.2 Data collection

Data were collected in Nicaragua from NLA members, their influenced partners, non-members of the NLA and their influenced partners, as well as from agribusiness private companies and universities. The data collection took place from July to September 2014 in Managua, and in the provinces of Matagalpa, Jinotega, Estelí, Madriz, Nueva Segovia, Masaya, and Chinandega (Fig. 3). These provinces were chosen, since most NLA members are working in these regions (e.g. CATIE in Matagalpa, CRS and FUNICA in Estelí and Jinotega).

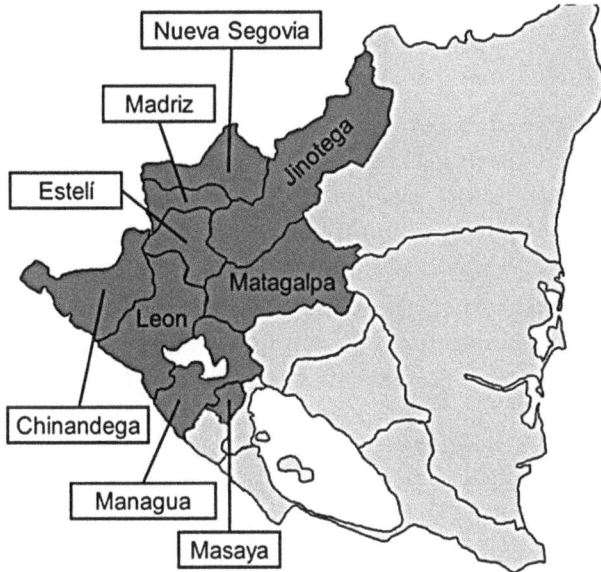

Fig. 3: *Map of Nicaragua identifying provinces where data collection took place. Source: Landmann & Cadilhon (2016)*

To collect qualitative data, key informant interviews and focus group discussions were held. Quantitative data were gathered through individual interviews using a structured questionnaire. Key informants (Table 1) were also interviewed with the aim of gaining a more profound understanding through a less structured conversation. The key informants in the different regions were chosen based on the different types of stakeholders involved in agricultural development: governmental organisations, non-governmental organisations, research institutes, universities, and private companies. The interviews conducted used a guide based on the individual survey and the conceptual framework.

Focus group discussions were held in the regions where the NLA is active. The characteristics of the villages and the composition of the members involved in the group discussions had to be similar to those of the villages representing the area of study. Focus group discussants and key informants were not included in the sample of individual questionnaires. The locations for the focus groups were randomly selected from the different regions in our study. All interview questionnaires and guidelines are accessible on the internet (Cadilhon & Landmann, 2015).

Focus group discussions followed the approaches of asking specific questions about definitions and background information, while also observing the direction taken by the focus group discussion. Focus group participants were members of first-level cooperatives chosen according to their membership and partners in the NLA network and their location. Three focus group discussions with different groups of producers and key informant interviews were held at the beginning of fieldwork as pre-tests of the individual

questionnaire. Results were considered before finalizing the questionnaire.

The individual questionnaire collected structural data about the organisation interviewed and used 53 Likert-scale statements to quantify the levels of conduct and performance. The respondents expressed their agreement with the statements proposed by the researcher through the Likert scale (Coding from 1 = 'strongly disagree' up to 5 = 'strongly agree' as well as N/A = not applicable).

In each location, preliminary individual interviews were held with managers of farmers' groups involved in NLA capacity development. To constitute the control group of farmers' organisations not involved in the NLA network, contact was made with other cooperatives and organisations with a similar structure in the same region. These organisations were identified by asking for references from the respondents within the NLA network, as well as by randomly interviewing numerous farmer organisations in the region where the fieldwork was undertaken. The resulting sample of interviews in all regions was double-checked against lists of all farmer organisations active in these areas. In total, 38 NLA-members or influenced partners and 52 members of other agribusiness development networks and organisations not influenced by the NLA were interviewed.

At the end of data collection, focus group discussions with NLA-influenced and non-influenced cooperatives were held to discuss unclear topics. Overall, data from six focus group discussions, 20 key informant interviews and 90 individual questionnaires were collected (Cadilhon & Landmann, 2015).

3.3 Data analysis

Graphical inspection and descriptive analysis of the structural data were undertaken first. Statistical differences were then identified between NLA members and their influenced groups as well as between the different levels of the network of the agricultural sector, compared with the reference group.

To avoid multicollinearity due to potential interrelationships between statements, factor analysis with orthogonal VARIMAX rotation reduced trust and capacity development statements to a smaller number of uncorrelated underlying factors. Reliability tests were carried out with all statements and afterwards with the calculated factors. The factors were also analysed with values of Cronbach's Alpha, Kaiser-Meyer-Olkin (KMO) measurement and Bartlett's Test of Sphericity. The acceptable factor loading chosen for this study (population of 90) is 0.564 (Field, 2009).

A multiple linear regression was undertaken with the factors developed from performance variables representing capacity development as the dependent variable. Independent variables were factors representing the trust component of the NLA members' conduct and additional individual structure and conduct variables as hypothesized by the conceptual framework (Fig. 2).

To affirm the validity and robustness of the regression models, common diagnostic tests were used: R-Squared showed the overall fit of the model and variance inflation factor (VIF) values were analysed (Field, 2009). Landmann (2015) describes the complete process of the factor and regression analyses.

The qualitative data relating to information sharing, communication, coordination, joint planning, and trust gathered from the focus group discussions and individual interviews were transcribed into a single document. Following best practices in mixed research methods (Patton, 2002), quotes from the stakeholders interviewed were selected to provide backing for the statistically significant results of the regression. Various other quotes were chosen because they enabled the interpretation of some of the non-significant results of the regression. The results presented below build upon the triangulation of both quantitative and qualitative data, therefore, go beyond an analysis produced by Landmann & Cadilhon (2016) in a case study form.

4 Results

4.1 Descriptive statistics of farmers' organisations

Coffee was the most produced crop for 41 farmers' organisations sampled; 33 reported basic grains (beans, corn and rice) and 16 declared other products (cattle, milk or dairy, vegetable, honey, cocoa). Twenty-six organisations focused on only one agricultural product.

In total, 12 respondents represent a national organisation (one NLA-member and 11 others). Six respondents represented regional organisations (three NLA-members and three others), two are from national-level cooperatives (one NLA-member and one other) and 14 represent regional-level cooperatives (seven NLA-partners and seven others). The sample included 54 village-level cooperatives (26 NLA-partners and 28 others).

Seventy respondents out of 90 mentioned their organisation was participating in more than one capacity development group. The majority of respondent organisations performed the function of farming, marketing or processing groups. Most groups were also providing capacity development services. Two-thirds of the organisations interviewed also had some financial role in providing credit to their members.

Table 1: *Institutions of key informants interviewed.*

Type of institution	Name of institution
Consultant	Kuan-Consultants & Associates
Foundation	FUNICA – Nicaraguan Foundation for Technological Development of Agriculture and Forestry
Governmental institution	CONICYT – Nicaraguan Council of Science and Technology
Governmental institution	MAGFOR – Ministry of Agriculture and Forestry
Governmental institution	MINED – Ministry of Education
NGO	CATIE – Tropical Agricultural Research and Higher Education Center
NGO	CRS – Catholic Relief Services
NGO	HEIFER International Nicaragua
NGO	LWR- Lutheran World Relief
NGO	SWISSAID
NGO	VECO MA – VredesEilanden Country Offices Mesoamerica
Private company	Exportadora Atlantic S.A.; ECOM Nicaragua
Private company	Ritter Sport
Producers organisation	APEN- Association of Producers and Exporters of Nicaragua
Producers organisation	FENACOOP – National Agricultural Cooperative Federation and Agroindustrial R.L.
Producers organisation	MAONIC – Movement of Nicaraguan Agroecology and Organic Producers
Producers organisation	UPANIC – Union of Agricultural Producers of Nicaragua
University	UCA – Centroamerican University
University	UCATSE – Agricultural Catholic University of the Dry Tropics
University	UNA – National Agrarian University

Source: Authors' own research.

Out of the 90 respondent organisations, 57 were established cooperatives, 14 associations, eight NGOs, five private companies, and six were related to government. The most important source of funding came from NGOs (37 cases) followed by cash from operations generated (25 cases), credit provided by the private sector (11 cases), membership fees (10 cases), while seven were government funded.

Of all organisations sampled, twenty-six organisations have fewer than 100 members as producers while 27 represent between 100 and 499 producers. Organisations speaking for 500 to 999 farmers are represented by nine respondents and organisations having 1,000 to 4,999 farmers as members are represented by 16 respondents. Only ten organisations represent more than 5,000 producers. The largest organisation interviewed counts 50,000 producers as members. In terms of gender balance of the producers, 69 % were men and 31 % women (Cadilhon & Landmann, 2015).

4.2 Regression analysis – structure and conduct influencing performance

The regression results presented in Table 2 derive from an econometric model to test selected parts of the conceptual framework depicted in Fig. 2. The variables chosen as explanatory variables are consistent with the literature review and conceptual elaboration by Mariami et al. (2015), who propose constructs that can be used to measure structure, conduct, and performance in the context of innovation platforms. Thus, Table 2 identifies, on the one hand, the influence of structure (characterized by the number of years the interviewee has worked for the organisation, the connection of the organisation with the NLA, and the position of the organisation inside the network) on the factor 'innovation'. Secondly, it describes the influence of the learning partners' conduct (represented by statements clustered into information sharing, coordination, joint planning and two trust factors) on the factor 'innovation'. The factor 'innovation' represents the capacities developed by the organisation in the last years measured by new products, knowledge and new techniques or machinery. 'Innovation' is based on the following statements: (1) We have developed new products in the last six years; (2) Our knowledge about our activity has improved in the past six years; (3) In the past six years, we have used new techniques or machinery in our production, production process or management. In the end, two structure variables, two joint planning statements and both trust factors show a significant impact on performance. The adjusted R-Squared of this regression is 40.4 %, and the whole regression is statistically significant at a level inferior to 0.1 %. All B-values are between one and minus one with only one exception. Respecting the conditions of the equa-

tion model meant that the influence of the independent variables on the dependent variable is relatively small (Field, 2009).

The number of years working for the organisation has a significance of 0.1 % and a Beta-value of 0.294, which shows that the amount of time the interviewee has worked for an organisation increases the factor innovation. The connection of the organisation with the NLA does not have a statistically significant influence on the factor innovation. On the other hand, the position of the organisation inside the network does have a significant influence (Sig. = 4.8 %; Beta-value = −0.178). The attributes of this variable are ordered in an ordinal scale related to the position inside the network: national organisations or institutions have a value of one, regional organisations a value of two, cooperatives at third level a value of three, cooperatives at second level a value of four and cooperatives at the first level closest to the farmers a value of five. The bigger the value, the more local the organisation. Being closer to the farmers' level rather than at national level decreases 'innovation'.

Two statements related to information sharing between the farmers and other stakeholders as well as between the NLA or their organisations and the farmers do not have a significant influence. The statement 'We plan our activities together with the NLA/ our organisation according to our production potential and customer demand' has a negative significant influence (Sig. = 2.6 %; Beta-value = −0.224). On the other hand, the statement 'Joint planning of activities with the NLA/ our organisation has improved in the last six years' has a positive and statistically significant influence (Sig. = 0.1 %; Beta-value = 0.378) on innovation. Yet, the statements 'our viewpoints are taken into account by the NLA/ our organisation when they plan their activities' as well as 'we prefer to have long-term relationships' do not have a significant influence.

Both trust factors labelled 'trustful relationships' (Sig. = 1.1 %; Beta-value = 0.248) and 'trustful contracts' (Sig. = 1.3 %; Beta-value = 0.231) have a positive significant influence on 'innovation'. The factor 'trustful relationships' is based on the following statements: (1) The NLA/our organisation always keep their promises; (2) The NLA/our organisation always give us correct information; (3) The NLA/our organisation actions and behaviours are very consistent; (4) The NLA/our organisation always try to inform us if problems occur. The factor 'trustful contracts' is based on the statements: (1) We only develop a relationship with business partners who are fair to us and (2) We only maintain a relationship with our business partners with clearly written terms and conditions.

Table 2: *Regression analysis of selected structure and conduct indicators on the factor 'innovation'.*

Dependent variable: Factor: Innovation		*Unstandardized Coef.*		*Standardized Coef.*			*Collinearity Statistics*	
		B	Std. Error	Beta	t	Sig.	Tolerance	VIF[†]
(Constant)		−1.709	0.907		−1.883	0.064		
Structure	*Years working for the organisation*[‡]	0.044	0.013	0.294	3.381	*0.001*	0.914	1.094
	Connection with NLA[§]	0.249	0.177	0.124	1.405	0.164	0.885	1.129
	Position of the organisation inside the network[¶]	−0.131	0.065	−0.178	−2.010	*0.048*	0.883	1.132
Conduct Information exchange	We usually share information about production with other stakeholders.[‖]	0.172	0.117	0.130	1.467	0.147	0.881	1.135
	The NLA/ our organisation exchange information about their on-going activities with us.[‖]	0.208	0.123	0.167	1.690	0.095	0.711	1.407
Conduct Coordination	*We plan our activities together with the NLA/ our organisation according to our production potential and customer demand.*[‖]	−0.260	0.115	−0.224	−2.265	*0.026*	0.707	1.415
Conduct Joint planning	Our viewpoints are into account by the NLA/ our organisation when they plan their activities.[‖]	0.028	0.142	0.022	0.201	0.842	0.558	1.791
	Joint planning of activities with the NLA/ our organisation has improved in the last six years.[‖]	0.447	0.126	0.378	3.541	*0.001*	0.607	1.646
Conduct Trust	We prefer to have long term relationships.[‖]	−0.174	0.125	−0.127	−1.387	0.169	0.828	1.208
	Factor: Trustful relationships	0.252	0.096	0.248	2.613	*0.011*	0.771	1.298
	Factor: Trustful contracts	0.230	0.091	0.231	2.532	*0.013*	0.834	1.200

Notes: Variables with significant influence on factor 'Innovation' are shown in italics;
R-Square = 0.480; Adjusted R-Square = 0.404; Significance = 0.000; level of significance $p < 0.05$
[†] VIF: Variance inflation factor;
[‡] Scale: Years in numbers;
[§] Scale: 0 = No; 1 = Yes;
[¶] Scale: 1 = National organisation; 2 = Regional organisation; 3 = Cooperative 3rd level; 4 = Cooperative 2nd level; 5 = Cooperative 1st level;
[‖] Scale: 1 = strongly disagree; 2 = disagree; 3 = undecided; 4 = agree; 5 = strongly agree
Source: Landmann & Cadilhon (2016)

4.3 Qualitative data

According to one key informant and professor at UCATSE – Agricultural Catholic University of the Dry Tropics being asked on the training activities of producers organisations: 'It will be given [the same] training that another organisation has conducted. It could be provided a service complementing the organisations offers: currently this is a little bit the problem in Nicaragua. There are sixty-eight organisations working in extension, but there is nobody who offers credit or the full range of extension service'. This statement underlines the findings from the regression model that joint planning of activities between farmers' cooperatives and training organisations is negatively related with innovations. Because there might be other organisations involved in training activities on similar subjects.

Information sharing is strongly linked to trust, which in return has a strong impact on fostering innovations. For example, a key informant and a board member of the national commission for coordination and management at MAONIC, the Movement of Agro-ecological and Organic Producers of

Nicaragua, declared: 'Each alliance depends to which extent transparency is achieved. If the organisation is transparent, trust is existing. [...] That is the condition we see, which is required. But we also ask the other partner, if he agrees on our point of view when possible'. The professor at UCATSE explained the discrepancy between information sharing and innovation by a mainly top-down flow of information in Nicaraguan agricultural development and extension: 'one problem of these alliances is the management of information. [...] Feedback is required, as a member but also from the bottom, meaning producer of the cooperative and the organisations. There should be a bottom-up information flow but according to our experience it doesn't work. The information does not flow with the same speed from one side to another [as easily as from the other]'.

Our key informant from MAONIC supports the fact that organisations and stakeholders working with producers are mostly aware of challenges and improve farmer's training. To 'focus on small producers as the core of the solution to solve problems' is one goal of MAONIC. According to the board member, the farmer himself is the 'leader of change to achieve the goal of change and improvement'. MAONIC is 'empowering [the farmer] to improve the information about his own farm [...]. If that fails, no project [...and] no government will help him out of poverty'.

A project manager at FUNICA – the Nicaragua Foundation for Agricultural and Forestry Technological Development, an NLA member and key informant, supports the point that trust influences the factor innovation, saying 'the level of trust has an influence on capacity development within the NLA'. He explains how trust was created within the NLA network and the beneficiary farmers' organisations: 'meeting their expected contributions is important. For example, if there is a project and they will comply, the organisation has to contribute a certain amount of resources, generate the planned results, provide information which is requested, basically'.

To illustrate the results of the 'trustful relationships' factor, we consider the declarations of a manager working for the private sector firm Exportadora Atlantic S.A. in Nicaragua and key informant. He states very generally that 'everything is based on trust. Trust and transparency have to be the axis of all organisations'. More specifically, he explains that farmers trust 'the economic solidity of the company' as well as 'the transparency' with which the company negotiates with farmers. 'We also trust in producers; we deposit 2,000, 3,000, 20 or 100 million dollars that he can work and grow his fruits'. The board member of MAONIC confirms the statement of the private firm by highlighting the importance of the relationship between transparency and trust but also says that, 'If there is trust then, even with a few resources, much more could be done'.

A consultant at CATIE – the Tropical Agricultural Research and Higher Education Centre – and key informant describes the structure of the NLA as one where '[all NLA-members] are at the same level. There is no hierarchical structure, [...]. There is a lot of trust, and quite some transparency in communication'.

During a focus group discussion with small-scale farmers on 5th August 2014 in Chinandega, the farmers said they trust NGOs and 'mistrust the governmental institutions'. They linked the former to the fulfilment of promises and financial support. A consultant of MINED – the Ministry of Education – and key informant supports this argument saying that: 'producers [...] distrust the government [...], because sometimes [...] the financial expectations of producers are not fulfilled'. Most stakeholders do not work together with governmental institutions. The farmers 'read [the contracts and agreements of the government] [...] but they do not have much confidence'. The farmers interviewed explain this as follows: 'we do not have all information [...] and do not know the intention of the gov-

ernment'. From the farmers' perspective, there is a lack of transparency in governmental activities.

5 Discussion of results and conclusions

5.1 Representativeness of the data

The Central American Bank for Economic Integration (BCIE) has undertaken a similar study of farmers' organisations in Central America using data from 63 representative Nicaraguan cooperatives (Lafortezza & Consorzio, 2009). The results from this study are similar to those of the BCIE study: the main products produced and exported are the same, as is the gender balance within the cooperatives. The main difference between the two studies lies in the size of the organisations interviewed: our study incorporates very large cooperatives with more than 10,000 members and active at regional and village levels. The BCIE study's sample also differs from the sample of this study as 35 percent of the BCIE sample had not received any training while all the farmers' organisations sampled in this study had benefited from some sort of training. Although there are some differences between the two studies' samples, the similarity in overall results allows to consider that this study's sample is also representative of farmers' organisations active in the provinces where the NLA is present.

5.2 Efficiency of the NLA's capacity development process

The old structures of cooperatives and the different levels within the cooperative network are still present and an important factor in agribusiness development. The private, public, and NGO sectors are familiar with farmers' cooperatives being widely spread out in Nicaragua and have adapted their methods to this structure. The NLA has used the wide network of agricultural cooperatives to snowball its training on agribusiness from the national level to a large number of individual farmers. However, the other support organisations in the Nicaraguan agricultural innovation system are doing the same so all farmers' cooperatives and their farmer members are connected one way or another to a source of training on agribusiness management. Every key stakeholder interviewed was practising capacity development in the study area. A majority, 77 percent, of respondents reported being supported by more than one organisation. This is the most reasonable explanation for the fact that a 'connection to the NLA' does not have a significant influence on 'innovation'.

The lack of significant influence of information sharing on innovation in the regression is surprising as qualitative data indicate that information sharing is strongly linked to trust, which in turn has a strong impact on fostering innovations. However, one possible explanation is that agricultural training is generally a top-down process. As a probable consequence, information sharing has no significant influence on innovation. The generalised top-down extension method also explains that only when joint planning is seen to be improving does it have a positive impact on innovation. Furthermore, the data show that the farmers' viewpoint is not taken into account. This supports the idea of rethinking and reorganizing the capacity development sector as well as the structure of agricultural organisations and institutions, just as the NLA has done. Although all providers of agribusiness training in Nicaragua are using the strong cooperative-oriented structure within the country's agricultural sector to improve farmers' knowledge of agribusiness and markets, yet, all stakeholders follow their own approaches. Cooperation and networks between different types of stakeholders are rare.

In fact, the qualitative data indicated that the organisations participating in the NLA network would greatly value an exchange of their experience and progress with other organisations. Thus, the NLA and other stakeholders working in the sector of capacity development should open their network to other stakeholder types, even though the government does not seem to show interest in such a cooperation. This could make the training method more efficient, sustainable, and successful. Using organised networks seems to be the right pathway for agricultural training given the structure of Nicaragua's agriculture, but interactions could be improved between actors of the same level and different levels.

On the other hand, there is no visible influence of the duration of a relationship on 'innovation'. One explanation could be farmers' easy access to cooperatives: out of 90 respondents, 70 were at least in two different cooperatives.

Several cooperatives consider that capacity development will be successful only if accompanied by financial support. The NLA has thus embedded financial support to put the agribusiness development skills learned during its learning cycles into practice. Financial support is necessary, but the main goal is to have successful producers who are not dependent on the financial support from NGOs (Lundy & Gottret, 2005). Indeed, successful IPs in the long-term are those that manage to renew their funding source to keep covering their costs, or those that manage to change their business model in order to become financially self-sustainable (Cadilhon *et al.*, 2016, Dror *et al.*, 2016).

The quantitative and qualitative data show on the one hand that the content and process of NLA's training are very good. On the other hand, the qualitative data show that the way the training is undertaken with farmers could be im-

proved if adjusted to the regional circumstances. This statement was confirmed during the focus group discussions, several key informant interviews, and some individual questionnaires. The current strategies of some NLA members to adapt the training to the local environment are successful, and this is a response to some criticism from the final beneficiaries. On the other hand, it is difficult to trace the success of the NLA training. If one member changes the method and uses the information on their own, it is no longer helpful for the other NLA members. Indeed, changing the learning method to adapt it to local context jeopardizes the approach of the LAs to build up a platform to share information and learn from each other. The institutional learning platform is no longer efficient and sustainable if improvements and changes are not shared with the other NLA members. Opportunities for communication and meetings with other knowledge networks to share and exchange information are also missed by some cooperatives. Lundy & Gottret (2005) describe an approach in the method of the LA to create networks at the micro, meso, and macro levels. These intertwined networks do not exist in Nicaragua and would be the answer to some current criticisms. The NLA itself already identified enlarging knowledge networks as one weakness and included this in the changes that are planned for the coming years (AdA-Nicaragua, 2012).

5.3   Trustful relationships improve the NLA learning process

The quantitative and qualitative results of this study show a clear impact of various components of trust within the NLA network and its capacity development outcomes. The regression model showed that 'trustful relationships' and 'trustful contracts' both had a positive significant influence on 'innovation'.

As described above, the farmers have a better perception of NGOs than of government. More worryingly, it seems that NGOs almost have no collaboration with governmental organisations. This, on the one hand, shows distrust of government action but on the other hand, also shows that the alliances between NLA members and governmental institutions could be strengthened.

The study's results identify the relationship existing between trust and capacity development in the case of the NLA. The regression and qualitative data triangulation indicate that structure and conduct have an influence on performance. The influence of trust on capacity development and innovation identified by this research contributes evidence for the conceptual framework of innovation platforms proposed by Mariami *et al.* (2015). Furthermore, the influence between IP structure and conduct is also observable, for example, in the fact that NGOs, as financial sources of the farmers' organisations, have a significant influence on 'trustful relationships'. The influence of conduct on performance is visible as well: results show that 'trustful relationships' and 'trustful contracts' have a positive influence on 'innovation' emerging from capacity development variables. The findings complement those of past research identifying linkages between how an IP is structured, the conduct of its members and its expected outcomes (Badibanga et al., 2013; Subedi et al., 2014; Kago et al., 2015; Mariami et al., 2015; Pham et al., 2015; Dror et al., 2016; Teno & Cadilhon, 2016).

Reverse causality is a limitation which could occur using regressions as an analysis tool (Field, 2009). In our study, we assume that this problem is relatively small as the conceptual framework is based on theories which are already used very often. Furthermore, this specific conceptual framework was already used in a similar way in different settings, moreover we triangulated our quantitative data with qualitative data. However, to be sure that reverse causality is not occurring, more detailed data would be helpful.

The method used to evaluate the impact of the NLA could also be improved, by including financial and business figures in the questionnaire for direct comparison of economic impacts. This data could also be collected through the auto-evaluation mechanism developed by the NLA whereby farmers' organisations evaluate their own learning progress. This would simplify making adjustments to the training method and the conceptual framework used to evaluate impact.

Acknowledgements

Authors wish to thank Falguni Guharay and colleagues at the International Center of Tropical Agriculture (CIAT) in Nicaragua for helping out with the fieldwork, Magdalene Trapp for translating and cleaning the qualitative data, and Susanna Thomas for her careful proof-reading.

Funding

This work was undertaken as part of the CGIAR research program on Policies, Institutions, and Markets (PIM) led by the International Food Policy Research Institute (IFPRI). Funding support for this study was provided by the CGIAR research program Humidtropics that aims to develop new opportunities for improved livelihoods in a sustainable environment, and the CGIAR research program PIM. The opinions expressed here belong to the authors and do not necessarily reflect those of PIM, IFPRI, or CGIAR.

References

AdA (2014). Alianzas de Aprendizaje (AdA). CIAT. Available at: http://www.alianzasdeaprendizaje.org/portal/ index.php (accessed on: 15 June 2017).

AdA-Nicaragua (2012). *Alianza de Aprendizaje Nicaragua: Plan Estratégico 2013–2016*. Alianza de Aprendizaje Nicaragua, Managua: 13.

Anderson, J. C. & Narus, J. A. (1984). A Model of the Distributor's Perspective of Distributor-Manufacturer Working Relationships. *Journal of Marketing*, 48 (4), 62–74. doi:10.2307/1251511.

Badibanga, T., Ragasa, C. & Ulimwengu, J. (2013). *Assessing the Effectiveness of Multistakeholder Platforms: Agricultural and Rural Management Councils in the Democratic Republic of the Congo*. International Food Policy Research Institute (IFPRI), Washington DC.

Bandura, A. (1971). *Social Learning Theory*. General Learning Press, New York.

Bolger, J. (2000). *Capacity Development: Why, What and How*. Capacity Development-Occasional Series, Vol. 1, No. 1, May 2000, CIDA, Policy Branch, Hull, Quebec.

Cadilhon, J.-J. (2013). A conceptual framework to evaluate the impact of innovation platforms on agrifood value chains development. Paper presented at the 138th EAAE Seminar on Pro-Poor Innovations in Food Supply Chains, Ghent, Belgium, 11–13 September 2013.

Cadilhon, J.-J. & Landmann, D. H. (2015). Nicaragua Learning Alliance monitoring and evaluation study 2014. ILRI – International Livestock Research Institute, Nairobi, Kenya. Available at: http://data.ilri.org/ portal/dataset/nicaragualearningalliance (accessed on: 18 September 2016).

Cadilhon, J.-J., Pham, N. D. & Maass, B. L. (2016). The Tanga Dairy Platform: fostering innovations for more efficient dairy chain coordination in Tanzania. *International Journal on Food System Dynamics*, 7 (2), 81–91.

Chambers, R. (1994). Participatory rural appraisal (PRA): Challenges, potentials and paradigm. *World Development*, 22 (10), 1437–1454. doi:10.1016/0305-750X(94)90030-2.

Claro, D. P., Hagelaar, G. & Omta, O. (2003). The determinants of relational governance and performance: How to manage business relationships? *Industrial Marketing Management*, 32 (8), 703–716. doi: 10.1016/j.indmarman.2003.06.010.

Coghlan, D. & Brydon-Miller, M. (eds.) (2014). *The SAGE encyclopedia of action research*. SAGE Publications Ltd.

CRS (2009). *Working Together, Learning Together – Learning Alliances in Agroenterprise Development*. Edited by: Best, R., Ferris, S. & Mundy, P., Catholic Relief Services (CRS), Baltimore, USA.

Dror, I., Cadilhon, J.-J., Schut, M., Misiko, M. & Maheshwari, S. (eds.) (2016). *Innovation Platforms for Agricultural Development. Evaluating the mature innovation platforms landscape..* Routledge, Oxon (United Kingdom) and New York (USA).

FAOSTAT (2014). Nicaragua. FAOSTAT, Statistics Division, Food and Agriculture Organization of the United Nations, Rome, Italy.

Field, A. (2009). *Discovering statistics using SPSS. Introducing Statistical Methods.* SAGE Publications, Los Angeles [i.e. Thousand Oaks, Calif.]; London.

Furubotn, E. G. & Richter, R. (2010). Introduction. *In:* Furubotn, E. G. & Richter, R. (eds.), *The New Institutional Economics of Markets.* Edward Elgar Publishing, Cheltenham, UK.

Hall, A. (2005). Capacity development for agricultural biotechnology in developing countries: an innovation systems view of what it is and how to develop it. *Journal of International Development*, 17 (5), 611–630. doi: 10.1002/jid.1227.

Hall, A. (2007). *Challenges to Strengthening Agricultural Innovation Systems: Where Do We Go From Here?.* United Nations University – UNU-MERIT, The Netherlands.

Homann-Kee Tui, S., Adekunle, A., Lundy, M., Tucker, J., Birachi, E., Schut, M., Klerkx, L., Ballantyne, P., Duncan, A., Cadilhon, J. & Mundy, P. (2013). *What are innovation platforms?.* Innovation Platforms Practice Brief 1, ILRI, Nairobi, Kenya.

Horton, D., Alexaki, A., Bennett-Lartey, S., Brice, K. N., Campilan, D., Carden, F., de Souza Silva, J., Duong, L. T., Khadar, I., Boza, A. M., Muniruzzaman, I. K., Perez, J., Chang, M. S., Vernooy, R. & Watts, J. (2003). *Evaluating capacity development: experiences from research and development organizations around the world..* International Service for National Agricultural Research (ISNAR), the Netherlands; International Development Research Centre (IDRC), Canada; ACP-EU Technical Centre for Agricultural and Rural Cooperation (CTA), the Netherlands.

Kago, K. M., Cadilhon, J.-J., Maina, M. & Omore, A. (2015). Influence of innovation platforms on information sharing and nurturing of smaller innovation platforms: a case study of the Tanzania Dairy Development Forum. 29th International Conference of Agricultural Economists, Milan, Italy.

Klerkx, L., Aarts, N. & Leeuwis, C. (2010). Adaptive management in agricultural innovation systems: The interactions between innovation networks and their environment. *Agricultural Systems*, 103 (6), 390–400. doi: 10.1016/j.agsy.2010.03.012.

Klerkx, L., Leeuwis, C. & van Mierlo, B. (2012). Evolution of systems approaches to agricultural innovation: concepts, analysis and interventions. *In:* Darnhofer, I., Gibbon, D. P. & Dedieu, B. (eds.), *Farming Systems Research into the 21st Century: The New Dynamic.* pp. 457–483, Springer, Dordrecht, New York.

Kumar, N. (1996). *The power of trust in manufacturer-retailer relationship.* Harvard Business Review.

Laeequddin, M., Sahay, B. S., Sahay, V. & Waheed, K. A. (2010). Measuring trust in supply chain partners relationships. *Measuring Business Excellence*, 14 (3), 53–69.

Lafortezza, D. & Consorzio, E. S. C. (2009). Inventario de las cooperativas productivas: Honduras. Banco Centroamericano de Integración Económica (BCIE), Honduras.

Landmann, D. H. (2015). *The influence of trust in the Nicaraguan Learning Alliance on capacity development of members and other influenced groups.* Master's thesis, Georg-August-Universität Göttingen, Germany.

Landmann, D. H. & Cadilhon, J.-J. (2016). With trust and a little help from our friends: how the Nicaragua Learning Alliance scaled up training in agribusiness. *In:* Dror, I., Cadilhon, J.-J., Schut, M., Misiko, M. & Maheshwari, S. (eds.), *Innovation Platforms for Agricultural Development. Evaluating the mature innovation platforms landscape.* pp. 16–41, Routledge, New York.

Lundy, M. & Gottret, M. V. (2005). *Learning Alliances: An Approach for Building Multi-stakeholder Innovation Systems.* International Center for Tropical Agriculture (CIAT), Cali, Colombia.

Lusthaus, C., Adrien, M.-H. & Perstinger, M. (1999). Capacity development: definitions, issues and implications for planning, monitoring and evaluation. Universalia Occasional Paper, No. 35, Universalia, Québec, Canada.

Mariami, Z. A., Cadilhon, J.-J. & Werthmann, C. (2015). Impact of innovation platforms on marketing relationships: The case of Volta Basin integrated crop-livestock value chains in Ghana. *African Journal of Agricultural and Resource Economics*, 10 (4), 328–342.

Nederlof, S., Wongtschowski, M. & van der Lee, F. (eds.) (2011). *Putting heads together.* Bulletin 396, KIT Publishers, Amsterdam, The Netherlands.

Neely, C. L. (2010). *Capacity Development for Environmental Management in the Agricultural Sector in Developing Countries.* OECD Environment Working Papers, No. 26, OECD Publishing.

Patton, M. Q. (2002). *Qualitative research and evaluation methods*. (3rd ed.). SAGE Publications, Inc. ISBN: 0-7619-1971-6

Pham, N. D., Cadilhon, J.-J. & Maass, B. L. (2015). Field Testing a Conceptual Framework for Innovation Platform Impact Assessment: The Case of MilkIT Dairy Platforms in Tanga Region, Tanzania. *East African Agricultural and Forestry Journal*, 81 (1), 58–63. doi: 10.1080/00128325.2015.1041257.

Pretty, J. N. (1995). Participatory learning for sustainable agriculture. *World Development*, 23 (8), 1247–1263. doi: 10.1016/0305-750X(95)00046-F.

Sanzo, M. J., Santos, M. L., Vázquez, R. & Alvarez, L. I. (2003). The Role of Market Orientation In Business Dyadic Relationships: Testing an Integrator Model. *Journal of Marketing Management*, 19 (1), 73–107. doi: 10.1362/026725703763771971.

Subedi, S., Cadilhon, J.-J., Ravichandranc, T. & Teufel, N. (2014). Impact evaluation of innovation platforms to increase dairy production: A case from Uttarakhand, Northern India. 8th International Conference of Asian Society of Agricultural Economists (ASAE) on Viability of Small Farmers in Asia 2014; Saver, Banladesh. International Livestock Research Institute (ILRI), Nairobi, Kenya.

Teno, G. & Cadilhon, J.-J. (2016). Innovation platforms as a tool for Improving agricultural production: the case of Yatenga province, northern Burkina Faso. *Field Actions Science Reports*, 9. Available at: http://factsreports.revues.org/4239

Trienekens, J. H. (2011). Agricultural Value Chains in Developing Countries A Framework for Analysis. *International Food and Agribusiness Management Review*, 14 (2), 51–82.

Ubels, J., Acquaye-Baddoo, N.-A. & Fowler, A. (eds.) (2010). *Capacity development in practice*. Earthscan, London; Washington, DC.

UNDP (2008). *Capacity Assessment Methodology: User's Guide*. Edited by: J. Colville. United Nations Development Programme (UNDP), New York, USA.

World Bank (2012). *Agricultural Innovation Systems*. World Bank, Washington D.C. doi: 10.1596/978-0-8213-8684-2

World Bank (2017). *Nicaragua: Paving the way to faster growth and inclusion. Systematic country diagnostic*. Washington D.C.

5

Dietary protein and energy requirements of Venda village chickens

Thomas Raphulu [a,b,*], Christine Jansen van Rensburg [a]

[a] *Department of Animal and Wildlife Sciences, University of Pretoria, Pretoria 0002, South Africa*
[b] *Limpopo Department of Agriculture and Rural Development, Mara Research Station, P/bag x 2467, Makhado, 0920, South Africa*

Abstract

The objective of this study was to determine the dietary protein and apparent metabolisable energy (AME) requirements of local chickens. Freshly laid eggs of scavenging chickens collected in rural villages were hatched and randomly distributed to 27 floor pens, 10 chicks per pen. Chicks were fed 9 experimental diets that were combinations of three CP levels (140, 170 and 190 g kg^{-1} DM) and three AME levels (11.0, 11.7 and 12 MJ kg^{-1}) during the starter phase (0–6 weeks) and combinations of three CP levels (120, 150 and 180 g kg^{-1} DM) and three ME levels (11.3, 12.0 and 12.4 MJ kg^{-1}) during the grower phase (7–17) weeks. Significant differences within means on CP $\times$ AME interaction effect were observed in all parameters measured, except feed intake during starter period and dressing percentage (%) and breast yield of 17 weeks old chickens. The results of the present study indicated that during the starter and grower phases, unsexed chickens would require dietary combinations of 170 g CP kg^{-1} and 11.0 AME MJ kg^{-1} and 150 g CP kg^{-1} and 12 AME MJ kg^{-1} in their diets to optimise weight gain and FCR, and 150 g CP kg^{-1} and 11.3 MJ kg^{-1} to optimise ash content of muscles, protein content of the breast and fat content of the leg muscle. Supplementation of 27 g CP kg^{-1} feed to grower scavenging chickens would be enough to improve chicken production in the rural villages.

Keywords: starter phase, grower phase, weight gain, feed conversion ratio

1 Introduction

Villages in the rural areas of the Vhembe District, Venda, South Africa, have relative high numbers of chickens which in most instances have to scavenge for feed. These chickens are mainly kept for religious and cultural reasons, income generation and a supply of high quality protein in the form of meat and eggs (Swatson *et al.*, 2001; Raphulu *et al.*, 2015a). The typical chicken population found in these areas are characterised by slow growth rate and late maturity, low egg production and small egg size (Melesse, 2000). Low husbandry input levels cause numerous nutritional and parasitic problems in these chickens. In a stress-free environment, given adequate access to essential nutrients, growth rate will increase until a genetically determined upper limit

is reached (Campbell & Taverner, 1988). A few studies have been conducted to determine the nutrient requirements of indigenous chickens in Africa. Mbajiorgu (2010) reported that a diet containing a crude protein content level of 178 g kg^{-1} DM and energy level of 14 MJ ME kg^{-1} DM allowed for optimal utilisation of absorbed protein and energy for growth of Venda chickens between one and six weeks old. Alabi *et al.* (2013) fed Venda Indigenous chickens diets that were isonitrogenous with different energy levels and concluded that dietary energy levels of 12.42 and 12.66 MJ ME kg^{-1} DM, in a diet of 180 g CP kg^{-1} DM, supported optimum growth rates at starter (1–7 weeks) and grower phases (8–13 weeks) of Venda chickens, respectively. The dietary protein requirement of indigenous chickens in Kenya were found to be between 130 g kg^{-1} (Chemjor, 1998) and 160 g kg^{-1} (Kingori *et al.*, 2003) during 14–21 weeks of age and on average 170 g kg^{-1} during a 19 weeks growing period (Ndegwa *et al.*, 2001).

*Corresponding author – thomas.raphulu@gmail.com

The improvement of chicken production in the rural areas of Venda can increase the availability of quality protein in the form of meat and eggs to the communities, which will result in alleviation of malnutrition, increased household income and job creation. Because of the limited information on the nutrient requirements of the chickens in Venda, it is difficult to come up with nutrient supplementation strategies to assist the rural communities. The objective of this study was to determine the genetic growth potential of these chickens (eggs collected from the rural villages) under good environmental conditions and management, and to determine the nutrient requirements in terms of dietary protein and metabolisable energy to optimise its growth performance.

2 Materials and methods

The experiment was conducted at the University of Pretoria, Gauteng Province; South Africa (coordinates 25°15′28.9″ E, 25°45′03.6″ S). All procedures were approved by the University of Pretoria, Animal Use and Care Committee (EC008-08). Thousand freshly laid eggs of scavenging chickens collected in 6 rural villages of Venda in the Vhembe District, South Africa, were hatched at the University of Pretoria Experimental farm. Two hundred and

seventy (270) chicks were randomly distributed to 27 floor pens, 10 chicks per pen. Chicks were fed 9 experimental diets that were combinations of three CP (140, 170, 190 and 120, 150 and 180 g kg^{-1}) and three AME (11.0, 11.7, 12.0 and 11.3, 12.0 and 12.4 MJ kg^{-1}) levels during the starter (0–6 weeks) and grower (7–17 weeks) phases, respectively (Tables 1 and 2).

Chickens were raised in an environmentally controlled house and temperature maintained at 30 to 33 °C and 23 to 25 °C during the starter and grower phase, respectively. Lighting was provided continuously. The chicks were vaccinated against Newcastle virus disease and infectious bronchitis. Feed and water were available ad libitum throughout the experiment. Feed intake, weight gain, and feed conversion ratio (FCR) were measured per pen weekly. Weight gain was calculated as the difference between the final and initial chicken body weight during each of the weighing periods. Feed conversion ratio was calculated by dividing feed intake by weight gain. Mortality was recorded as it occurred. At the end of the experiment, eight 17-week old chickens per treatment (4 males and 4 females) were randomly selected following a 12-hour fast and killed by bleeding from a neck cut. The chickens were scalded in water for 2–3 minutes to ease plucking. Head, neck, feet and the visceral organs were removed. The hot carcass was weighed

Table 1: *Composition of the experimental starter diets.*

Dietary CP/ME [†]	190/11	190/11.7	190/12.4	170/11	170/11.7	170/112.4	140/11.7	140/11.7	140/12.4
Ingredient (g kg^{-1} DM)									
White maize	605.4	677.6	612.3	626.6	702.8	694.9	658.3	734.6	808.4
Soya oilcake meal	226.8	281.3	98.9	158.8	180.5	119.9	56.9	78.6	120.3
Bran	76.8	0	0	119.7	21.4	0	184	85.7	0
Sunflower oilcake meal	50	0	0	50	50	0	50	50	20
Fullfat soya	0	0	247.5	0	0	140	0	0	0
Limestone	16.9	16.8	17.1	17.2	17.0	17.1	17.5	17.4	17.3
Monocalcium phosphate	10.3	11.2	11.3	10.7	11.5	12	11.2	12.1	13
Salt	3.9	4.0	4.0	3.9	4.0	4.0	3.9	3.9	4.0
Premix [‡]	3.0	3.0	3.0	3.0	3.0	3.0	3.0	3.0	3.0
Lysine	2.9	2.1	1.8	4.7	4.4	3.6	7.5	7.2	6.6
Methionine	1.7	1.8	1.9	2.0	2.0	2.1	2.5	2.4	2.4
Threonine	2.1	2.1	2.0	3.0	2.9	2.7	4.4	4.2	4.0
Tryptophan	0.11	0.13	0.24	0.39	0.4	0.5	0.8	0.8	0.8
Calculated analysis (g kg^{-1})									
Calcium	10	10	10	10	10	10	10	10	10
Phosphorus	4.5	4.5	4.5	4.5	4.5	4.5	4.5	4.5	4.5

[†] CP/ME, crude protein (g kg^{-1} DM) / metabolisable energy (MJ kg^{-1} DM)

[‡] Composition of the broiler starter premix. Each 2.5 kg contained vitamin A , 13,000,000 IU, vitamin D$_3$, 3,500,000 IU, B$_1$ (Thiamine) 2000 mg, vitamin B$_2$ (Riboflavin) 6000 mg, vitamin B$_6$ (Pyrodoxine) 6000 mg, folic acid 1500 mg, vitamin B$_{12}$ (Cobalamine) 20 mg, E 40,000 mg, betaine 500,000 mg, niacin 60,000 mg, pantothenic acid 15000 mg, Vitamin K$_3$ stab 4000 mg, biotin 150 mg, cobalt 500 mg, iodine 3000 mg, selenium 300 mg, manganese 70,000 mg, copper 20,000 mg, zinc 70,000 mg, iron 50,000 mg, antioxidant 200,000 mg.

Table 2: *Composition of the experimental grower diets.*

Dietary CP/ME[†]	180/11.3	180/12	180/12.4	150/11.3	150/12	150/12.4	120/11.3	120/12	120/12.4
Ingredient (g kg^{-1} DM)									
White maize	648.5	691.8	654.9	680.2	756.5	778.9	711.8	788.3	831.8
Soya oilcake meal	201.8	215.3	111.2	99.8	121.5	142.7	46.9	19.6	32.0
Fullfat soya	0	50.3	191.2	0	0	30.0	0	0	0
Bran	57.2	0	0	121.4	23.2	0	187	87.4	31.3
Sunflower oilcake meal	50	0	0	50	50	0	0	50	50
Limestone	17.0	17.0	17.1	17.4	17.2	17.2	17.8	17.6	17.5
Monocalcium phosphate	10.8	11.6	11.7	11.4	12.3	12.7	12.0	12.9	13.4
Lysine	3.7	2.9	2.7	6.5	6.2	5.5	9.3	9.0	8.8
Methionine	1.9	2.0	2.0	2.3	2.2	2.3	2.7	2.7	2.6
Threonine	2.5	2.3	2.4	3.9	3.7	3.5	5.2	5	4.9
Tryptophan	0.3	0.3	0.4	0.7	0.7	0.7	1.1	1.1	1.1
Salt (fine)	3.9	4.0	4.0	3.9	4.0	4.0	3.9	4.0	4.0
Premix[‡]	2.5	2.5	2.5	2.5	2.5	2.5	2.5	2.5	2.5
Calculated analysis (g kg^{-1})									
Calcium	10	10	10	10	10	10	10	10	10
Phosphorus	4.5	4.5	4.5	4.5	4.5	4.5	4.5	4.5	4.5

[†] CP/ME, crude protein (g kg^{-1} DM) / metabolisable energy (MJ kg^{-1} DM)

[‡] Composition of the broiler grower premix. Composition of each 2.5 kg contained vitamin A, 11,500,000 IU, vitamin D, 3,500,000 IU, vitamin E 40,000 mg, vitamin K_3, 35,000 mg, vitamin B_1, 2000 mg, vitamin B_2 and B_6, 55000 mg, vitamin B_{12}, 20 mg, niacin, 55,000 mg, calpan, 13,500 mg, folic acid 1250 mg, bictin, 100 mg, selenium, 300 mg, manganese, 700,000 mg, iron, 450,000 mg, iodine 3000 mg, zinc, 65,000 mg, copper, 20,000 mg, cobalt, 500 mg, phyzyme XP 10,000 TPT, 100,000 mg, choline, 400,000 mg, antioxidant, 200,000 mg and rovabio excel AP, 60,0000.

and dressing percentage calculated as weight of carcass divided by live body weight multiplied by 100. Carcass cuts (breast, thigh, drumstick and fat) were also weighed and expressed as proportions of the carcass weight. Dry matter, ether extract, crude protein and ash were evaluated in the breast and leg (thighs and drumstick) muscles according to the recommendation of AOAC Official Methods of Analysis (2000). Data collected were subjected to analysis of variance for a 3 × 3 factorial in a completely randomised design, using the General Linear Model procedure of SAS, version 9.3 (SAS Institute, 2016). Differences among means were determined by LSD's option of SAS.

3   Results

3.1   Performance

The influence of dietary protein and metabolisable energy on growth performance of the chickens is shown in Tables 3 and 4. During the starter period (0–6 weeks), average body weight at the start of the experiment was 28.67 g. There was no effect of CP and AME or their interaction on feed consumption ($p > 0.05$). Dietary CP, AME and CP × AME interaction had significant effect ($p < 0.05$) on weight gain and FCR of chickens during the starter phase.

Weight gain and FCR improved with increasing levels of dietary protein ($p < 0.05$). Crude protein above 140 g kg^{-1} feed at all dietary AME levels resulted in improved weight gain and FCR in the chickens. Chickens that received CP of 170 and 190 g kg^{-1} at all AME levels had similar weight gain and FCR. However, at CP level of 170 g kg^{-1} weight gain and FCR numerically, but not significantly, decreased with an increase in dietary AME, with 11.0 MJ kg^{-1} maximising weight gain and FCR. The dietary combination of CP 170 g kg^{-1} × AME 11 MJ kg^{-1} yielded best weight gain and FCR.

During the grower period (7–17 weeks) there was no significant CP × AME effect ($p \geq 0.05$) observed on any of the traits measured, although significant differences within means were observed. Feeding above 120 g CP kg^{-1} at all AME levels improved weight gain and FCR. There were no significant differences observed between chickens that received CP of 150 and 180 g kg^{-1} at all dietary AME levels on weight gain and FCR. The lightest and heaviest chickens were observed on dietary combinations of 120 g CP kg^{-1} × 12.4 MJ AME kg^{-1} and 150 g CP kg^{-1} × 12.0 MJ AME kg^{-1}, respectively. The feed intake decreased with increasing dietary AME levels ($p < 0.05$). The dietary combination of 150 g CP kg^{-1} feed × 12.0 MJME kg^{-1} feed maximised weight gain and FCR.

Table 3: *The influence of dietary protein and metabolisable energy levels on growth performance of Venda chickens (0 to 6 weeks old).*

Parameter	Feed intake (g/bird/day)	Weight gain (g/bird)	FCR (g/g)
Protein (g kg^{-1} DM)			
140	27.11	258.49[b]	4.44[a]
170	28.00	342.38[a]	3.48[b]
190	26.32	372.47[a]	3.00[c]
SEM[†]	1.05	7.46	0.13
AME (MJ kg^{-1} DM)			
11.0	27.95	328.49[a]	3.67[ab]
11.7	26.92	347.53[a]	3.32[b]
12.4	26.56	297.32[b]	3.92[a]
SEM[†]	1.05	7.46	0.13
CP × AME			
140 × 11.0	29.07	261.50[de]	4.67[a]
140 × 11.7	26.99	292.64[cd]	3.86[abc]
140 × 12.4	25.28	221.34[e]	4.79[a]
170 × 11	27.56	371.80[a]	3.12[bc]
170 × 11.7	26.56	357.40[ab]	3.14[bc]
170 × 12.4	29.86	297.94[ab]	4.20[ab]
190 × 11	27.20	352.16[abc]	3.25[bc]
190 × 11.7	27.21	392.56[a]	2.97[c]
190 × 12.4	24.55	372.70[a]	2.77[c]
SEM[†]	1.81	12.92	0.22
P values			
CP	0.5410	< 0.0001	< 0.0001
AME	0.6338	0.0006	0.0218
CP × AME	0.3855	0.0194	0.0111

[a,b,c,d,e] Means with different superscript within a column and a factor differ significantly ($p < 0.05$)
[†] Standard error of the mean; AME: apparent metabolisable energy; CP: crude protein

Table 4: *The influence of dietary protein and metabolisable energy levels on growth performance of Venda chickens (7–17 weeks old).*

Parameter	Feed intake (g/bird/day)	Weight gain (g/bird)	FCR (g/g)
Protein (g kg^{-1} DM)			
120	71.73	976.99[b]	5.76[a]
150	70.00	1262.53[a]	4.34[b]
180	67.32	1256.33[a]	4.19[b]
SEM[†]	2.08	38.68	0.09
AME (MJ kg^{-1} DM)			
11.3	72.59[a]	1138.03	4.98[a]
12.0	71.87[a]	1219.91	4.79[ab]
12.4	64.61[b]	1137.91	4.50[b]
SEM[†]	2.08	36.68	0.09
CP × AME			
120 × 11.3	75.47[b]	1009.96[bc]	5.77[ab]
120 × 12.0	77.41[a]	1010.85[bc]	6.16[a]
120 × 12.4	62.33[c]	910.16[c]	5.40[b]
150 × 11.3	71.43[abc]	1197.31[ab]	4.64[c]
150 × 12.0	73.93[ab]	1392.28[a]	4.24[cde]
150 × 12.4	64.65[bc]	1198.00[ab]	4.16[cde]
180 × 11.3	70.87[abc]	1206.83[ab]	4.52[cd]
180 × 12.0	64.26[bc]	1256.61[a]	3.98[e]
180 × 12.4	66.85[abc]	1305.57[a]	4.05[de]
SEM[†]	3.60	66.99	0.16
P values			
CP	0.3408	< 0.0001	< 0.0001
AME	0.0257	0.2506	0.0149
CP × AME	0.2023	0.3062	0.0593

[a,b,c,d,e] Means with different superscript within a column and a factor differ significantly ($p < 0.05$)
[†] Standard error of the mean; AME: apparent metabolisable energy; CP: crude protein

3.2 Carcass parameters

There was no effect of CP and ME or their interaction on dressing per cent and breast relative weight (RW) ($p > 0.05$), except thigh RW that decreased with increasing levels of dietary ME ($p < 0.05$) (Table 5). Dietary AME of 11.3 MJ kg^{-1} showed higher thigh RW and lower fat RW. The differences within means on CP × AME interaction effect were observed on thigh, drumstick and fat relative weights, but there was no convincing trend.

3.3 Carcass chemical content

The influence of dietary protein and metabolisable energy levels on the carcass chemical content of the 17-week-old Venda chickens is shown in Table 6. There was CP × AME interaction effect on dry matter of the breast and leg muscles, ash and protein of the leg muscle ($p < 0.05$). The highest ash content of the leg muscle was recorded on dietary combination of 180 g CP kg^{-1} × 12.0 MJ AME kg^{-1}. Means of CP × AME interaction on ash, fat and protein of the breast muscle and fat of the leg muscle differed significantly, but there was no convincing trend. The protein and ash content of the breast muscle improved with increasing levels of dietary CP ($p < 0.05$). The fat content of the leg muscle decreased with increasing levels of dietary CP ($p < 0.05$). The breast muscle showed more protein content and less fat than leg muscle at all CP and ME levels.

Table 5: *The effect of dietary crude protein and metabolisable energy levels on some carcass parameters (%) of 17 weeks old Venda chickens.*

Parameter	Dressing	Breast RW	Thigh RW	Drumstick RW	Fat RW
Protein (g kg^{-1} DM)					
120	69.05	24.01	16.93[a]	14.75	4.40
150	70.10	23.47	17.08[ab]	14.7	3.18
180	69.04	23.92	17.45[a]	14.29	3.64
SEM[†]	0.53	0.50	0.21	0.25	0.57
AME (MJ kg^{-1} DM)					
11.3	68.89	23.65	17.65[a]	14.82	2.98[a]
12.0	70.21	23.83	17.30[a]	14.29	4.47[b]
12.4	69.59	23.90	16.63[b]	14.64	3.48[ab]
SEM[†]	0.53	0.50	0.21	0.25	0.58
CP × AME					
120 × 11.3	68.73	23.95	17.16[ab]	14.77[a]	3.23[a]
120 × 12.0	69.87	24.50	16.94[ab]	14.53[a]	5.81[b]
120 × 12.4	68.54	23.57	16.69[ab]	14.95[a]	4.13[ab]
150 × 11.3	69.31	23.72	17.98[a]	14.92[a]	2.18[a]
150 × 12.0	70.70	22.48	17.32[ab]	15.12[a]	5.16[b]
150 × 12.4	70.28	24.21	15.95[b]	14.20[ab]	2.20[a]
180 × 11.3	68.62	23.33	17.84[a]	14.78[a]	3.49[ab]
180 × 12.0	70.06	24.52	17.63[ab]	13.25[b]	3.33[ab]
180 × 12.4	69.95	23.94	17.26[ab]	14.78[a]	4.09[ab]
SEM[†]	0.91	0.86	0.36	0.43	0.98
P values					
CP	0.3767	0.7102	0.0805	0.3024	0.2773
AME	0.2165	0.9408	0.0036	0.3293	0.0977
CP × AME	0.9306	0.4769	0.2055	0.0747	0.3375

[a,b] Means with different superscript within a column and a factor differ significantly ($p < 0.05$)
[†] Standard error of the mean; AME: apparent metabolisable energy; CP: crude protein

4 Discussion

There was a highly significant interaction of dietary CP and AME on weight gain and FCR of six weeks old local chickens. Significant interaction between dietary CP and AME showed the importance of balanced protein to energy ratio to achieve optimum performance (Wang & Liu, 2002). Makinde & Egbekun (2016) reported a CP × AME interaction with high protein and energy levels resulting in higher body weight. In the present study, starter ration containing a CP content of 140 g kg^{-1} at different AME levels yielded the lowest weight gain as compared to 170 or 190 g kg^{-1} at all AME levels. The results suggest that Venda village chickens would require a dietary combination of 170 g CP kg^{-1} and 11.0 MJ kg^{-1} in the starter diet to maximise weight gain and FCR. These results imply that excess

CP (190 g CP g kg^{-1} feed) has no advantage in village chickens of 0–6 weeks old. The results are in agreement with the findings of Nguyen & Bunchasak (2005) who observed that 170–180 g CP kg^{-1} gave optimal results in Betong chickens of similar age. Previous studies however, reported higher optimal AME values than the present findings. Nakkazi *et al.* (2015) reported diet containing 180 g CP kg^{-1} and 11.7 MJ ME kg^{-1} were sufficient for rearing local chickens during the early growth phase (0–6 weeks), whereas Payne (1990) and Nguyen & Bunchasak (2005) recommended diets containing 11.46 and 12.56 MJ ME kg^{-1} respectively. Mbajiorgu (2010) observed that slightly higher dietary CP level of 178 g kg^{-1} DM and higher energy level of 14 MJ kg^{-1} DM allowed for optimal nutrient utilisation for growth in Venda chickens between one and six weeks of age. Alabi *et al.* (2013) found that 12.42 MJ ME kg^{-1} DM

Table 6: *The influence of dietary protein and metabolisable energy levels on the carcass chemical content of the 17 weeks old Venda chickens.*

Parameter	Breast muscle				Leg muscle			
	Dry matter	Ash	Fat	Protein	Dry matter	Ash	Fat	Protein
Protein (g kg^{-1} DM)								
120	270.29^a	45.27^b	37.81	892.44^b	253.47^a	39.86^b	141.61^a	813.85
150	270.75^a	52.31^a	35.67	907.03^a	246.56^b	40.89^b	121.65^b	824.60
180	265.33^b	52.91^a	44.67	901.84^a	240.20^b	42.22^a	117.31^b	824.72
SEM	1.34	1.34	2.97	3.85	2.46	0.62	5.32	5.43
AME (MJ kg^{-1} DM)								
11.3	267.60	52.90^a	38.35^a	902.55^a	243.98	40.96^{ab}	132.75	816.26
12.0	269.75	50.80^a	48.04^b	891.33^b	249.23	41.76^a	128.52	819.28
12.4	269.02	46.80^b	31.75^a	907.43^a	247.03	40.24^b	119.30	827.65
SEM	1.39	1.37	2.75	3.62	2.44	0.45	5.29	5.22
CP × AME								
120 × 11.3	270.89^{abc}	46.43^e	34.50^{bc}	894.32^{abc}	243.83^{bc}	40.76^b	139.64^b	824.55^{ab}
120 × 12.0	273.74^{ab}	44.57^e	48.82^{ab}	882.87^c	263.70^a	39.20^b	142.06^a	813.41^{ab}
120 × 12.4	266.28^{bcd}	44.79^e	30.09^c	900.00^{abc}	252.89^{ab}	39.60^b	143.14^a	803.60^b
150 × 11.3	264.14^{cd}	58.83^a	38.77^{bc}	904.00^{abc}	246.71^{bc}	41.43^b	135.32^c	805.73^b
150 × 12.0	274.18^a	52.52^{abcd}	39.46^{bc}	902.39^{abc}	242.98^{bc}	40.37^b	118.20^{abc}	826.86^{ab}
150 × 12.4	273.36^{abc}	45.58^{de}	28.76^c	914.71^a	250.01^{bc}	40.87^b	111.44^{bc}	841.12^a
180 × 11.3	267.77^{abcd}	53.42^{abc}	41.78^{abc}	909.33^{ab}	241.39^c	40.70^b	123.31^{abc}	818.37^{ab}
180 × 12.0	260.80^d	55.29^{ab}	55.83^a	888.73^{bc}	241.00^c	45.71^a	125.30^{abc}	817.56^{ab}
180 × 12.4	267.41^{abcd}	50.03^{bcde}	36.40^{bc}	907.47^{ab}	238.21^c	40.23^b	103.32^c	838.24^a
SEM	2.41	2.36	4.76	6.22	4.22	0.78	9.15	9.04
P values								
CP	0.0098	0.0002	0.0901	0.0309	0.0015	0.0034	0.0044	0.2751
ME	0.5224	0.0074	0.0011	0.0141	0.3254	0.2349	0.1967	0.3110
CP × AME	0.0009	0.0845	0.6745	0.7680	0.0470	0.0109	0.4449	0.0446

^{a,b,c,d,e} Means with different superscript within a column and a factor differ significantly ($p < 0.05$)
AME: apparent metabolisable energy; CP: crude protein

and 12.66 MJ ME kg^{-1} DM at 180 g CP kg^{-1} DM supported optimum growth rate and FCR, respectively, in Venda chickens. In the present study, chickens obtained a FCR of 3.2, which was slightly better than the FCR value of 3.5 noticed by Alabi *et al.* (2013). Differences in responses to dietary CP and AME interactions by local chickens might be attributed to different dietary protein and energy levels in the diet used during experimentation and also breed differences.

Feed intake of the local chickens during both the starter and grower phases was not influenced by the dietary CP level, agreeing with findings of Nguyen & Bunchasak (2005) who also reported no differences in feed intake of Betongs chickens with varying CP levels. Similarly, Nde-gwa *et al.* (2001) reported that indigenous growing chickens fed diets containing 170–230 g CP kg^{-1} had similar feed intake. On the contrary, Melesse *et al.* (2013) observed an increase in the level of feed consumption of Koekoek chickens with increasing levels of protein supplementation. The effect of dietary protein on feed intake in poultry species is inconsistent due to genotype, age, body weight, stage of maturity and sex of the bird.

The decrease of feed intake with increasing dietary AME levels may suggest that the birds regulated their intake according to dietary energy. Several authors have shown that chickens eat to satisfy their energy requirements (Scott *et al.*, 1982; Leeson *et al.*, 1996; Velkamp *et al.*, 2005; Nahashon *et al.*, 2006). Onwudike (1983) and Nawaz *et al.* (2006) also found that feed consumption in broilers was lower with higher energy diets. These observations seem to be applicable to the grower phase of the local chickens aged 7–17 weeks in the present study, as chicks aged less than six weeks old failed to adjust their feed intake to match AME in the diet. It can be suggested that the adjustment of feed intake according to dietary ME may be related to age and energy needs for maintenance, growth and production of birds.

In the grower phase, there was no significant interaction of dietary CP and AME levels on weight gain and FCR, but significant differences within means were observed. Weight gain improvement with increasing levels of dietary CP is in agreement with findings of several researchers that increased dietary protein content resulting in improved growth performance (Jackson *et al.*, 1982; Nguyen & Bunchasak, 2005). There were no significant differences observed between chickens that received CP levels of 150 and 180 g kg^{-1} at all dietary AME levels on weight gain and FCR. However, at 150 g CP kg^{-1} feed weight gain decreased with increasing dietary AME levels, with 12 MJ ME kg^{-1} supporting optimum weight gain and FCR. Optimal use of protein is imperative for any feeding system because protein supplements are usually more expensive than energy feeds and wasteful usage increases the cost of production besides leading to environmental degradation (Church & Kellens, 2002). Bikker *et al.* (1994) reported that feeding above the protein requirements did not result in an increase in protein deposition, but nitrogen excretion through the urine increased. Feeding beyond 150 g CP kg^{-1} feed DM, irrespective of AME levels, yielded no improvement in weight gain and FCR, suggesting 150 g CP kg^{-1} feed DM as a threshold level for optimum production of Venda chickens. It can be concluded that the dietary combination of 150 g CP kg^{-1} × 12.0 MJ ME kg^{-1} maximised weight gain and FCR during the grower phase of the local chickens used in the present study. Nguyen *et al.* (2010) reported no significant interaction effect between protein and energy on any performance parameters measured during the growing phase of Betong chickens and subsequently recommended 190 g CP kg^{-1} and 12.56 MJ AME kg^{-1}, which are higher than the present study findings. Makinde & Egbekun (2016) observed no CP × AME interaction on feed intake and weight gain for 6–12 weeks old Fulani ecotype chickens, but recommended 200 g CP kg^{-1} and 12.56 MJ ME kg^{-1}. The recommended AME in the grower phase is in accordance with those recommended by NRC (1994) (12.14 MJ kg^{-1}) and Tadelle & Ogle (2000) (11.99 MJ kg^{-1}). Variations in age, genotype of the chickens and the environment may explain observed differences in nutritional requirements between studies.

A study on the assessment of the nutrient adequacy from the crop contents of free-ranging indigenous chickens in the rural villages of the Vhembe District, South Africa observed CP content of 123 g kg^{-1} and 118 g kg^{-1} for growers and adults, respectively (Raphulu *et al.*, 2015b). According to NRC (1994), the recommended levels of CP in diets for growing chickens (not broilers) range from 150 g CP kg^{-1} DM to 200 g CP kg^{-1} DM and for mature chickens from 100 g CP kg^{-1} DM to 160 g CP kg^{-1} DM. It appears therefore as if the birds raised under village condition did not receive adequate levels of dietary protein to support efficient production. Protein deficit of 27 g kg^{-1} for growers between feed resource base for scavenging chickens in the rural villages and the required nutrients observed in the present study has to be compensated with supplemental feed. No information is available on the performance of adult indigenous chickens fed different protein and energy levels raised in closed confinement.

The improvement of FCR with increasing levels of energy is in close agreement with results of Nawaz *et al.* (2006) and Holsheimer & Veerkamp (1992). The obtained mean FCR during the grower phase in the present study was slightly better than that reported by Kingori *et al.* (2003) and Chemjor (1998), who found FCR values of 5.8 and 5.2, respectively, for indigenous chickens of 14–21 weeks old.

The shares of major basic carcass parts (breast, drumsticks and thighs) and the presence of certain tissues in them, as well as the chemical composition of the muscular tissue, are regarded as vital parameters determining broiler meat quality (Holcman *et al.*, 2003). Carcass relative weights of the local birds were not affected by dietary protein or AME levels, with the exception of thigh yield that decreased with increasing levels of dietary AME. The results of the present study confirm other reports where dietary protein did not affect carcass yields (Nguyen & Bunchasak, 2005; Iheukwumere *et al.*, 2007; Melesse *et al.*, 2013). Makinde & Egbekum (2016) observed no significant effect of dietary ME on carcass yield in Fulani Ecotype chickens. The observed decrease of thigh RW with increase in dietary AME is similar to the results of Nguyen *et al.* (2010) where high energy levels decreased wing and leg relative weights. Feeding above 120 g CP kg^{-1} DM improved ash content, protein content of the breast and decreased fat content of the muscle and these parameters were not affected by dietary AME levels. Furlan *et al.* (2004) also reported that carcass protein was affected by the level of protein in the diet. The increase in dietary protein improves carcass protein by reducing lipid deposition and increasing protein content (Gous *et al.*, 1990). An increase in protein intake induces a decrease in the protein to energy ratio that causes a reduction in energy intake relative to the protein intake, resulting in decreasing body fat percentage (Ghahri et al., 2010). Feeding 150 or 180 g CP kg^{-1} feed DM at any AME level yielded similar values. Results in the present study imply that dietary combination of 150 g CP kg^{-1} feed DM and 11.3 MJ ME kg^{-1} feed DM may optimise carcass chemical composition of 17 weeks old chickens, since there was no dietary AME effect observed. The fat content of the leg muscle was influenced by the changes in dietary protein. Higher protein diets induce higher meat protein content, while reducing the fat content of muscles (Si et al., 2001; Bogosav-Boskovic et al., 2010). The influence of dietary protein on the carcass chemical content seems to be dependent on the specific portion. In the present study, the breast muscle showed higher protein and less fat content when compared to leg muscle at all CP and AME levels. Ferket & Sell (1990) reported that breast meat contains more protein than leg meat. Diaz et al. (2010) related the differences in CP effect on the very structure of these portions, with breast being mostly composed of white fibres, as opposed to drumstick made up of muscles that contain more red fibres, which differ in metabolic function.

5   Conclusions

During 0–6 and 7–17 weeks, unsexed local Venda chickens would require dietary combinations of 170 g CP kg^{-1} feed DM and 11.0 MJ ME kg^{-1} feed DM and 150 g CP kg^{-1} feed DM and 12 MJ ME kg^{-1} feed DM, respectively, in their diets to optimise body weight gain. Neither dietary CP and AME nor their interactions influenced carcass relative weights, except for thigh yield that decreased with increasing AME levels. The dietary combination of 150 g CP kg^{-1} feed and 11.3 MJ ME kg^{-1} feed optimised ash content of muscles, protein content of the breast and fat content of the leg muscle of the 17 weeks old local chickens. It can be recommended that supplementation of 27 g CP kg^{-1} feed to grower scavenging chickens, would be enough to improve chicken production in the rural villages. Locally available feed resources high in protein like groundnuts, beans, meat and bone scraps, and insects should be used as supplement to compensate the nutrient intake deficit of scavenging chickens and also to reduce input costs.

Acknowledgements

We are grateful to the National Research Foundation (NRF) for financial support of the present study, indigenous chicken's farmers and the Limpopo Department of Agriculture and Rural Development officials for their assistance in this study.

References

Alabi, O. J., Ng'ambi, J. W. & Norris, D. (2013). Dietary energy level for optimum productivity and carcass characteristics of indigenous Venda chickens raised in closed confinement. *South African Journal of Animal Science*, 43, S75–S80.

AOAC (2000). *Official Methods of Analysis*. (17th ed.). The Association of Official Analytical Chemists, Alington, VA, USA.

Bikker, P., Verstegen, M. W. A. & Tamminga, S. (1994). Partitioning of dietary nitrogen between body components and waste in young pigs. *Netherlands Journal of Agricultural Science*, 42, 37–45.

Bogosav-Boskovic, S., Mitrovic, S., Djokovic, R., Ladimir Doskovic, V. & Djermanovic, V. (2010). Chemical composition of chicken meat produced in extensive indoor and free range rearing systems. *African Journal of Biotechnology*, 53, 9069–9075.

Campbell, R. G. & Taverner, M. R. (1998). Genotype and sex effects on the relationship between energy intake and protein deposition in growing pigs. *Journal of Animal Science*, 66, 676–686.

Chemjor, W. (1988). *Energy and protein requirements of indigenous chickens of Kenya*. Master's thesis, Egerton University, Kenya.

Church, D. C. & Kellens, R. O. (2002). *Livestock feeds and feeding*. Prentice Hall, New Jersey.

Díaz, O., Rodríguez, L., Torres, A. & Cobos, A. (2010). Chemical composition and physico-chemical properties of meat from capons as affected by breed and age. *Spanish Journal of Agricultural Research*, 8 (1), 91–99.

Ferket, P. R. & Sell, J. L. (1990). Effect of early protein and energy restriction of large turkey tomes fed high or low fat realimentation diets. 2. Carcass characteristics. *Poultry Science*, 69, 1982–1990.

Furlan, R. L., Faria, F., Rosa, P. S. & Macari, M. (2004). Does Low-Protein Diet Improve Broiler Performance Under Heat Stress Conditions? *Brazilian Journal Poultry Science*, 6, 71–79.

Ghahri, H., Gaykani, R. & Toloie, T. (2010). Effect of dietary crude protein level on performance and lysine requirements of male broiler chickens. *African Journal of Agricultural Research*, 5 (11), 1228–1234.

Gous, R. M., Emmans, G. C., Broadbent, L. A. & Fisher, C. (1990). Nutritional effects on the growth and fatness of broilers. *British Poultry Science*, 31, 495–505.

Holcman, A., Vadnjal, R., Žlender, B. & Stibiji, V. (2003). Chemical composition of chicken meat from free range and extensive indoor rearing. *Archiv für Geflügelkunde*, 67, 120–124.

Holsheimer, J. P. & Veerkamp, C. H. (1992). Effect of dietary energy, protein, and lysine content on performance and yields of two strains of male broiler chicks. *Poultry Science*, 71, 872–879.

Iheukwumere, F. C., Ndubuisi, E. C., Mazi, E. A. & Onyekere, M. U. (2007). Growth, blood chemistry and carcass yield of broilers fed cassava leaf meal (*Manihot esculenta* Crantz). *International Journal of Poultry Science*, 6, 555–559.

Jackson, S., Summer, J. D. & Leeson, S. (1982). Effect of dietary protein and energy on broiler performance and production costs. *Poultry Science*, 61, 2232–2240.

Kingori, A. M., Tuitoek, J. K., Muiruri, H. K. & Wachira, A. M. (2003). Protein requirements of growing indigenous chickens during the 14–21 weeks growing period. *South African Journal of Animal Science*, 33, 78–82.

Leeson, S., Caston, L. J. & D., S. J. (1996). Broiler response to diet energy. *Poultry Science*, 75, 529–535.

Makinde, O. A. & Egbekun, C. P. (2016). Determination of optimum dietary energy and protein levels for confined early–stage Fulani Ecotype chickens. *Livestock Research for Rural Development*, 28 (9), #164. Available at: http://www.lrrd.org/lrrd28/9/maki28164.html (accessed on: 3 March 2018).

Mbajirogu, C. A. (2010). *Effect of dietary energy to protein ratio levels on growth and productivity of indigenous Venda chickens raised in close confinement from one day up to 13 weeks of age*. Ph.D. thesis, Animal Production Department, University of Limpopo, South Africa.

Melesse, A. (2000). *Comparative studies on performance and physiological responses of Ethiopian indigenous ("Angete-melata") chicken and their F1 crosses to long term heat stress*. Ph.D. thesis, Martin-Luther University, Halle-Wittenberg, Berlin, Germany. pp. 4–5.

Melesse, A., Dotamo, E., Banerjee, S., Berihun, K. & Beyan, M. (2013). Studies on carcass traits, nutrient retention and utilization of Kokoeck chickens fed diets containing different protein levels with Iso-Caloric Ration. *Journal of Animal Science Advances*, 10, 532–543.

Nahashon, S. N., Adefope, N., Amenyenu, A. & Wright, D. (2006). Effect of varying metabolizable energy and crude protein concentrations in diets of pearl gray guinea fowl pullets 1. Growth performance. *Poultry Science Journal*, 85, 1847–1854.

Nakkazi, C., Kugonza, D. R., Kayitesi, A., Mulindwa, H. E. & Okot, M. W. (2015). The effect of diet and feeding system on the on-farm performance of local chickens during the early growth phase. *Livestock Research for Rural Development*, 27 (10), #204. Available at: http://www.lrrd.org/lrrd27/10/nakka27204.html (accessed on: 7 March 2018).

Nawaz, H., Mushtag, T. & Yaqoob, M. (2006). Effect of varying levels of energy and protein on live performance and carcass characteristics of broiler chicks. *Journal of Poultry Science*, 43, 388–393.

Ndegwa, J. M., Mead, R., Norrish, P., Kimani, K. W. & Wachira, A. M. (2001). The performance of indigenous Kenya chickens fed diets containing different protein levels during rearing. *Tropical Animal Health Production*, 33, 441–448.

Nguyen, T. V. & Bunchasak, C. (2005). Effects of dietary protein and energy on growth performance and carcass characteristics of Betong chicken at early growth stage. *Songklanakarin Journal of Science and Technology*, 27, 1171–1178.

Nguyen, T. V., Bunchasak, C. & Chantsavang, S. (2010). Effects of dietary protein and energy on growth performance and carcass characteristics of betong chickens (*Gallus domesticus*) during growing period. *International Journal of Poultry Science*, 9, 468–472.

NRC (1994). *Nutrient requirements of poultry. Ninth revised edition*. National Academy Press, Washington DC, USA. pp. 61–79

Onwudike, O. C. (1983). Energy and protein requirements of broilers in humid tropics. *Tropical Animal Production*, 8, 39–44.

Payne, W. J. A. (1990). *An Introduction to Animal Husbandry in the Tropics*. (4th ed.). Longman Scientific and Technical, Essex and New York. pp. 684–744

Raphulu, T., Jansen van Rensburg, C. & Coertze, R. J. (2015a). Carcass composition of Venda indigenous scavenging chickens under village management. *Journal of Agriculture and Rural Development in the Tropics and Subtropics*, 116 (1), 27–35.

Raphulu, T., Jansen van Rensburg, C. & van Ryssen, J. B. J. (2015b). Assessing nutrient adequacy from the crop contents of free-ranging indigenous chickens in rural villages of the Venda region of South Africa. *South African Journal of Animal Science*, 45, 143–152.

SAS Institute (2016). *SAS® Statistics Users Guide, Statistical Analysis System, 9.2 version*. SAS Institute Inc., Cary, NC, USA.

Scott, M. L., Neshei, M. C. & Young, R. J. (1982). *Poultry Nutrition*. (3rd ed.). ML.SCTT and Associates, Ithca, New York, USA.

Si, J., Fritts, C. A., Burnham, D. J. & Waldroup, P. W. (2001). Relationship of dietary lysine level to the concentration of all essential amino acids in broiler diets. *Poultry Science*, 80, 1472–1479.

Swatson, H. K., Nsahlai, I. V. & Byebwa, B. K. (2001). The status of smallholder poultry production in the Alfred District of KZN (South Africa): priorities for intervention. *In:* The Proceedings of Association of Institutions for Tropical Veterinary Medicine. 10th International Conference on "Livestock, Community and Environment" 20–23rd August 2001, Copenhagen, Denmark. pp. 143–149.

Tadelle, D. & Ogle, B. (2000). Nutritional status of village poultry in the central highlands of Ethiopia as assessed by analyses of crop contents. *Ethiopian Journal of Agricultural Science*, 17, 47–57.

Velkamp, T., Kwakkel, R. P., Ferket, P. R. & Verstegen, M. W. A. (2005). Growth response to dietary energy and lysine at high and low ambient temperature in male turkeys. *Poultry Science*, 84, 273–232.

Wang, S. Y. & Liu, H. Y. (2002). Effect of different energy and protein on production performance of broilers. *Shandong Agricultural Science*, 4, 43–44.

6

Roles and optimisation rate of potassium fertiliser for immature oil palm (*Elaeis guineensis* Jacq.) on an Ultisol soil in Indonesia

Sudradjat [a,*], Oky Dwi Purwanto [a], Ega Faustina [a], Feni Shintarika [b], Supijatno [a]

[a]*Department of Agronomy and Horticulture, Faculty of Agriculture, Bogor Agricultural University, Bogor, Indonesia*
[b]*State Polytechnic of Lampung, Lampung, Indonesia*

Abstract

Potassium (K) is an essential macronutrient needed in large amounts by oil palm as it is directly involved in physiological processes. This research focused on the influence of K fertiliser on the vegetative growth of oil palm and determined the optimum rate of K fertiliser for immature plants (aged 1 to 3 years). The study was conducted at IPB-Cargill Oil Palm Teaching Farm, Jonggol, Bogor, Indonesia, from March 2013 to March 2016. The application rates were 0, 196, 392, 588, and 784 g K_2O plant^{-1} year^{-1} during the first year; 0, 384, 768, 1152, and 1536 g K_2O plant^{-1} year^{-1} during the second year; and 0, 450, 900, 1350, and 1800 g K_2O plant^{-1} year^{-1} during the third year of the immature oil palms. This experiment used a randomized complete block design with three replications. The optimum fertiliser rate was calculated by differentiating the regression equation of the quadratic response curve for variable growth. Potassium fertiliser significantly affected plant morphology, and increased plant height, stem girth, frond number, frond length, and leaf area of frond 17. Potassium application also significantly increased chlorophyll content, stomatal density, and K nutrient content of the leaves of immature oil palm. The optimum K fertilizer rate for 1-, 2-, and 3-year-old immature oil palm was 512, 966, and 1430 g K_2O plant^{-1} year^{-1}, respectively. Application of K fertiliser provided the amount of K needed to support oil palm growth.

Keywords: fertiliser, fronds number, growth, macronutrient, potassium, stem girth

1 Introduction

Potassium plays important roles in photosynthesis, regulation of osmotic balance, and phloem transport in plants (Tripler *et al.*, 2006; Harris & Nazari, 2011). It also activates enzymes, regulates transpiration, transports assimilates, controls transport through the cell membrane, aids in the formation of proteins and carbohydrates, and strengthens plant tissues (Marschner, 2012). Potassium also supports resistance to abiotic and biotic stresses, such as drought, pests, and diseases (Wang *et al.*, 2013).

Potassium deficiency is usually found in plants grown on peat soil, sandy soil, and acid soils with low cation exchange capacity. This deficiency is due to low exchangeable K in the soil and the lack of K fertiliser application. In contrast, excessive K can cause nutrient imbalances and deficiencies of Ca and Mg (Reddy *et al.*, 1997; Roggatz *et al.*, 1999). Potassium is a highly mobile element in plants, and it is translocated from older to younger tissue (Selvaraja *et al.*, 2013). The presence of adequate levels of nutrients in plants is important for normal growth and high production. Fertiliser application can help to meet the nutrient needs of plants. The rate, manner, and timing of application, as well as the type of fertiliser, should be based on the age of the plants and soil characteristics.

Oil palm (*Elaeis guineensis* Jacq.) is an agricultural commodity that has a high economic value. Indonesia is the largest palm oil producer in the world, and it accounts for 45 % of global crude palm oil production (Pusdatin Pertanian, 2014). From 2004 to 2014, the oil palm plantation area in Indonesia grew by 7.7 % annually on average, and palm oil production increased by an average of 11.1 % per year (Ditjenbun, 2015). However, the available land for

* Corresponding author
Sudradjat (sudradjat_ipb@yahoo.com)

the cultivation of oil palm generally has low fertility levels. These lands need to be restored through the application of appropriate technical cultivation (intensification of land use) to enable plant growth.

However, fertilisation of oil palm is expensive, and it accounts for about 30 % of the production costs and 60 % of the maintenance costs. Therefore, appropriate recommendations for fertiliser application are needed to keep the cost of fertiliser to a minimum, while providing the maximum benefit (Sugiyono et al., 2005). Optimising the rate of fertilisation is intended to ensure that enough fertiliser is applied to meet crop needs, while limiting the amount of residues. The basic theory for determining the optimum rate relies on a quadratic function. This function can help determine the state of nutrients in deficient, replete, and toxic conditions (Corley & Tinker, 2003; Webb, 2009). The Indonesian Oil Palm Research Institute has formulated a general standard fertilisation rate for oil palm (Sutarta et al., 2007). However, this standard rate is not appropriate for ensuring maximum productivity at each location or for every soil type. Today, Indonesia has more than 11.9 million ha of oil palm plantations grown on various soil types, including Ultisols, Andisols, Entisols, Vertisols, Inceptisols, Spodosols, and Oxisols. Ultisols are classified as marginal soils, especially in tropical regions, and these soils have a low pH (pH < 5.5), low cation exchange capacity, low base saturation, and deep profiles (high leaching losses). As Ultisols are mostly found in high rainfall areas with more than 200 mm month^{-1}, high leaching losses and erosion occur. Marginal soils need special techniques or best management practices to reach maximum productivity (Paramananthan, 2013). Therefore, appropriate fertilisation rates for specific locations need to be identified so that oil palm can grow and reach maximum production, thus improving the potential of existing land. Efforts to increase the productivity of oil palm should be done as early as possible, ideally when the plants are still immature so that potential production will be high when the plants reach maturity. Therefore, the effects of K fertiliser on a marginal soil, Ultisol, were studied to identify the optimum fertiliser rate for immature oil palms during their first 3 years of development.

2 Materials and methods

Research was conducted at IPB-Cargill Oil Palm Teaching Farm, Jonggol, Bogor, Indonesia, on an Ultisol at an altitude of ca. 113 m asl. Analysis of fertiliser, soil, leaf tissue, and frond tissue was carried out at the Laboratory of Chemistry and Soil Fertility, Department of Soil Science and Land Resources, Bogor Agricultural University. The

study was conducted from March 2013 to March 2016 (year 1–3).

Plant materials used in this study were immature oil palms aged 1 month of the Tenera type, Dami Mas variety. These palms were planted at a density of 136 plants hectare^{-1} with a spacing of 9.2 m × 9.2 m × 9.2 m; the planting system was an equilateral triangle. Holes of 60 cm × 60 cm × 60 cm were dug before planting, and basic fertiliser was added to each hole at the following amounts: 60 kg of manure, 500 g of rock phosphate, and 500 g of dolomite. This experiment used a randomized complete block design with one factor: five levels of K fertiliser. Each treatment consisted of five plants, and was repeated thrice leading to a total of 75 plants.

Fertiliser was broadcast spread in a circle around the oil palm twice a year, once at the beginning and once at the end of the rainy season. Fertiliser application in the first year was performed at 4 and 9 months after planting (MAP), in the second year at 14 and 20 MAP, and in the third year at 25 and 32 MAP. Each application was one half of the yearly fertiliser rate. In addition to the K fertiliser treatments (Table 1), each oil palm received standard fertiliser with nitrogen (N) and phosphorus (P) at 250 g N and 255 g P_2O_5 plant^{-1} year^{-1} in the first year; 630 g N and 450 g P_2O_5 plant^{-1} year^{-1} in the second year, and 960 g N and 720 g P_2O_5 plant^{-1} year^{-1} in the third year of the experiment.

Table 1: K fertiliser treatments on immature oil palm plants during the 3 years of the experiment (n = 15 per treatment).

K rate	Rate of fertilizers (g K_2O plant^{-1} year^{-1})		
	First year	Second year	Third year
Level 0 (control)	0	0	0
Level 1	196	384	450
Level 2	392	768	900
Level 3	588	1152	1350
Level 4	784	1536	1800

Plant morphological responses were evaluated based on the variables plant height, stem girth, frond number, frond length, and leaf area. Plant height was measured from the base of the stem to the youngest unfolded leaf and stem girth was determined at approximately 20 cm above the ground. The number of fronds that were still fresh and the leaflets of fronds that had completely opened were counted. The youngest leaf that was completely open was designated as the first leaf following Legros et al. (2009).

Frond lengths were measured from the frond base that was attached to the stem to the tip of fronds 9 (first-year immature oil palm) and fronds 17 (second-year and third-year immature oil palm). Leaf area measurements were made at the leaf midrib of frond 9 (first-year immature oil palm) and frond 17 (second-year and third-year immature oil palm). Leaf area was calculated using the following formula: leaf area $= \left((\sum_{(i=1)}^{2} l_i \times w_i)/2 \right) \times 2n \times k$, where l_i was the length of leaflets (m), w_i was the width of the leaflets (m), n was the number of leaflets left or right, and k was a constant (0.57 for immature oil palm) (Hardon *et al.*, 1969; Sutarta *et al.*, 2007).

Evaluations of physiological plant response were based on the variables chlorophyll content, stomatal density, and K nutrient content in the leaves. Analysis of chlorophyll content and nutrient content of K in the leaves was done on leaf samples of frond 9 in first-year immature oil palm and leaf samples of frond 17 in second-year and third-year immature oil palm (Chapman & Gray, 1949; Ochs & Olivin, 1977; von Uexkull & Fairhurst, 1991). Stomatal density was determined by applying cellulose acetate to the lower surface of the cut leaves and leaving it to dry; then these leaves were affixed with transparent insulation and placed on object glass. Stomata were observed using a microscope with a magnification of 10×40.

Study of K dynamics is essential because K is mobile in soil. The measurement of K dynamics in the soil was carried out at three soil-layer depths (0–20, 20–40, and 40–60 cm) within the radius of fertiliser application around each oil palm at 12 MAP (first year), 24 MAP (second year), and 36 MAP (third year). These measurements were done for the level 3 K rate only (Table 1), as this was assumed to be the optimum rate.

Nutrient balance calculation was done using the following method:

(1) *Potassium source*

K total in the soil before fertilisation (g) =
K content in the soil (%) × soil dry weight.
K in fertiliser (g) =
K content in the fertiliser (%) × the amount of fertiliser applied.

(2) *K Recovered*

K in the soil one year after fertiliser application (g) =
K content in the soil after fertilisation (%) × soil dry weight.
Plant uptake (g) = K content in the rachis and leaflets of frond 17 × dry weight of plant tissue.

The plant nutrient contents used in calculating the nutrient balance only in the rachis and leaflets referred to the formula of Aholoukpe *et al.* (2013):
Dry weight (DW) frond =
$1.147 + 2.135 \times$ DW rachis (kg)

(3) *Fertiliser efficiency (%) =*

$$\frac{\text{plant uptake}}{\text{the amount of K in the fertiliser}} \times 100\%$$

(4) *Unmeasured fertiliser (%) =*

$$\frac{\text{the amount of K in the fertiliser–plant uptake}}{\text{the amount of K in the fertiliser}} \times 100\%$$

Data were analysed by analysis of variance (ANOVA) at the level $P < 0.05$, using SAS Proprietary Software 9.4 (TS 1M3). If a significant effect was found, then analysis proceeded with a test of the orthogonal polynomial at level $P < 0.01$ and $P < 0.05$. In addition, regression analyses were done to determine the optimum rate of fertilisers (Mattjik & Sumertajaya, 2006).

3 Results

3.1 Soil characteristics and climatic conditions

Soil from the research site underwent initial analysis in March 2013 for the circle of fertiliser application and the area between two oil palm rows (dead pathway; Table 2).

The average annual rainfall during the experimental period (April 2013 to March 2016) was 2734 mm year^{-1} with an average of 227.8 mm month^{-1}. Rain occurred on 140 days year^{-1} on average, the average temperature ranged from 26 to 32 °C, and the average humidity was 73.8 %.

Based on the Schmidt-Ferguson climate criteria, the rainfall distribution in year 1 was evenly distributed throughout the year without dry months, and there was no water deficit. During the second experimental year, two dry months occurred (September and October 2014). Extended droughts occurred in the third year, with a dry period of six months from May to October 2015.

3.2 Morphological responses

Potassium fertiliser had a significant effect on plant height of oil palm at 12 MAP, but not at 24 MAP and 36 MAP (Table 3). In addition, the long dry period of six months in the third year resulted in a water deficit that affected plant morphology, especially plant height. Plant growth in the second year was approximately 193 cm but only 149 cm in the third year (mean overall treatments).

Application of K fertiliser yielded a quadratic response for stem girth of oil palm at 6 and 12 MAP, 24 MAP, and 30 and 36 MAP, but not at 18 MAP. Application of K fertiliser at level 2 increased the stem girth up to 23 % at 12 MAP, 7.9 % at 24 MAP, and 9.7 % at 36 MAP compared to the control.

Application of K fertiliser increased the frond number of oil palm at 12, 18, 24, 30 and 36 MAP. At level 2, it increased the number of fronds of oil palm up to 59.4 % at

Table 2: *Analysis of physical and chemical soil properties before the start of the experiment.*

Soil characteristics	Value [†]	Criteria [*]	Value [‡]	Criteria [*]
Texture				
Sand (%)	17.34		18.63	
Dust (%)	28.67		22.75	
Clay (%)	53.99		58.63	
pH_{H_2O}	4.55	Acid	4.55	Acid
pH_{KCl}	3.85		3.85	
C-organic (%)	2.15	Moderate	2.19	Moderate
N-total	0.19	Low	0.20	Low
P-available (ppm)	10.15	Low	11.15	Low
P-total (ppm)	88.30	Very low	92.75	Very low
Cation exchange value				
Ca (meq/100 g)	4.14	Low	3.68	Low
Mg (meq/100 g)	1.86	Moderate	1.56	Moderate
K (meq/100 g)	0.38	Moderate	0.32	Moderate
Na (meq/100 g)	0.28	Low	0.24	Low
CEC (meq/100 g)	34.38	High	34.18	High
Alkali saturation (%)	19.39	Very low	16.81	Very low
Acidity				
Al (meq/100 g)	10.84		10.65	
H (meq/100 g)	2.34		3.03	

[†]: analysis sample on oil palm circle. [‡]: analysis sample on oil palm dead pathway.
[*]: based on general criteria of soil chemical properties (Balittanah, 2005).

12 MAP, 30.0 % at 24 MAP, and 40.0 % at 36 MAP as compared to the control.

Potassium fertiliser application affected quadratically the leaf area in the second-year immature oil palm (18 and 24 MAP) and the third-year immature oil palm (30 and 36 MAP), but it had no effect in the first year. Potassium fertiliser application at level 3 increased the leaf area of oil palm up to 25.8 % at 24 MAP and 11.1 % at 36 MAP compared to the control.

3.3 Physiological response

Potassium fertiliser application linearly affected the chlorophyll content at 6 and 12 MAP, but it did not affect the chlorophyll content in the second-year and third-year immature oil palm (Table 4). Potassium fertiliser at the level 3 rate was associated with an increase in the chlorophyll content of 9.3 % compared to controls at 12 MAP. Potassium fertiliser application quadratically affected the stomatal density at 12 and 24 MAP, but linearly affected it at 36 MAP. Potassium fertiliser at the level 3 rate increased the stomatal density of leaves by 7.4 % compared to controls at 12 MAP, while the rate level 2 increased the density of leaf stomata by 15.2 % compared to controls at 24 MAP.

Potassium fertiliser application linearly affected the nutrient content of K in the leaves at 6, 12, and 18 MAP, but not at 24 and 36 MAP. This result suggests that the K nutrient at 6, 12, and 18 MAP was absorbed and was needed by the plants for growth, although it is possible that the K was not yet optimally absorbed. The K nutrient content in the leaf tissue was suspected to be below the critical nutrient levels.

3.4 Potassium nutrient content in the soil

Measurement of K nutrient content in the soil in the first year (Fig. 1a) was performed at 12 MAP, in the second year (Fig. 1b) at 24 MAP, and in the third year at 36 MAP (Fig. 1c). Figure 1 shows that K nutrient content in the soil for oil palm at 12, 24, and 36 MAP decreased with increasing soil depth. The movement of the K nutrient accumulated at a depth of 0–20 cm, while the content was the lowest at a depth of 40–60 cm. This finding suggests that some of the K from fertilisers did not reach deeper soil layers. Potassium

Table 3: *The effect of K fertiliser application on the morphological responses of immature oil palm aged 1 to 3 years.*

Rate level of K	Plant height (cm)						
	0 MAP	6 MAP	12 MAP	18 MAP	24 MAP	30 MAP	36 MAP
Level 0 (control)	158.7	240.3	292.0	431.4	496.4	533.6	623.4
Level 1	162.5	246.7	310.2	431.7	500.0	556.4	652.1
Level 2	165.8	248.1	308.8	431.5	497.5	561.5	667.5
Level 3	167.7	257.9	318.0	434.4	513.8	568.2	654.2
Level 4	166.3	248.1	313.0	431.6	500.2	557.6	656.9
Response pattern $^{\text{¢}}$	*ns*	*ns*	Q^*	*ns*	*ns*	*ns*	*ns*
	Stem girth (cm)						
Level 0 (control)	25.2	44.2	79.0	134.4	167.0	222.8	239.6
Level 1	28.9	47.7	86.4	137.2	170.0	233.7	256.3
Level 2	28.7	48.9	90.3	140.1	176.6	238.0	258.2
Level 3	29.7	51.8	97.2	140.2	180.1	240.2	262.8
Level 4	30.5	50.2	84.1	137.1	178.7	236.3	257.5
Response pattern $^{\text{¢}}$	*ns*	Q^{**}	Q^{**}	*ns*	Q^*	Q^*	Q^*
	Frond number						
Level 0 (control)	14	24	32	52	70	45	60
Level 1	16	26	42	70	88	65	82
Level 2	16	27	47	66	86	63	80
Level 3	17	29	51	71	91	66	84
Level 4	16	29	48	69	89	63	80
Response pattern $^{\text{¢}}$	*ns*	*ns*	Q^{**}	Q^{**}	Q^*	Q^{**}	Q^{**}
	Frond length (cm)						
Level 0 (control)	118.1	147.3	200.9	301.1	374.2	384.9	439.1
Level 1	121.3	152.1	207.3	280.4	374.4	411.9	468.0
Level 2	123.5	153.4	205.7	287.8	374.9	416.5	464.5
Level 3	123.4	154.9	215.8	306.5	377.2	420.2	470.8
Level 4	122.9	153.5	207.2	291.9	375.4	416.6	464.9
Response pattern $^{\text{¢}}$	*ns*	*ns*	*ns*	Q^{**}	*ns*	Q^*	Q^*
	Leaf area (m^2)						
Level 0 (control)	0.38	1.04	1.85	2.46	2.87	3.80	4.69
Level 1	0.39	1.17	1.93	2.38	3.49	4.12	5.42
Level 2	0.43	1.13	1.94	2.26	3.51	4.18	5.34
Level 3	0.45	1.17	2.11	2.72	3.61	4.27	5.21
Level 4	0.43	1.26	1.93	2.48	3.56	4.08	5.26
Response pattern $^{\text{¢}}$	*ns*	*ns*	*ns*	L^*	Q^{**}	Q^{**}	Q^*

$^{\text{¢}}$: based on contrast test. **: significance of treatment effect at level $P < 0.01$. *: significance of treatment effect at level $P < 0.05$. *ns*: not significant. *L*: linear response. *Q*: quadratic response. MAP: month after planting.

Table 4: *The effect of K fertiliser on physiological response on immature oil palm aged 1 to 3 years.*

Rate level of K	Chlorophyll content ($mg\,cm^{-2}$)					
	6 MAP	12 MAP	18 MAP	24 MAP	30 MAP	36 MAP
Level 0 (control)	0.041	0.043	0.043	0.044	0.034	0.043
Level 1	0.043	0.044	0.044	0.045	0.036	0.044
Level 2	0.043	0.044	0.043	0.047	0.037	0.043
Level 3	0.045	0.046	0.044	0.047	0.039	0.044
Level 4	0.046	0.047	0.044	0.046	0.036	0.043
Response pattern $^{¢}$	L^{**}	L^{**}	*ns*	*ns*	*ns*	*ns*
	Stomatal density (unit mm^{-1})					
Level 0 (control)	188.90	210.60	217.69	245.75		189.6
Level 1	207.80	216.70	227.04	259.35		215.0
Level 2	213.40	218.70	249.15	283.16		226.9
Level 3	216.90	226.20	250.00	279.76		205.0
Level 4	211.60	215.40	238.95	270.41		225.3
Response pattern $^{¢}$	*ns*	Q^{*}	*ns*	Q^{*}		L^{*}
	Leaf nutrient content of K (%)					
Level 0 (control)	0.82	0.85	0.96	0.65		0.49
Level 1	0.97	0.88	0.93	0.59		0.51
Level 2	1.10	1.08	1.07	0.63		0.60
Level 3	0.99	0.91	1.08	0.67		0.57
Level 4	1.06	1.01	1.05	0.65		0.52
Response pattern $^{¢}$	L^{*}	L^{*}	L^{*}	*ns*		*ns*

$^{¢}$: based on polynomial orthogonal test. **: significance of treatment effect at level $P < 0.01$.
*: significance of treatment effect at level $P < 0.05$. *ns*: not significant.
L: linear response. *Q*: quadratic response. MAP: months after planting.

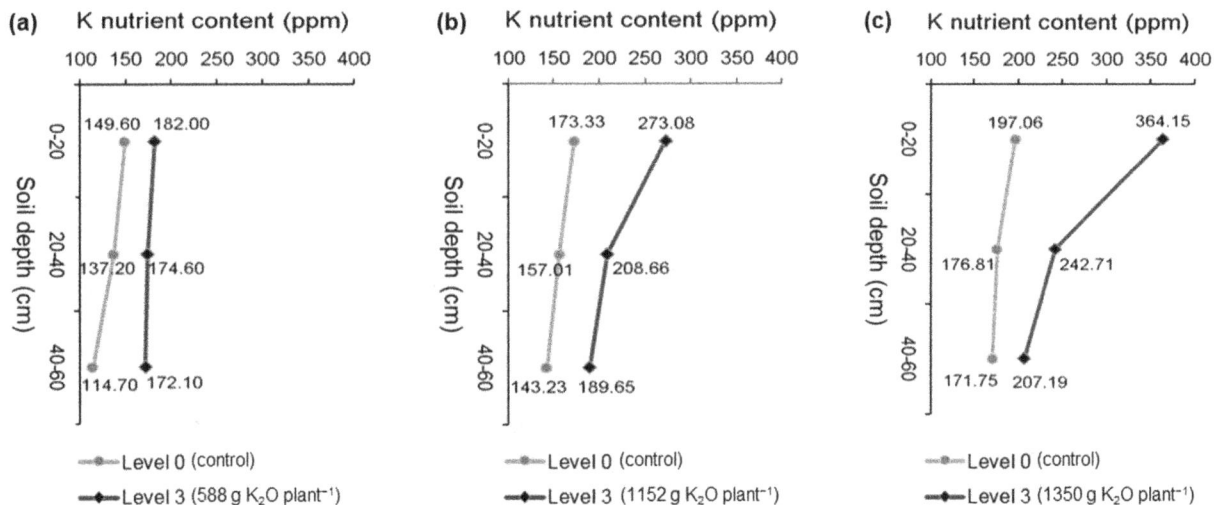

Fig. 1: *Potassium nutrient content of palm circle at various depths of the soil layers for first-year (a), second-year (b), and third-year (c).*

availability in the soil is influenced by weathering of minerals, K fixation by clay, leaching, erosion, and plant uptake. When the temperature increases, the movement of K also increases through diffusion (Korb *et al.*, 2002).

3.5 Optimization rate of K fertiliser on immature oil palm aged 1, 2, and 3 years

The effect of fertiliser optimisation was assessed for the variables stem girth and frond number. Siallagan *et al.* (2014) stated that the optimum rate could be obtained by lowering the quadratic curve regression equation on variables of morphology that were significantly affected, while Sudradjat *et al.* (2014a) stated that the optimum rate could be seen from the morphology variable that was the most responsive. The results of optimum rate in the first-year im-

mature oil palm, the second-year immature oil palm, and the third-year immature oil palm were $512, 966$, and 1431 g K_2O ha^{-1} year^{-1}, respectively (Table 5).

3.6 Potassium nutrient balance

Nutrient balance of K in this experiment was investigated to determine the optimum rate range. Fertiliser efficiency, which shows that nutrients derived from fertilisers can be absorbed by plants, was 26 to 27 % in this research (Table 6), while the fertiliser efficiency of previous research (Busyra, 2010) was 17 to 39 %. Unmeasured fertiliser in the current study reached 73 %. This fertiliser may have been in other plant tissues (fruits, seeds, roots) or partially lost because of leaching and adsorption to soil (clay type $1:2$) (Havlin *et al.*, 2005).

Table 5: *Optimum rate of K fertiliser based on regression function on immature oil palm.*

Variable	Age of plants (MAP[†])	Regression function	R^2	Optimum rate (g K_2O plant^{-1} year^{-1})
Stem girth	6	$y = -0.00002x^2 + 0.020x + 44.13$	0.95	500
	12	$y = -0.00007x^2 + 0.065x + 77.80$	0.77	464
Fronds number	10	$y = -0.00001x^2 + 0.0118x + 22.116$	0.99	590
	12	$y = -0.00003x^2 + 0.0297x + 28.625$	0.96	495
		Average on the first-year immature oil palm		512 ± 54
Stem girth	20	$y = -0.00001x^2 + 0.0221x + 146.72$	0.91	1105
	24	$y = -0.000006x^2 + 0.0173x + 166.09$	0.94	1442
Fronds number	14	$y = -0.00002x^2 + 0.0347x + 41.891$	0.83	868
	16	$y = -0.00002x^2 + 0.0355x + 47.598$	0.81	888
	18	$y = -0.00002x^2 + 0.034x + 53.598$	0.81	850
	20	$y = -0.00002x^2 + 0.0331x + 58.264$	0.84	828
	22	$y = -0.00002x^2 + 0.0344x + 64.154$	0.84	860
	24	$y = -0.00002x^2 + 0.0354x + 72.176$	0.82	885
		Average on second-year immature oil palm		966 ± 211
Stem girth	26	$y = -0.000009x^2 + 0.0234x + 215.18$	0.94	1300
	28	$y = -0.00001x^2 + 0.0254x + 219.63$	0.97	1270
	30	$y = -0.00001x^2 + 0.0277x + 222.93$	0.99	1385
	36	$y = -0.00001x^2 + 0.0357x + 240.49$	0.95	1785
Fronds number	26	$y = -0.00001x^2 + 0.0323x + 41.054$	0.83	1615
	28	$y = -0.00001x^2 + 0.0338x + 44.932$	0.85	1690
	30	$y = -0.00001x^2 + 0.0339x + 47.224$	0.85	1695
	32	$y = -0.00001x^2 + 0.0342x + 50.761$	0.84	1710
	34	$y = -0.00002x^2 + 0.037x + 56.805$	0.85	925
	36	$y = -0.00002x^2 + 0.0375x + 62.542$	0.85	938
		Average on third-year immature oil palm		1431 ± 319

[†]MAP: months after planting.

Table 6: *Nutrient balance of potassium in immature oil palm aged 1 to 3 years.*

Analysis	per plant		
	First-year	*Second-year*	*Third-year*
Potassium source			
Soil prior to fertiliser (g)	147.7	672.1	1194.8
Fertiliser (g)	488.0	955.9	1120.2
Recovered nutrient			
Soil after fertiliser (g)	224.5	545.7	1991.2
Plant uptake (g)	132.4	259.9	285.5
Fertiliser efficiency (%)	27.0	27.2	26.2
Unmeasured fertiliser (%)	73.0	72.8	73.8

4 Discussion

For 3-year-old oil palm, K has a significant effect on the morphological variables stem girth, frond number, frond length, and leaf area, but it has no effect on plant height (Table 3). The stem girth is closely correlated with production (Corley & Tinker, 2003, 2016) likely because it is very closely related to the stem function of oil palm. The stem contains carrier systems for water, plant nutrients, and the results of photosynthesis, and it is also the largest site of food storage in plants (Corley & Tinker, 2003). Muhdi *et al.* (2015) explained that the oil palm stem has the highest proportion of biomass compared with other organs, such as the midrib and leaves. The oil palm stem is also suspected to be an active sink, which would explain the increase in stem girth in response to K fertiliser (Goh & Hardter, 2003). One of the roles of K is to support transfer of the results of photosynthesis to the tissues that need these, and sufficient K content in plant tissues assists in this transfer process within stems.

Morphological variables such as girth stem, frond number, frond length, and leaf area were strongly affected by water shortage (Table 3). Water deficit occurred during the period 26–31 MAP, with an average rainfall of only 19 mm month^{-1} (against 325 mm per month during the rest of the experimental period). Sudradjat *et al.* (2014b) explained that a long dry period can affect the amount of oil palm fronds as the uptake of nutrients is blocked. In the current study, young fronds did not unfold (leaf spear) during the water deficit period (30 MAP), but the number of fronds significantly increased at 36 MAP when rainfall was adequate. Young midribs (leaf spears) were inhibited and growth was restrained during drought stress, but the midribs began to reopen when rainfall returned to normal.

The rate of growth of oil palm is related to water availability, especially rainfall during periods of growth. Taiz & Zeiger (1998) explained that drought reduces growth and photosynthesis. The process of growth of oil palm needs nutrients that are absorbed by the soil, such as K which plays a role in cell osmotic regulation and regulation of enzymes. Water deficit conditions decrease the effectiveness of mineral nutrient uptake and translocation of photosynthate into plant tissue and cause plants to exhibit stunted growth (Darmosarkoro *et al.*, 2001).

Application of K fertiliser had no effect on the frond length at 12 and 24 MAP, but had a significant effect on the frond length at 36 MAP. Greater frond length could increase the leaf area index, which could in turn potentially increase yield components such as fruit weight (Prayitno *et al.*, 2008). Table 3 shows that K has a linear effect at 18 MAP and a quadratic effect at 24, 30, and 36 MAP on the leaf area of oil palm. Leaf area increases with increasing age of oil palm, and the ability of photosynthesis is related to leaf area (Cha-um *et al.*, 2013). Arsyad *et al.* (2012) explained that improving nutrient uptake requires a sufficient quantity of groundwater as well as optimum leaf surface area. The more photosynthate produced, the higher the effectiveness of photosynthesis.

Potassium is directly involved in several physiological processes (Table 4), and it plays important roles in the biochemical and physiological processes that are vital to plant growth (Cakmak & Kirkby, 2008). The chlorophyll content in first-year immature oil palm seemed to increase along with the amount of K fertiliser applied. High chlorophyll content indicates that photosynthesis is efficient and the plants have energy for growth (Suharno *et al.*, 2007). In this study, K only affected leaf chlorophyll content in first-year immature oil palm. Leaf chlorophyll content in second-year and third-year immature oil palm was assumed to have reached the maximum.

This study showed that stomatal density had an inconsistent response to application of K fertiliser from first-year to third-year immature oil palm. Table 4 shows that K affected the stomatal density at 12 to 36 MAP, with the exception of 18 MAP. The inconsistent stomatal density response might be closely related to K mobility in plant tissue. Potassium is involved in maintaining the osmotic potential of plants, including setting the opening and closing of stomata, which allow the CO_2 needed for photosynthesis to enter plants (Corley & Tinker, 2003). Stomata have several characteristics that control the rate of photosynthesis, such as density, size, and conductance. Stomatal density can increase the activity of the gas exchange through the stomata, which can increase photosynthesis, transpiration, and respiration in the leaves (Khazaei *et al.*, 2010).

Potassium content in leaf tissue at 24 and 36 MAP was lower compared to previous time points. This decrease was presumably because of the phase change from vegetative to generative plants at 24 and 36 MAP, which resulted in nutrients being transferred to the seeds and fruit that were forming (Schwab *et al.*, 2007). The critical nutrient potassium level of leaf frond 9 is 1.25 % (Ochs & Olivin, 1977) and that of the leaf midrib frond 17 is 1 % in young oil palm less than 6 years after transplanting (Ochs & Olivin, 1977; von Uexkull & Fairhurst, 1991; IFIA, 1992). Table 4 indicates that the leaf K content (average over five treatments) at 12, 24, and 36 MAP was 0.95, 0.64, and 0.54 %, respectively. Therefore, the value of leaf K content was 24 % (the first-year), 36 % (the second-year), and 46 % (the third-year) lower as compared to this critical level.

5   Conclusion

Potassium consistently increases morphological variables such as girth stem, frond number, frond length, and leaf area at the age of one to three years of oil palm. Based on the response of morphological variables on potassium fertilisation we obtained the quadratic equation to determine the optimum rate according to the age of the plant. This study has determined the optimum rate for one, two and three years old of oil palm planted on Ultisol soils, i.e. 512, 966 and 1430 g K_2O plant^{-1} year^{-1}. Potassium is needed in large quantity by oil palm for vegetative growth during immature period (0 to 36 months). The optimum rate of K fertiliser obtained from this research can be recommended to accelerate the productive period and to increase the overall productivity of oil palms cultivated in Indonesia.

Acknowledgements

This research was supported by PT Cargill Indonesia and the Department of Agronomy and Horticulture of Bogor Agricultural University.

References

Aholoukpe, H., Dubos, B., Flori, A., Deleporte, P., Amadji, G., Chotte, J. L. & Blavet, D. (2013). Estimating aboveground biomass of oil palm: Allometric equations for estimating frond biomass. *Forest Ecology Management*, 292, 122–129.

Arsyad, A. R., Junedi, H. & Farni, Y. (2012). Pemupukan kelapa sawit berdasarkan potensi produksi untuk meningkatkan hasil TBS pada lahan marginal Kumpeh. *Jurnal Penelitian Universitas Jambi Seri Sains*, 14 (1), 29–36.

Balittanah (2005). Analisis Kimia Tanah, Tanaman, Air, dan Pupuk. Balai Penelitian Tanah, Badan Penelitian dan Pengembangan Pertanian, Departemen Pertanian, Jakarta.

Busyra, B. S. (2010). Phosphorus requirement referred to phosphate status of rice field in Jambi Province. *Jurnal Agronomi Indonesia*, 8 (1), 60–74.

Cakmak, J. & Kirkby, E. A. (2008). Role of magnesium in carbon partitioning and alleviating photo oxidative damage. *Physiology Plant*, 133 (4), 692–704.

Cha-um, S., Yamada, N., Takabe, T. & Kirdmanee, C. (2013). Physiological features and growth characters of oil palm (*Elaeis guineensis* Jacq.) in response to reduced water-deficit and rewatering. *Australian Journal of Crop Sciences*, 7 (3), 432–439.

Chapman, G. W. & Gray, H. M. (1949). Leaf analysis and the nutrition of the oil palm (*Elaeis guineensis* Jacq.). *Annals of Botany*, 13 (52), 415–433.

Corley, R. H. & Tinker, P. B. (2003). *The Oil Palm*. (4th ed.). Blackwell Science Ltd, Oxford.

Corley, R. H. V. & Tinker, P. B. (2016). *The Oil Palm*. (5th ed.). Blackwell Science Ltd, Oxford.

Darmosarkoro, W., Harahap, I. Y. & Syamsuddin, E. (2001). Pengaruh kekeringan pada tanaman kelapa sawit dan upaya penanggulangannya. *Warta Pusat Penelitian Kelapa Sawit*, 9 (3), 83–96.

Ditjenbun (2015). Statistik Perkebunan Indonesia Komoditas Kelapa Sawit 2014–2016. Direktorat Jenderal Perkebunan, Jakarta.

Goh, K. J. & Hardter, R. (2003). General oil palm nutrition. *In:* Fairhurst, T. H. & Hardter, R. (eds.), *Oil Palm Management for Large and Sustainable Yield*. pp. 191–230, Potash and Phosphate Institute of Canada, Norcross.

Hardon, J. J., Williams, C. N. & Watson, I. (1969). Leaf area and yield in the oil palm in Malaya. *Experimental Agriculture*, 5, 25–32.

Harris, A. & Nazari, Y. A. (2011). Kajian status hara tanah dan jaringan tanaman kelapa sawit di kebun kelapa sawit Tungkap. *Agroscientiae*, 18 (3), 122–128.

Havlin, J. L., Beaton, J. D., Tisdale, S. L. & Nelson, W. L. (2005). *Soil Fertility and Fertilizer*. Pearson Prentice Hall, Upper Saddle River, New Jersey.

IFIA (1992). Agricultural Research Station. International Fertilizer Industry Association, Limburgerhof, Germany.

Khazaei, H., Monneveux, P., Hongbo, S. & Mohammady, S. (2010). Variation for stomatal characteristics and water use efficiency among diploid, tetraploid and hexaploid Iranian wheat landraces. *Genetic Resources and Crop Evolution*, 57, 307–314.

Korb, N., Jones, C. & Jacobsen, J. (2002). Potassium Cycling, Testing, and Fertilizer Recommenda-tions. Communications Coordinator, Communications Services, Montana State University, Bozeman, Montana.

Legros, S., Mialet-Sera, I., Caliman, J. P., Siregar, F. A., Clement-Vidal, A., Fabre, D. & Dingkuhn, M. (2009). Phenology, growth and physiological adjustments of oil palm (*Elaeis guineensis* Jacq.) to sink limitation induced by fruit pruning. *Annals of Botany*, 104, 1183–1194.

Marschner, P. (2012). *Marschner's Mineral Nutrition of Higher Plants*. (3rd ed.). Academic Press, London.

Mattjik, A. A. & Sumertajaya, I. M. (2006). *Perancangan Percobaan dengan Aplikasi SAS dan Minitab*. PT Penerbit IPB Press, Bogor.

Muhdi, Risnasari, I., Bayu, E. S., Hanafiah, D. S., Hutasoit, A., Sitanggang, G. N. & Silaban, D. S. (2015). Kuantifikasi biomassa perkebunan kelapa sawit di Langkat, Sumatera Utara. *Jurnal Pertanian Tropik*, 2 (1), 17–20.

Ochs, R. & Olivin, J. (1977). Leaf analysis for the control of nutrition in oil palm plantations: taking of leaf samples. *Oleagineux*, 32 (5), 211–216.

Paramananthan, S. (2013). Managing marginal soils for sustainable growth of oil palms in the Tropics. *Journal of Oil Palm &The Environment*, 4, 1–16.

Prayitno, S., Indradewa, D. & Sunarminto, B. H. (2008). Produktivitas kelapa sawit (*Elaeis guineensis* Jacq.) yang dipupuk dengan tandan kosong dan limbah cair pabrik kelapa sawit. *Ilmu Pertanian*, 15, 37–48.

Pusdatin Pertanian (2014). Outlook Komoditi Kelapa Sawit. Pusat Data dan Sistem Informasi Pertanian, Kementerian Pertanian, Jakarta.

Reddy, K. R., Hodges, H. F. & McKinion, J. M. (1997). Modeling temperature effects on cotton internode and leaf growth. *Crop Science*, 37, 503–509.

Roggatz, U., McDonald, A. J. S., Stadenberg, I. & Schurr, U. (1999). Effects of nitrogen deprivation on cell division and expansion in leaves of *Ricinus communis* L. *Plant Cell and Environment*, 22, 81–89.

Schwab, G. J., Lee, C. D. & Pearce, R. (2007). *Sampling Plant Tissue for Nutrient Analysis*. University of Kentucky College of Agriculture, Lexington, Kentucky.

Selvaraja, S., Balasundram, S. K., Vadamalai, G. & Husni, M. H. A. (2013). Use of spectral reflectance to discriminate between potassium deficiency and orange spotting symptoms in oil palm (*Elaeis guineensis* Jacq.). *Life Science Journal*, 10 (4), 947–951.

Siallagan, I., Sudradjat & Hariyadi (2014). Optimizing rate of organic and NPK compound fertilizers for immature oil palm. *Jurnal Agronomi Indonesia*, 42 (2), 166–172.

Sudradjat, Darwis, A. & Wachjar, A. (2014a). Optimizing of nitrogen and phosphorus rates for oil palm (*Elaeis guineensis* Jacq.) seedling in the main nursery. *Jurnal Agronomi Indonesia*, 42 (3), 222–227.

Sudradjat, Sukmawan, Y. & Sugiyanta (2014b). Influence of manure, nitrogen, phosphorus and potassium fertilizer application on growth of one-year-old oil palms on marginal Soil in Jonggol, Bogor, Indonesia. *Journal of Tropical Crop Science*, 1 (2), 18–24.

Sugiyono, Sutarta, E. S., Darmosarkoro, W. & Santoso, H. (2005). Peranan perimbangan K, Ca, dan Mg tanah dalam penyusunan rekomendasi pemupukan kelapa sawit. *In:* Sutarta, E. S., Siregar, H. H., Erningpraja, L., Darnoko, Winarna, Yudanto, B. G. & Listia, E. (eds.), *Prosiding Peningkatan Produktivitas Kelapa Sawit melalui Pemupukan dan Pemanfaatan Limbah PKS*. Medan, Indonesia.

Suharno, I., Mawardi, Setiabudi, N., Lunga, S. & Tjitrosemito (2007). Efisiensi penggunaan nitrogen pada tipe vegetasi yang berbeda di stasiun penelitian Taman Nasional Gunung Halimun Jawa Barat. *Biodiversitas*, 8 (4), 287–294.

Sutarta, E. S., Rahutomo, S., Darmosarkoro, W. & Winarna (2007). Peranan unsur hara dan sumber hara pada pemupukan tanaman kelapa sawit. *In:* Darmosarkoro, W., Sutarta, E. S. & Winarna (eds.), *Lahan dan Pemupukan Kelapa Sawit, Ed ke-1*. pp. 79–90, Indonesian Oil Palm Research Institute, Medan, Indonesia.

Taiz, L. & Zeiger, E. (1998). *Plant Physiology*. (1st ed.). Sinauer Association Inc, London.

Tripler, C. E., Kaushal, S. S., Likens, G. E. & Walter, M. T. (2006). Patterns in potassium dynamics in forest ecosystems. *Ecology Letters*, 9, 451–466.

von Uexkull, H. R. & Fairhurst, T. H. (1991). *Fertilizing for High Yield and Quality: The Oil Palm*. International Potash Institute, Bern.

Wang, M., Zheng, Q., Shen, Q. & Guo, S. (2013). The critical role of potassium in plant stress response. *International Journal of Molecular Sciences*, 14, 7370–7390.

Webb, M. J. (2009). A conceptual framework for determining economically optimal fertiliser use in oil palm plantations with factorial fertiliser trials. *Nutrient Cycling in Agroecosystems*, 83, 163–178.

7

Motivations to consume agroecological food: An analysis of farmers' markets in Quito, Ecuador

Cristian Vasco [a,*], Carolina Sánchez [a], Karina Limaico [a], Víctor Hugo Abril [b]

[a] Facultad de Ciencias Agrícolas, Universidad Central del Ecuador, Quito, Ecuador
[b] Universidad de las Fuerzas Armadas-ESPE, Sangolquí, Ecuador

Abstract

This paper examines the motivations to consume agroecological foods in Quito, Ecuador. Using data from a survey ($n = 254$) conducted among the customers of farmers' markets, the results reveal that agroecological consumers are substantially different from the rest of the population in terms of education, income and life style. The perceived healthiness of agroecological food is by far the main motivation to buy at farmers' markets, with environmental concern as the least important motivation. In terms of who spends the most on agroecological produce, the results of a multilevel regression model indicate that these are single, educated, wealthy individuals who exercise regularly and are part of a social/environmental organisation. These results reflect that agroecological produce is mainly consumed by a segment of wealthy and educated individuals who are not really concerned of the positive effects for the environment that agroecological production involves, so that, additional efforts are needed to make agroecological food accessible to the general population.

Keywords: agroecological food, farmers' markets, motivations, expenditure, Quito

1 Introduction

In many parts of Latin America, social movements and farmers' organisations have established agroecological farmers' markets (*ferias agroecológicas*) as a space where producers can showcase and sell the produce and consumers can access healthy food at reasonable prices (McKay & Nehring 2014). These initiatives have been supported by governmental (principally at local and regional level) and non-governmental organisations, as a suitable strategy to reach social equity, food security, food sovereignty and environmental awareness among subsistence and small-scale family farms (McKay & Nehring, 2014; Gomes *et al.*, 2015; Heifer International, 2015). Agroecological farmers' markets also arose as a criticism to the organic movement which according to some (Nelson *et al.*, 2010; Buck *et al.*, 1997; Guthman, 2002) has done little to limit the participation of large companies in the organic market, protect small-

scale farmers (principally in the developing world), protect the rights of agricultural labourers and restrict monocropping. This has resulted in a "conventionalized organic sector" (Nelson *et al.*, 2010), the "elitization of healthy produce", and the distinction between "markets for the rich or for the poor" (Intriago *et al.*, 2017), with costs of organic certification being prohibiting for small-scale family farms and the prices of certified products being unaffordable for most consumers in the developing world.

Since the future of ecological production depends, to a large extent, on its ability to satisfy consumers' motivations, preferences, needs and desires (Lockie *et al.*, 2002), a significant body of research has examined peoples' motivations to consume organic foods. In this vein, a number of studies have been conducted in more developed countries including Germany (Bravo *et al.*, 2013), Switzerland (Tanner & Wölfing Kast, 2003), Norway (Kvakkestad *et al.*, 2018), Ireland (O'Donovan & McCarthy, 2002), the Netherlands (Schifferstein & Ophuis, 1998), Italy (Pino *et al.*, 2012), Spain (López Galán *et al.*, 2013), the United States (Lee & Yun, 2015), and Australia (Lockie *et al.*, 2004). Nevertheless,

* Corresponding author – clvasco@uce.edu.ec

to the best of our knowledge, no previous empirical study has been conducted to establish the determinants of agroecological consumption in less developed countries.

Prior literature on organic food consumption classifies motivations to consume organic food into *"personal good"* and *"public good"* factors (Bravo *et al.*, 2013; OECD, 2014). The first term, *personal good*, refers to individuals who consume organic food mainly motivated by the personal benefits associated to the consumption of organic food. These include, for instance, the perceived healthiness, safety and better taste associated to organic products. In the case of *public good* or *altruistic* motivations, individuals consume organic foods because they are perceived as environmentally friendly, respectful of animal welfare, and, to a lesser extent, socially fair (Kvakkestad *et al.*, 2018). Overall, health is found to be the principal motivation for consumers to buy organic food, with environmental concern as the second most important motivation (Gracia & De Magistris, 2008; Shepherd *et al.*, 2005; Lockie *et al.*, 2002).

Environmental motivations are associated with higher levels of buying (Gracia & De Magistris, 2008; Durham & Andrade, 2005; OECD, 2014). Concerning the determinants of expenditures on organic food, several studies have analysed the role of individual and household characteristics. To illustrate, prior research has reported mixed effects of income on expenditures on organic food, with some studies reporting a positive effect (Bellows *et al.*, 2008; Allender & Richards, 2010) and others finding no effect (Zepeda & Li, 2007). Education is reported to have a positive effect on expenditures on organic foods (Zepeda & Li, 2007; Bellows *et al.*, 2008). Mixed effects are also reported for the presence of children in a household. Some studies (Loureiro *et al.*, 2001; Thompson & Kidwell, 1998) found that households with children are more likely to buy organic food while others (Riefer & Hamm, 2011) reported that organic food consumption decreases when children become adolescent. But do these findings apply for agroecological food and the concept of agroecological farmers' markets in developing countries? Furthermore, have farmers' markets achieved the goal of making healthy food available to those unable to afford certified organic food? These and related questions are addressed in this study.

With data from a survey conducted among consumers of farmers' markets in Quito, the capital of Ecuador, this paper examines the motivations of agroecological consumption and the determinants of the expenditures on agroecological foods. The rest of this paper is structured as follows: the next section describes the context of agroecological production and farmers' markets in Ecuador, the subsequent section defines the data collection process and the statistical methods utilised. Next, the most salient results are presented and discussed, while the final section concludes.

1.1 The context: agroecology and agroecological farmers' markets in Ecuador

As in most Latin American countries, agroecology in Ecuador arose in the 1980s as a response to the environmental degradation resulting from the green revolution. In this context, several NGOs and farmers' associations have emerged to rescue local farmers' knowledge focusing not only on the technical and economic dimensions but also incorporating the social, cultural, and environmental dimensions of agricultural production (Macas & Echarry, 2009; Heifer International, 2015). While the definition of agroecology may change from one organisation to another, most of them are focused on: mitigating environmental problems resulting from the use of pesticides and chemical fertilisers in conventional agriculture; integrating the social, economic and cultural dimensions of agriculture; and recognizing and recovering local/indigenous knowledge (Macas & Echarry, 2009; Intriago *et al.*, 2017; Heifer International, 2015).

Since most agroecological producers have little land and have no access to supermarkets, commercialisation of agroecological produce normally occurs in the framework of alternative commercialisation schemes, including farmers markets, food basket programs, barter markets and agroecological shops, with farmers' markets accounting for most (78%) agroecological trade in Ecuador (Intriago *et al.*, 2017; Heifer International, 2015). The direct contact between producer and customer drives middlemen out of the business and allows farmers to keep a larger profit. Another advantage of farmers' markets is that they are normally organised by farmers' associations so that all members can participate regardless the amounts they are able to supply. Since this form of commercialisation is more flexible than conventional markets in terms of stability of supply and product characteristics, it allows the participation of those farmers unable to maintain a permanent supply of produce throughout the year (Macas & Echarry, 2009).

In 2002, the Municipality of Quito started the project AGRUPAR (standing for Participative Urban Agriculture). This initiative focuses on reducing poverty and improving food security among the poor residents of urban, peri-urban and rural neighbourhoods of the city of Quito, prioritising the participation of marginalized groups (i.e., women, single mothers, the elderly, recently arrived migrants, indigenous people, the jobless) (Rodríguez Dueñas & Proaño Rivera, 2016; Anguelovski, 2009), with most of the farmers being women (84%) with farms smaller than 600 m^2 (Rodríguez Dueñas & Proaño Rivera, 2016). As a part of the project,

AGRUPAR trained farmers in agroecological production, entrepreneurship, and management and commercialisation. While initially the project aimed at enhancing food security, promoting environmental consciousness and the production and consumption of healthy food among poor residents of Quito, the surpluses are marketed in fourteen farmers' markets, locally labelled as *Bioferias*, which are organised by AGRUPAR on a weekly basis (Quinga Guallichico, 2016; Mena Pérez, 2012). The transparency of the production process and the quality of the produce marketed on *Bioferias* are guaranteed by a Participative Guarantee System (PGS), based upon relationships of trust between producers and consumers (Rodríguez Dueñas & Proaño Rivera, 2016).

Income from the surpluses marketed on *Bioferias* helps to balance household income of agroecological producers, with an average monthly income from sales of US $ 130, an amount that accounts for 36 % of Ecuador's minimum wage[1] (Oviatt, 2016). Another objective of the project AGRUPAR is promoting the consumption of agroecological food at affordable prices among the urban population, principally among those who, otherwise, would not be able to buy healthy produce (Rodríguez Dueñas & Proaño Rivera, 2016; Mena Pérez, 2012).

2   Methods

2.1   The survey

Data came from a survey conducted among buyers of the farmers' markets as organised by AGRUPAR in the Metropolitan District of Quito, the capital of Ecuador, a city of 2.2 million inhabitants (INEC, 2010). The questionnaire included information on buyers' demographic characteristics (age, sex, schooling-level, personal income, type of occupation), household characteristics (household composition, household total income, home ownership, zone of residence), personal habits (physical activity, membership in an organisation). Another section asked respondents about the principal reason to consume agroecological products, with health, characteristics of the product (appearance and taste), environmental concern and support to small-scale farmers as the response choices. Finally, the survey included questions concerning the monthly expenditures and the frequency of buying of agroecological produce.

The survey was conducted in April–May 2017 in farmers' markets organised by AGRUPAR. Undergraduate students were trained to administer the questionnaire to 254 customers of *Bioferias*. Interviewees were selected using a two-stage sampling method. First, in order to ensure a broad

diversity of customers, nine out of fourteen farmers' markets were included in the sample. While initially we intended to include all the farmers' markets organised by AGRUPAR, we had to drop five of them from the final sample because of several reasons. In two cases we did not find any farmer showcasing agroecological produce in the places advertised in the webpage of AGRUPAR. In two other cases, we noticed that conventional produce was being sold next to agroecological produce without clear distinction. Finally, in one case we noted that almost all the buyers were employees of the municipality of Quito which could have become a source of bias. Nevertheless, we were able to obtain a balanced sample in terms of income, schooling-level and ethnicity of the customers.

Since markets are scattered throughout the city, including wealthy and poor neighbourhoods, at the centre and the outskirts of the city, variability in terms of income, schooling-level and the ethnicity of the customers was ensured. Next, the survey was randomly administered to customers within each selected market, that is, all adult customers had the same probability to be part of the sample, which ensured variability in terms of age, schooling-level and gender. We defined customers as those persons who declared having bought agroecological produce in the farmers' market where they were approached. Clearly this is not a strictly random sample, i.e., interviewees were approached directly in the farmers' markets. Although there is no way of knowing if the individuals in the sample are representative for the general population, the survey is expected to provide important insights on the motivations to consume agroecological produce in Quito.

2.2   Statistical methods

In order to find the determinants of the expenditures on agroecological produce, we used a multivariate regression approach in which the monthly amount spent on fruits and vegetables[2] agroecologically produced was a function of a number of individual and household characteristics. Before proceeding though, a methodological issue must be addressed. As mentioned earlier in the text, *Bioferias* are scattered throughout the city, so the sample includes individuals with different endowments of education and income. Such differences may be distinctive to individuals residing in a specific neighbourhood/ part of the city. Failing to control for the hierarchical nature of the data, that is the effect that the place of residence may exert on expenditure patterns, may lead to misleading interpretations. An alternative to cope with these kinds of data is the use of multilevel

---

[1] Note that since 2000, Ecuador adopted the US Dollar as its official currency.

[2] While we inquired about expenditures on other products (e.g., meat and dairy products, candies, jams, among others), these values were negligible compared to the expenditures on vegetables and fruits.

models, which, in this case, control for the effect of the marketplace where the survey was conducted. Thus, we used a multilevel regression model of the following form:

$$Y_{ij} = \beta \mathbf{X}_{ij} + \varepsilon_{ij} + v_j, \qquad (1)$$

where Y is the natural logarithm of the monthly expenditures on agroecological produce[3] by individual i in market j, $\mathbf{X}$ is a vector of individual and household characteristics to be described later on, β is a vector of coefficients the direction and magnitude of which are of interest in this study, ε_{ij} is the error term, and v_j is the market-level error term.

In terms of the explanatory variables, we included sets of individual and household characteristics. The first group comprised demographic characteristics including the age, gender, schooling-level and marital status of the buyer. We also included the squared age of the buyer in order to capture any possible non-linearity between age and the expenditure on agroecological products. The natural logarithm of the individual's income is used as a proxy of wealth. Additionally, three dichotomous variables taking the value of 1 if the interviewee declares: exercising regularly, being member of an organisation and being vegetarian, respectively, control for the effect of the buyer's life style on the expenditure on agroecological products. At household-level, we included the household size and the number of children (individuals younger than 15). Finally, three dichotomous variables taking the value of 1 if the individual bought agroecological food principally motivated by either the characteristics of the product (appearance and taste), wish to support small-scale producers or environmental concern, respectively, are included in the model. The group of individuals who buy agroecological produce because it is associated with healthy produce -the most numerous in the sample- are left as the comparison group.

Most of these variables have been found to shape decisions concerning the consumption of organic produce (OECD, 2014; Thompson & Kidwell, 1998; Loureiro *et al.*, 2001; Zhang *et al.*, 2008) in more developed countries. Thus, we expected that these predictors also have a significant effect on the expenditures on agroecological produce in Ecuador.

3 Results

3.1 Characteristics of agroecological consumers

In Table 1 we compare some demographic and socioeconomic characteristics of the consumers of agroecological produce to those of the general population. We use data

from the National Census–2010 (INEC, 2010), the Employment and Unemployment Survey–2016 (INEC, 2016), the Time Use Survey–2012 (INEC, 2012), and the Living Standards Survey 2013–2014 (INEC, 2014) to produce figures for the general population of Quito.

Table 1: *Comparison between agroecological consumers and the general population of Quito.*

Variable	Individuals in the sample	General population of Quito
Schooling level (%)		
Illiterate	0.0	2.5[a]
Primary education	5.0	22.9[a]
Secondary education	32.8	23.7[a]
University education	48.9	32.1[a]
Postgraduate education	13.3	3.4[a]
Household size	3.2	3.5[a]
Number of children	0.6	0.3[a]
Income	1081	381[b]
Organisation (%)	19	1.1[c]
Exercise (%)	29	16.5[d]

[a] Computed with data from the National Census–2011
[b] Computed with data from the Employment and Unemployment Survey–2016
[c] Computed with data from the Living Standards Survey–2013
[d] Computed with data from the Time Use Survey–2011

On average, the customers of *Bioferias* are better educated than the rest of the population of Quito, with the share of agroecological consumers holding a university degree (48.9 %) substantially higher than that estimated for the general population of Quito (32.1 %). Similarly, the fraction of consumers holding a postgraduate degree (13.3 %) is four times as high as that of the average population of Quito. The average household of an agroecological buyer is slightly smaller than the average household in Quito. Nevertheless, households consuming agroecological products have twice as many children (individuals younger than 15) as the typical household in Quito. The average monthly income of agroecological consumers (US $ 1081) is three times as high as that of an average citizen of Quito in age to work (US $ 381), as estimated from the Employment and Unemployment Survey–2016. The share of agroecological buyers that are part of a social organisation is considerably higher (19 %) than that of the general population (1 %). Finally, in terms of life style, the fraction of the customers of farmers' markets that exercise regularly is twice as high as that of the average citizen of Quito.

3.2 Motivations to buy agroecological produce

Figure 1 shows the motivations to buy agroecological produce as estimated from a sample of 254 buyers of agroecological farmers' markets in Quito. The results show that

---

[3] We took natural logarithms in order to avoid the effect of outliers.

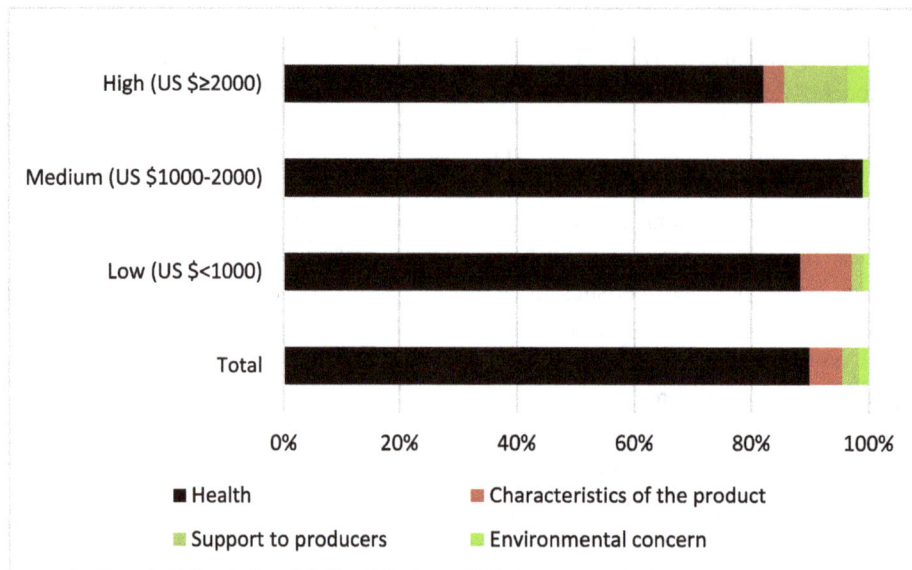

Fig. 1: *Motivations to consume agroecological produce by category of income.*

personal good motivations appear to be more important than *public good* ones in shaping consumption decisions. Health is the principal motivation to buy in *Bioferias*, with most respondents (90 %) buying agroecological products because they are perceived as healthy if compared to conventional produce. A small share of the respondents (5.6 %) buy agroecological produce because of its characteristics (taste and appearance). Moving on to altruistic motivations, a very small fraction (4.6 %) of the interviewees consume agroecological food because of *public good* (support to producers and environmental concern) reasons.

Although health continues to be the main motivation to eat agroecological foods, when analysing motivations by income category, there are some changes that warrant to be mentioned. Altruistic reasons seem to be more important for the group of buyers with high incomes (more than US $ 2000 a month), with 10.7 % of respondents answering that "support to producers" is their principal motivation to buy in agroecological farmers' markets. In contrast, almost all respondents with medium incomes (US $ 1000–2000 a month) buy agroecological foods because they are perceived as healthy. In the case of buyers with low incomes (< US $ 1000 a month), they are more concerned about the characteristics of the product than their peers with medium and high incomes, with 8.7 % of the respondents in this category buying agroecological products due to their appearance and taste.

Figure 2 shows the motivations to buy in farmers' markets by the monthly expenditure on agroecological produce. Respondents with low expenditures (< US $ 100 a month) seem to be the most concerned about the harmlessness of

the food they consume, with 94 % of the individuals in this group buying agroecological products because they are associated with healthy food. In contrast, 8.3 % of the individuals with medium expenditures (US $ 100–300 a month) eat agroecological food because of its appearance and taste.

3.3 Determinants of expenditure on agroecological products

The definitions and descriptive statistics used for the analysis are presented in Table 2. As referred to earlier in the text, we used the natural logarithm of the overall monthly expenditure on agroecological products as the outcome variable.

Table 3 presents the results of the multilevel linear regression model, with the natural logarithm of the monthly expenditures on agroecological products as the outcome variable. The intra-class correlation, which is the proportion of the error variance due to differences across marketplaces is of 42 % and highly significant, which proves that using random-effects was a sensible decision. The results show that, on average, women spend 27 % less on agroecological produce than men[4]. Education has a positive effect on the consumption of agroecological food. Every additional year of formal education increases the expenditures on agroecological produce by 4.9 %. Everything else held equal, married individuals spend 43 % less than their single counterparts. On average, respondents who are part of a social organisation spend 37 % more than those that are not.

---

[4] The percent change of a coefficient c multiplying a dummy variable in a semi logarithmic model is given by $100[\exp(c) - 1]$ (see Halvorsen & Palmquist, 1980).

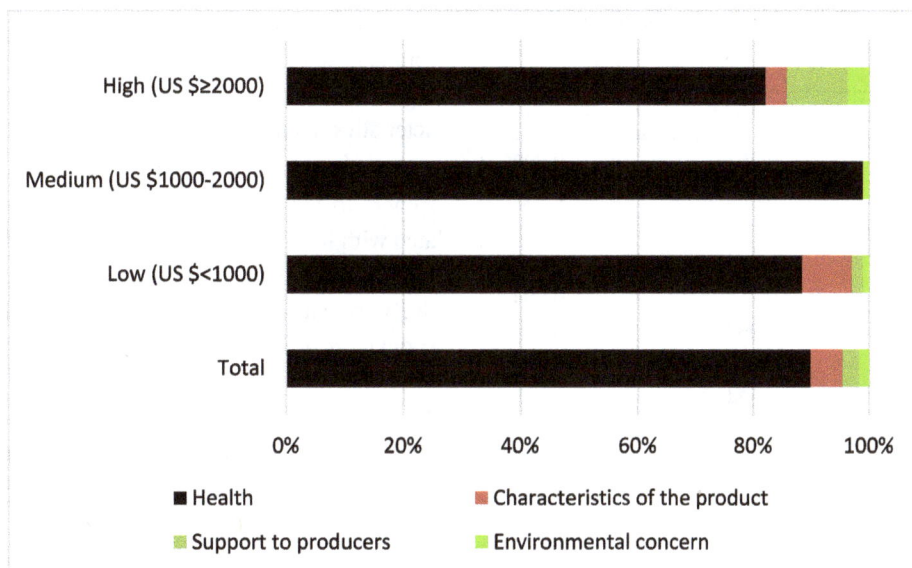

Fig. 2: *Motivations to buy in agroecological farmers' markets by monthly expenditure on agroecological produce.*

Table 2: *Descriptive statistics and definitions of variables.*

Variable	Description	Mean	St. Dev.
Dependent variable			
Expenditure	Overall monthly expenditures on agroecological products (US $)	196.2	182.6
Independent variables			
Age	Age of the individual (years)	45.3	14.8
Woman (0/1)	Individual is woman	0.6	–
Education	Completed years of formal education of individual (years)	16.5	3.9
Married (0/1)	Individual is married	0.46	–
Income	Monthly income (US $)	1081.8	1027.5
Exercise (0/1)	Individual exercises regularly	0.33	–
Vegetarian (0/1)	Individual is vegetarian	0.11	–
Association (0/1)	Individual is member of an organisation	0.22	–
Household size	Household size (n)	3.15	1.35
Children	Number of children in the household (n)	0.54	0.80
Health (0/1)	Health is the main motivation to consume agroecological food	0.90	–
Characteristics (0/1)	Product characteristics are the main motivation to consume agroecological food	0.06	–
Support (0/1)	Supporting farmers is the main motivation to consume agroecological food	0.03	–
Environment (0/1)	Environmental concern is the main motivation to consume agroecological food	0.01	–

Note: (0/1) identifies dummy variables.

Moving on to household variables, individuals from households with more children spend more on agroecological food. Each new child increases expenditures by 26 %. In terms of the motivations to buy in agroecological farmers' markets, only the dummy accounting for support to farmers has a significant effect. On average, those whose principal motivation to buy agroecological food is supporting farmers spend 60 % more on agroecological food than those who have health as their principal motivation.

Table 3: *Determinants of the expenditures on agroecological products.*

	Coefficients	p-values
Age	0.003	0.890
Age squared	0.000	0.595
Gender (0/1)	−0.244	0.046
Education	0.049	0.001
Married (0/1)	−0.356	0.004
Income	0.068	0.044
Exercise (0/1)	−0.060	0.624
Vegetarian (0/1)	0.324	0.115
Organisation (0/1)	0.316	0.045
Household size	−0.001	0.978
Children	0.262	0.000
Characteristics (0/1)	0.060	0.784
Support (0/1)	0.468	0.050
Environment (0/1)	0.081	0.851
Intra-class correlation	0.423	0.000
Prob > F	69	0.000
Number of observations	254	
R^2	0.38	

Notes: (0/1) identifies dummy variables. The model is estimated with robust standard errors.

4 Discussion

Although agroecological farmers' markets in Latin America have emerged to widespread healthy consumption among the local population, that is, among those who otherwise would not be able to afford healthy produce (Intriago *et al.*, 2017; Heifer, International 2015), our results indicate that agroecological consumers belong to a group of individuals that are substantially different from the average population in terms of socioeconomic characteristics. They are wealthier, better educated, and come from smaller households with more children than the average person in Quito. Additionally, as distinct from the average resident of Quito, many individuals in this group belong to a social organisation and exercise regularly.

That agroecological consumers are better educated than the average population of Quito may reflect that they are more aware of the dangers of eating conventional produce and the benefits associated with the consumption of agroecological food (Zepeda & Li, 2007). Similarly, that agroecological consumers have higher incomes than the average inhabitant of Quito may indicate that they can afford the price premium of agroecological produce (Bellows *et al.*, 2008; Allender & Richards 2010). While these results are fairly consistent with those found for organic consump-

tion (OECD, 2014), they reflect that far from being available for the general population, agroecological produce is mostly consumed by a segment of the population with special characteristics in terms of income, education and life style.

Our results indicate that the main motivation for consumers to buy agroecological produce is that it is associated with healthy produce. This finding is not at all surprising since health has been found to be the principal motivation to consume organic food in developed countries as well (Lockie *et al.*, 2002; Schifferstein & Ophuis, 1998; Gracia & De Magistris, 2008; O'Donovan & McCarthy, 2002; Tregear *et al.*, 1994; Kvakkestad *et al.*, 2018; Squires *et al.*, 2001; Durham & Andrade, 2005). Nevertheless, the share of respondents buying agroecological produce because it is considered healthier than conventional produce (90 % of our sample) is substantially higher than that reported for organic consumers in OECD countries (50 %) (OECD, 2014; Boccaletti, 2009). In distinction from OECD countries, where preserving the environment is the second most important motivation to buy organic produce, environmental concern appears not to be influential in the decisions of buying agroecological produce in Quito, with only a small fraction (1.7 %) of the respondents having environmental concern as their main motivation to consume agroecological food.

In terms of who spends the most on agroecological farmers' markets, this is a single, well-educated man with high income, who has more children. This individual is more likely to exercise regularly, to be part of a social or environmental organisation and to buy agroecological produce due to the desire of supporting small-scale farmers. That wealthier and educated individuals spend more on agroecological produce may reflect that these persons know more about the risks of the use of pesticides in conventional food production and have the resources to buy more agroecological food compared to poorer persons (Andrade & Flores 2008). The positive and significant effect of the dichotomous variable accounting for "exercise regularly" may indicate that people who strive a healthy life style are willing to spend more on food that they perceive as healthy. Similarly, membership in a social/environmental organisation may be associated with higher concern about social and environmental issues and so with higher willingness to pay a price premium for what is considered environmentally friendly and socially fair (OECD, 2014). Our results show that individuals from households with more children spend more on agroecological produce. This is consistent with prior research on the determinants of organic food consumption (Loureiro *et al.*, 2001; Kiesel & Villas-Boas, 2007) and may reflect that buyers with children pay more attention to the healthiness and harmlessness of the food they give to their children. Another interesting finding is that those whose principal motivation to consume agroecological produce is the support to small-scale farmers spend more than those whose main motivation is the perceived healthiness of agroecological produce.

Do our results support the statement that promoting farmers' markets may be an effective strategy to widespread healthy food consumption among the local population unable to afford certified organic food? Well, they do not. With only a selective group of educated and wealthy individuals buying agroecological produce, farmers' markets are far from reaching the goal of wide spreading healthy food consumption. In this sense, policy makers intending to increase the production and consumption of healthy food should focus on making agroecological produce accessible to a larger segment of the population.

Policies should be oriented at advertising more the benefits of consuming agroecological food among the population, principally among those with low levels of education and income. The findings presented here should be a source of concern to policy makers aimed at raising environmental awareness among citizens, since environmental concern ranks as the least important motivation to buy agroecological food. In this sense, campaigns promoting agroecological consumption should go beyond the health attribute of agroecological produce. Instead, potential consumers should also be informed that by buying agroecological produce (in farmers' markets) they are also contributing to promote the sustainable use of natural resources and to enhance the living condition of small-scale farmers.

5 Conclusion

This paper has examined the motivations to consume agroecological produce in Quito, the capital of Ecuador. Consumers of agroecological produce differ substantially from the rest of the population of Quito in terms of education, income and life style. Among those buying agroecological produce, the perceived healthiness of agroecological foods is by far the main motivation to buy in farmer's markets. In contrast, environmental concerns appear not to be influential in agroecological consumption decisions. Concerning who spends more on agroecological produce, these are single, educated and wealthy individuals who exercise regularly, are members of a social/environmental organisation and whose principal motivation to buy is supporting small-scale farmers.

Some argue that farmers' markets have the potential to promote healthy food consumption while improving the living conditions of farmers with little land, and promoting sustainable agriculture. Our findings indicate, however, that far from being accessible to all, agroecological products are principally bought by a group of educated and wealthy individuals who are not really concerned about the potential benefits for the environment that agroecological production entails. So, additional efforts are necessary to make agroecological production accessible to all.

Acknowledgements

This research was based upon work funded by Universidad Central del Ecuador.

References

Allender, W. J. & Richards, T. J. (2010). Consumer Impact of Animal Welfare Regulation in the California Poultry Industry. *Journal of Agricultural and Resource Economics*, 35 (3), 424–442.

Andrade, D. & Flores, M. (2008). *Consumo de productos orgánicos/agroecológicos en los hogares ecuatorianos*. VECO, Quito, Ecuador.

Anguelovski, I. (2009). Building the Resilience of Vulnerable Communities in Quito: Adapting local foodsystems to climate change. *Urban Agriculture*, 22, 25–26.

Bellows, A. C., Onyango, B., Diamond, A. & Hallman, W. K. (2008). Understanding consumer interest in organics: production values vs. purchasing behavior. *Journal of Agricultural & Food Industrial Organization*, 6 (1), 1–31.

Boccaletti, S. (2009). Organic Food Consumption: Results and Policy Implications. Paper presented at OECD Conference on "Household Behaviour and Environmental Policy", 3-4th June 2009, Paris.

Bravo, C. P., Cordts, A., Schulze, B. & Spiller, A. (2013). Assessing determinants of organic food consumption using data from the German National Nutrition Survey II. *Food Quality and Preference*, 28 (1), 60–70.

Buck, D., Getz, C. & Guthman, J. (1997). From farm to table: The organic vegetable commodity chain of Northern California. *Sociologia ruralis*, 37 (1), 3–20.

Durham, C. A. & Andrade, D. (2005). Health vs. environmental motivation in organic preferences and purchases. Paper presented at the American Agricultural Economics Association Annual Meeting, Providence, Rhode Island, July 24–27, 2005.

Gomes, R. A., Matias, T. L. & Paulino, J. S. (2015). Articulações interinstitucionais na realização de feiras agroecológicas na Microrregião de Campina Grande-PB. *Revista Brasileira de Desenvolvimento Regional*, 3 (1), 103–126.

Gracia, A. & De Magistris, T. (2008). The demand for organic foods in the South of Italy: A discrete choice model. *Food Policy*, 33 (5), 386–396.

Guthman, J. (2002). Commodified meanings, meaningful commodities: Re–thinking production–consumption links through the organic system of provision. *Sociologia ruralis*, 42 (4), 295–311.

Halvorsen, R. & Palmquist, R. (1980). The Interpretation of Dummy Variables in Semilogarithmic Equations. *American Economic Review*, 70 (3), 474–475.

Heifer International (2015). *La agroecología está presente: Mapeo de productores agroecológicos y del estado de la agroecología en la sierra y costabecuatoriana.* Heifer-Ecuador, Quito, Ecuador.

INEC (2010). Censo de Población y Vivienda–2010. Instituto Nacional de Estadística y Censos (INEC), Quito, Ecuador.

INEC (2012). Encuesta de Uso del Tiempo–2012. Instituto Nacional de Estadística y Censos (INEC), Quito, Ecuador.

INEC (2014). Encuesta de Condiciones de Vida 2014–2015. edited by I. N. d. E. y. Censos, Quito, Ecuador.

INEC (2016). Encuesta de Empleo, Subempleo y Desempleo 2016. Instituto Nacional de Estadística y Censos (INEC), Quito, Ecuador. Available at: http://www.ecuadorencifras.gob.ec/enemdu-2016/ (accessed on: 12 August 2018).

Intriago, R., Amézcua, R. G., Bravo, E. & O'Connell, C. (2017). Agroecology in Ecuador: historical processes, achievements, and challenges. *Agroecology and Sustainable Food Systems*, 41 (3–4), 311–328.

Kiesel, K. & Villas-Boas, S. B. (2007). Got organic milk? Consumer valuations of milk labels after the implementation of the USDA organic seal. *Journal of Agricultural & Food Industrial Organization*, 5, #4.

Kvakkestad, V., Berglann, H., Refsgaard, K. & Flaten, O. (2018). Citizen and consumer evaluation of organic food and farming in Norway. *Organic Agriculture*, 8 (2), 87–103.

Lee, H.-J. & Yun, Z.-S. (2015). Consumers' perceptions of organic food attributes and cognitive and affective attitudes as determinants of their purchase intentions toward organic food. *Food Quality and Preference*, 39, 259–267.

Lockie, S., Lyons, K., Lawrence, G. & Grice, J. (2004). Choosing organics: a path analysis of factors underlying the selection of organic food among Australian consumers. *Appetite*, 43 (2), 135–146.

Lockie, S., Lyons, K., Lawrence, G. & Mummery, K.

(2002). Eating 'green': motivations behind organic food consumption in Australia. *Sociologia Ruralis*, 42 (1), 23–40. doi:10.1111/1467-9523.00200.

López Galán, B. S., Royo, A. G. & Hurlé, J. B. (2013). ¿Conocimiento, medio ambiente o salud? Una investigación sobre los determinantes del consumo de alimentos ecológicos en España. *ITEA*, 109, 86–106.

Loureiro, M. L., McCluskey, J. J. & Mittelhammer, R. C. (2001). Assessing consumer preferences for organic, eco-labeled, and regular apples. *Journal of Agricultural and Resource Economics*, 26 (2), 404–416.

Macas, B. & Echarry, K. (2009). Caracterización de mercados locales agroecológicos y sistemas participativos de garantía que se construyen en el Ecuador. Coordinadora Ecuatoriana de Agroecología, Quito, Ecuador.

McKay, B. & Nehring, R. (2014). Sustainable agriculture: An assessment of Brazil's family farm programmes in scaling up agroecological food production. Working Paper number 123, International Policy Centre for Inclusive Growth (IPC - IG) United Nations Development Programme, Brazil.

Mena Pérez, V. E. (2012). Evaluación integral del programa AGRUPAR de Conquito correspondiente al Distrito Metropolitano de Quito. Trabajo de Graduación previo la Obtención del Título de Economista. Universidad Central del Ecuador, Quito, Ecuador.

Nelson, E., Tovar, L. G., Rindermann, R. S. & Cruz, M. Á. G. (2010). Participatory organic certification in Mexico: an alternative approach to maintaining the integrity of the organic label. *Agriculture and Human Values*, 27 (2), 227–237.

O'Donovan, P. & McCarthy, M. (2002). Irish consumer preference for organic meat. *British Food Journal*, 104, 353–370.

OECD (2014). Greening Household Behaviour. OECD.

Oviatt, K. (2016). *El impacto de la Agricultura Urbana como método de empoderamiento de las poblaciones pobres.* CONAUITO, Quito, Ecuador.

Pino, G., Peluso, A. M. & Guido, G. (2012). Determinants of regular and occasional consumers' intentions to buy organic food. *Journal of Consumer Affairs*, 46 (1), 157–169.

Quinga Guallichico, T. M. (2016). Evaluación y determinación del estado del proyecto de agricultura urbana participativa Agrupar de Conquito en el distrito metropolitano de Quito. Tesis para optar al título de Ingeniero Agrónomo, Universidad Central del Ecuador.

Riefer, A. & Hamm, U. (2011). Organic food consumption

in families with juvenile children. *British Food Journal*, 113 (6), 797–808.

Rodríguez Dueñas, A. & Proaño Rivera, I. (2016). Quito siembra agricultura urbana. CONQUITO-AGRUPAR, Quito, Ecuador.

Schifferstein, H. N. & Ophuis, P. A. O. (1998). Health-related determinants of organic food consumption in the Netherlands. *Food Quality and Preference*, 9 (3), 119–133.

Shepherd, R., Magnusson, M. & Sjödén, P.-O. (2005). Determinants of consumer behavior related to organic foods. *AMBIO: A Journal of the Human Environment*, 34 (4), 352–359.

Squires, L., Juric, B. & Cornwell, T. B. (2001). Level of market development and intensity of organic food consumption: cross-cultural study of Danish and New Zealand consumers. *Journal of Consumer Marketing*, 18 (5), 392–409.

Tanner, C. & Wölfing Kast, S. (2003). Promoting sustainable consumption: Determinants of green purchases by Swiss consumers. *Psychology & Marketing*, 20 (10), 883–902.

Thompson, G. D. & Kidwell, J. (1998). Explaining the choice of organic produce: cosmetic defects, prices, and consumer preferences. *American Journal of Agricultural Economics*, 80 (2), 277–287.

Tregear, A., Dent, J. & McGregor, M. (1994). The demand for organically grown produce. *British Food Journal*, 96 (4), 21–25.

Zepeda, L. & Li, J. (2007). Characteristics of organic food shoppers. *Journal of Agricultural and Applied Economics*, 39 (1), 17–28.

Zhang, F., Huang, C. L., Lin, B. H. & Epperson, J. E. (2008). Modeling fresh organic produce consumption with scanner data: a generalized double hurdle model approach. *Agribusiness*, 24 (4), 510–522.

8

Bionomics of the sweet potato weevil, *Cylas puncticollis* (Coleoptera: Brentidae) on four different sweet potato varieties in sub-Saharan Africa

Médétissi Adom [a,*], David D. Wilson [a,b], Ken O. Fening [a,c],
Anani Y. Bruce [d], Kwadwo Adofo [e]

[a]*African Regional Postgraduate Programme in Insect Science (ARPPIS), College of Basic and Applied Sciences, University of Ghana, P. O. Box LG 68, Accra, Ghana*
[b]*The Department of Animal biology and Conservation Science, College of Basic and Applied Sciences, University of Ghana, Accra, Ghana*
[c]*Soil and Irrigation Research Centre, School of Agriculture, College of Basic and Applied Sciences, University of Ghana, Accra, Ghana*
[d]*International Wheat and Maize Improvement Centre (CIMMYT), Nairobi, Kenya*
[e]*Council for Scientific and Industrial Research-Crops Research Institute (CSIR-CRI), Kumasi, Ghana*

Abstract

Sweet potato (*Ipomoea batatas* (L.) Lam.) is an important staple food crop in sub-Saharan Africa. A limiting production factor is infestation by the sweet potato weevil *Cylas puncticollis* (Boheman) (Coleoptera: Brentidae). The use of host plant resistance is an essential component of Integrated Pest Management (IPM). Therefore, the bionomic parameters of *C. puncticollis* were studied under laboratory conditions ($30 \pm 1\,°C$ and $75 \pm 5\,\%$ RH) on four sweet potato varieties commonly grown in Ghana (Apumoden, Ligri, Bohye, and Okumkom) to determine their susceptibility to the pest. There were significant varietal differences between some of the parameters. Egg to adult development time was highest on Okumkom while larval survival, emergence rate, intrinsic rate of increase, and the net reproductive rate were lowest on Bohye. The results of this study indicate that Bohye was the least susceptible variety compared to the other varieties tested and can be used in host plant resistance as part of an IPM programme against *C. puncticollis* in Ghana.

Keywords: host plant resistance, integrated pest management, life table parameters, susceptibility

1 Introduction

Sweet potato (*Ipomoea batatas* [L.] Lam.) is a staple food for a large proportion of the population in many parts of sub-Saharan Africa (Bouwkamp, 1985). Recent projections by the International Food Policy Research Institute indicate that sweet potato production in Africa will be doubled by 2020. However, productivity in sub-Saharan Africa has remained static at around $5.1\,\mathrm{t\,ha^{-1}}$ compared to a world average of $12.2\,\mathrm{t\,ha^{-1}}$ (FAOSTAT, 2016). It is a tolerant crop that, once established, will reliably produce adequate yield under marginal conditions with no inputs and minimum or intermittent care under a wide range of rainfall patterns (Ewell, 2002). The most important production constraint worldwide is plant damage caused by sweet potato weevils, viz. *Cylas* spp. (Rees *et al.*, 2003). *Cylas* spp. can damage every harvestable part of the crop, with devastating consequences especially for poor farmers (Chalfant *et al.*, 1990; Nottingham & Kays, 2002) leading to lower income and reduced food security (Smit, 1997; Magira, 2003).

In Africa, the three Cylas species, *C. formicarius* (Fabricius), *C. puncticollis* Boheman and *C. brenneus* (Fabricius) (Coleoptera: Apionidae) (Talekar, 1987; Ames *et al.*, 1997) are known to attack sweet potato in the field and in storage. In Ghana, *C. puncticollis* can cause yield losses of up

*Corresponding author
Médétissi Adom (adomsons1@gmail.com)

to 50 % (Darko, 2000). Infestation on stored tubers not only reduces quantity, but also reduces the marketability of the commodity because of the unpalatable terpenoids produced by the plant in response to infestation caused by the weevil (Ndunguru *et al.*, 1998; Stathers *et al.*, 2003a). Presently, no viable technology exists to combat sweet potato weevil in Africa. However, host plant resistance could be an essential component of Integrated Pest Management (IPM) against sweet potato weevil (Kabi *et al.*, 2001; Stathers *et al.*, 2003a; Stevenson *et al.*, 2009).

Life table parameters are important to draw more reliable conclusions on susceptibility status of a given crop plant to a specific insect pest (Braendle *et al.*, 2006; Mirmohammadi *et al.*, 2009; Goodarzi *et al.*, 2015; Mahmoudi *et al.*, 2015). This work assessed the susceptibility of four different sweet potato varieties commonly grown in Ghana to *C. puncticollis*.

2   Materials and methods

This work was carried out at the entomology laboratory of the African Regional Postgraduate Programme in Insect Science (ARPPIS), University of Ghana, Accra.

2.1   Insect culture

A colony of sweet potato weevils was established in the laboratory (30 ± 1 °C and 75 ± 5 % RH) from infested stored tubers collected from farmers in Akatsi located at 06°07′29.10″ N, 00°47′33.21″ E (Volta region, Ghana). The tubers were incubated in rectangular $17.0 \times 17.0 \times 9.5$ cm plastic containers fitted with a mesh cover at the top to allow ventilation and placed a paper towel at the bottom to suck excessive moisture resulting from transpiration or putrefaction of stored tubers . Upon emergence, male and female adults were removed from the containers and were kept in rectangular $40 \times 26 \times 26$ cm plastic containers where they were provided with un-infested tubers for infestation. The infested tubers were removed and replaced weekly with new ones. The infested tubers were kept in incubation boxes for adult emergence.

2.2   Determination of developmental period and survival rate of C. puncticollis immature stages

Four un-infested stored tubers of each sweet potato variety were washed, wiped dry and offered to gravid naive female weevils for 24 h for oviposition. Twenty cubes each containing one egg were cut from each of the infested stored tubers. A small slit was made on each cube just to make the egg visible inside the oviposition hole. The experiment was set up as a completely randomized design with four replications where cubes from one storage root represented one replication. The cubes were kept in containers and were observed daily under stereomicroscope until all eggs hatched or collapsed to determine the egg incubation period and egg survival.

Upon hatching, each 1st instar larva (≤ 24 h old) was transferred to an about 3 mm deep well made on a new fresh sweet potato cube ($1.5 \times 1.5 \times 2.0$ cm) of the same variety from which the egg hatched. This was done by using a needle nosed forceps and a fine brush. The well was then covered with a glass slide and the cube held between two slides with rubber bands. On the second day, dead larvae which didn't show any sign of feeding before they died were discarded; it was assumed that death was caused by the transfer procedure and not due to varietal effect. The remaining cubes were observed daily to determine larval and pupal developmental period and survival until adult emergence. Cubes were changed every week to avoid contamination by fungi or drying out till adult emergence. The sex of emerged adults was also determined.

2.3   Survival and fecundity of C. puncticollis adult

Newly emerged (< 24 h old) adults obtained from incubated pupae were paired (sexed by the shape of distal antennal segments, as described by Smit *et al.*, 2001) and kept in a transparent plastic jar ($\varnothing = 6.3$ cm; $h = 10.5$ cm). A fresh piece of un-infested tuber was cut almost flat (5 cm by 2 cm) and placed in the plastic jar with the periderm facing upwards until the last individual died. The jar was closed with muslin cloth. For each variety, fifteen pairs of weevils were daily monitored and continued until all adults died. The pre-oviposition period, oviposition period, post-oviposition period, and adult longevity and fecundity were determined.

2.4   Data analysis

Time from oviposition to adult emergence, survival, daily fecundity, and sex ratio (proportion of females) of *C. puncticollis* on different sweet potato varieties were used to construct the net female maternity, $l_x m_x$ (where l_x is the fraction of females alive at age x and m_x is the number of daughters born to surviving females at age x) life tables from which demographic growth parameters were calculated. The effects of sweet potato variety on developmental period and survival of immature life stages, sex-ratio, adult longevity and fecundity were analysed with one-way analysis of variance (ANOVA) and the means were separated by Student–Newman–Keuls (SNK) test. Survival of all immature stages was arcsine $\sqrt{(p/100)}$ transformed prior to analysis. Life table parameters including the intrinsic rate of increase (r_m), the net reproductive rate (R_0), the finite rate of increase (λ), the mean generation time (G) and the doubling time (t) were calculated according to Hulting *et al.* (1990), using the jack-knife programme and subjected to ANOVA, using the general linear model (GLM) procedure of SAS for PC (SAS Institute, 1997). Differences in intrinsic rate of increase (r_m) values among populations were calculated following the protocol by Dixon (1987) and compared with Newman–Keuls sequential tests (Sokal & Rohlf, 1995) based on jack-knife estimates of variance for r_m values (Meyer *et al.*, 1986).

3 Results

3.1 Developmental time, survival of immature life stages and sex-ratio of C. puncticollis

Eggs were laid singly in holes and covered with a root plug. The egg incubation period did not vary significantly with sweet potato variety ($F_{3,15} = 0.44$; $P = 0.726$) (Table 1). On all the varieties, egg hatching reached its peak on 4th day with about 60 % of total number of eggs laid. Larval developmental time ($F_{3,15} = 3.33$; $P = 0.056$) and the pupal developmental time ($F_{3,15} = 1.56$; $P = 0.251$) also did not vary among the varieties .The total developmental period of *C. puncticollis* was significantly longest ($F_{3,15} = 4.90$; $P = 0.019$) on Okumkom.

The egg and pupal survival rate did not differ significantly among the varieties ($F_{3,15} = 0.21$; $P = 0.888$; $F_{3,15} = 0.95$; $P = 0.251$, respectively). Whereas the larval survival was lowest on Bohye ($F_{3,15} = 7.18$; $P = 0.005$). Adult emergence varied significantly ($F_{3,15} = 4.90$; $P = 0.019$) among

the varieties. The sex ratio at emergence was similar on all the tested varieties ($F_{3,15} = 0.20$; $P = 0.897$) (Table 1).

3.2 Adult longevity and female fecundity

The pre–oviposition period did not vary significantly with variety ($F_{3,59} = 0.54$; $P = 0.656$) but the oviposition and post-oviposition periods varied significantly among varieties ($F_{3,59} = 12.73$; $P < 0.0001$ and $F_{3,59} = 3.75$; $P = 0.016$, respectively). The shortest oviposition periods were recorded on Bohye and Okumkom while the shortest post-oviposition period was recorded only on Okumkom (Table 2). During the oviposition period, the female fecundity differed significantly ($F_{3,59} = 14.85$; $P < 0.0001$) among the varieties with the lowest recorded on Bohye. Female longevity was higher on Apumoden and Ligri than on Bohye and Okumkom ($F_{3,59} = 17.00$; $P < 0.0001$) but male longevity did not vary significantly ($F_{3,59} = 0.46$; $P = 0.709$) among the varieties. Generally, *C. puncticollis* females tended to live longer than the male. The differences were however significant only on Apumoden ($t_{28} = 5.6$; $P < 0.0001$) and Ligri ($t_{28} = 3.99$; $P < 0.0001$) and statistically similar on Bohye ($t_{28} = 0.87$; $P = 0.390$) and Okumkom ($t_{28} = 0.44$; $P = 0.208$). The female age specific survival (Fig. 1) showed that all females lived up to 10 weeks on Ligri and Apumoden but only 6 and 5 weeks on Okumkom and Bohye, respectively. The age-specific reproduction of the females followed almost a similar pattern with the lowest rate at the early time of oviposition period and increased to reach at peak between 3 and 5 weeks, and decreased gradually.

Table 1: *Developmental period and survival rate of immature life stages (mean ± SEM) of Cylas puncticollis on different sweet potato varieties.*

Life stages	Sweet potato variety			
	Apumoden	Ligri	Bohye	Okumkom
Development period (days)				
Egg – larva	4.0 ± 0.00^a	3.8 ± 0.13^a	3.7 ± 0.14^a	3.9 ± 0.24^a
Larva – pupa	14.8 ± 0.33^a	15.1 ± 0.42^a	15.5 ± 0.38^a	16.6 ± 0.23^a
Pupa – adult	4.1 ± 1.87^a	3.7 ± 0.27^a	4.4 ± 0.33^a	3.9 ± 0.18^a
Developmental period	22.8 ± 0.54^a	22.6 ± 0.35^a	23.2 ± 0.25^a	24.7 ± 0.46^b
Survival rate (%)				
Egg	93.8 ± 4.73^a	96.3 ± 3.70^a	92.5 ± 4.78^a	90.0 ± 7.07^a
Larva	79.7 ± 2.24^a	82.7 ± 1.20^a	66.7 ± 4.04^b	72.3 ± 2.92^{ab}
Pupa	94.7 ± 3.25^a	93.7 ± 4.18^a	94.1 ± 3.71^a	100.0 ± 0.00^a
Emergence rate (%)	70.8 ± 3.24^a	74.9 ± 3.40^a	57.9 ± 1.69^b	64.5 ± 4.11^{ab}
Female sex – ratio	$0.49. \pm 0.22^a$	0.47 ± 0.24^a	0.49 ± 0.24^a	0.51 ± 0.51^a

Means within a row followed by the same letter are not significantly different
(Student–Newman–Keuls, $P < 0.05$).

Table 2: *Mean (± SE) pre-oviposition, oviposition, post-oviposition periods, fecundity and longevity of Cylas puncticollis on the different sweet potato varieties.*

Variables	Sweet potato variety			
	Apumoden	*Ligri*	*Bohye*	*Okumkom*
Pre-oviposition period (day)	5.0 ± 0.37^{a}	6.0 ± 0.70^{a}	5.3 ± 0.63^{a}	5.2 ± 0.51^{a}
Oviposition period (day)	88.8 ± 4.22^{a}	81.4 ± 4.56^{a}	57.7 ± 4.40^{b}	63.4 ± 3.09^{b}
Post-oviposition period (day)	12.9 ± 1.58^{a}	13.2 ± 2.31^{a}	8.6 ± 1.74^{ab}	6.6 ± 0.72^{b}
Fecundity (egg/ female)	233.3 ± 13.90^{a}	242.1 ± 17.72^{a}	125.7 ± 11.69^{b}	194.1 ± 10.65^{ab}
Longevity (day)[†]				
Female	106.8 ± 4.48^{Aa}	100.6 ± 4.38^{Aa}	71.6 ± 5.00^{Ab}	75.2 ± 3.07^{Ab}
Male	73.8 ± 4.03^{B}	69.4 ± 6.46^{B}	66.2 ± 3.77^{A}	68.9 ± 3.83^{A}

Means within the same row followed by the same lower case letter ([†] and within the same column followed by the same upper case letter) differ significantly (Student–Newman–Keuls, $P < 0.05$).

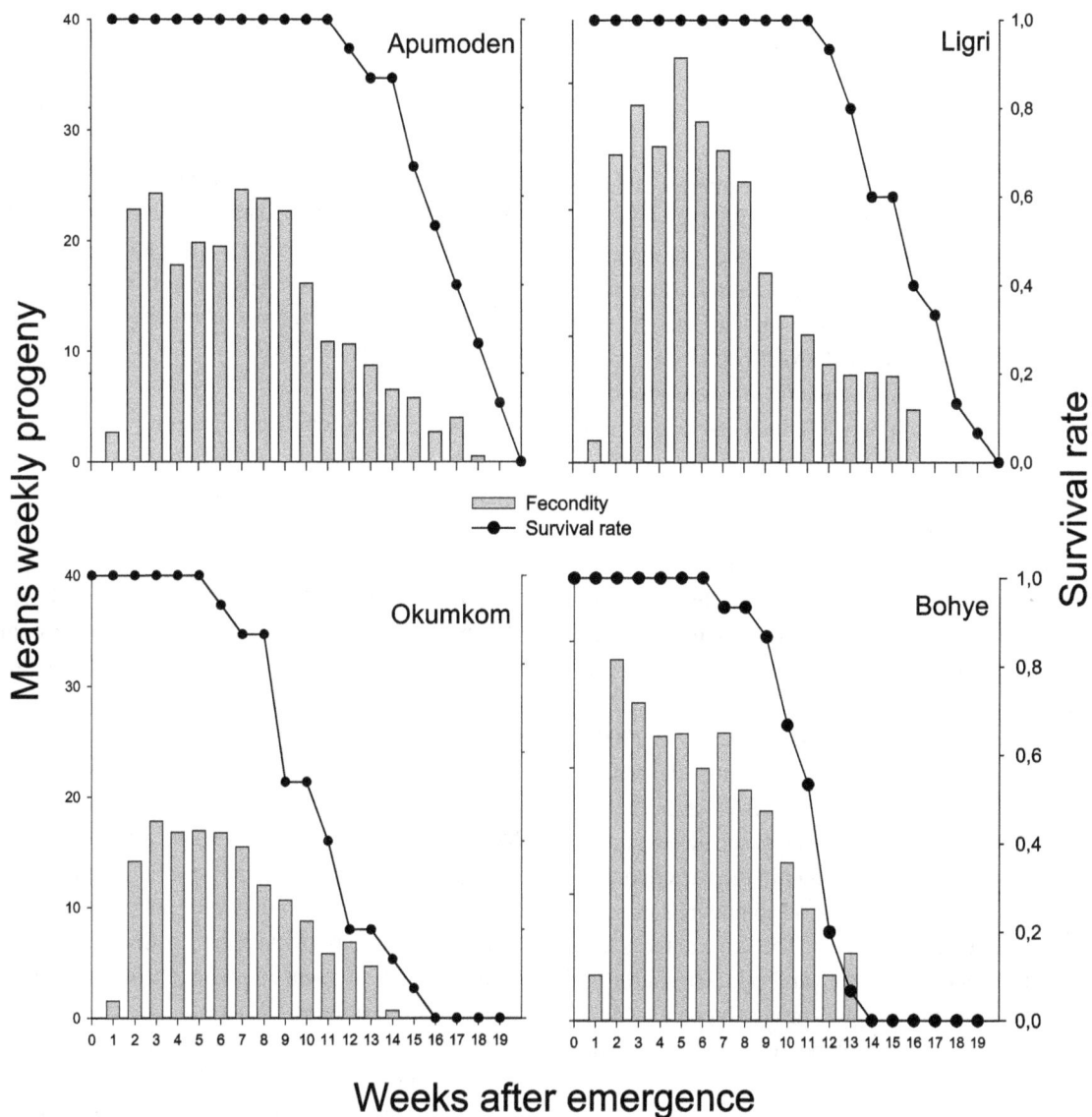

Fig. 1: *Survival and age-specific reproduction of females of Cylas puncticollis adults on different sweet potato varieties.*

Table 3: *Life table parameters (mean ± SEM) of Cylas puncticollis on different sweet potato varieties.*

Variety	Parameter				
	r_m	R_0	G (days)	T (days)	λ
Apumoden	0.082 ± 0.002[a]	90.5 ± 5.39[a]	54.7 ± 1.70[a]	8.4 ± 0.23[a]	1.08 ± 0.002[ab]
Ligri	0.087 ± 0.002[a]	86.1 ± 6.27[a]	50.9 ± 1.32[ab]	7.9 ± 0.18[ab]	1.09 ± 0.002[a]
Bohye	0.077 ± 0.004[b]	34.7 ± 6.27[c]	45.9 ± 2.22[b]	8.9 ± 0.64[a]	1.08 ± 0.004[b]
Okumkom	0.089 ± 0.002[a]	64.0 ± 7.54[b]	46.7 ± 1.26[ab]	7.8 ± 0.17[b]	1.09 ± 0.020[a]

Means within a column followed by different letters are significantly different (Student–Newman–Keuls sequential test; $P < 0.05$). r_m = intrinsic rate of increase, R_0 = net reproductive rate, G = mean generation time, T = doubling time, λ = finite rate of increase.

3.3 Life table parameters of C. puncticollis

All life table parameters of *C. puncticollis* were affected by the sweet potato variety. The r_m and R_0 were lower on Bohye than on the other three varieties. For the other parameters, there were no clear trends (Table 3).

4 Discussion

The determination of factors on which the fitness of an insect pest depends is the cornerstone for developing an environmentally friendly pest management strategy (Golizadeh & Razmjou, 2010; Reddy & Chi, 2015). In this study, the developmental period of *C. puncticollis* was affected by the sweet potato variety. The effects of sweet potato variety on the developmental period of *C. puncticollis* were reported earlier by Darko (2000) with two different Ghanaian varieties. The adult emergence rate differed significantly among tested varieties which collaborates with the findings of Kabi *et al.* (2001) and Stathers *et al.* (2003b). Several secondary metabolites occur in the sweet potato stored tubers and some of these possess diverse biological activities. Resin glycosides and caffeic acid in the latex of sweet potato may have an insecticidal effect on sweet potato weevil and affect survival and development of certain insects such as *Plutella xylostella* (Lepidoptera: Plutellidae) (Jackson & Peterson, 2000). In Tanzania, Stevenson *et al.* (2009) found that the resistance in sweet potato was governed by levels of hydroxycinnamic acid esters in the latex of one of the sweet potato varieties. Since these compounds occur in the latex and this latex occurs throughout the plant, these compounds may be responsible for the resistance effects. The same authors hypothesized that the reduction of *Cylas* spp. infestation is active, quantifiable and manageable through breeding. Variation in the content of these metabolites may explain the difference in developmental period and adult emergence rate found among the varieties in this study.

In addition, nutritional quality of an insect host plant can also affect growth and development of the insect (Das *et*

al., 1993; Golizadeh & Razmjou, 2010). In Nigeria, Anota & Odebiyi (1984) identified carotene as a major factor in tuber resistance to the sweet potato weevil. There was a lower survival rate in all life stages, smaller body weights and a longer developmental period of *C. puncticollis* raised on resistant cultivars (ibid). The proportion of females of *C. puncticollis* population emerged from all varieties in this study was close to 50 %, which corroborate results from other studies like Smit & Van Huis (1999), and Mohamed (2005). It appears that the environmental conditions coupled with the sweet potato varieties do not have any effect on the weevil sex-ratio. Insects regulate the sex ratio of its progeny based on the total number of eggs laid and the clutch size (Yu *et al.*, 2003).

Various studies found different pre-oviposition periods in female *C. puncticollis* and this could be attributed to differences in experimental conditions (Moyer *et al.*, 1989; Smit & Van Huis, 1999). However pre-oviposition period recorded in the current study on different sweet potato varieties are not far from the findings of Mohamed (2005). The oviposition period found varied from 88.8 on Apumoden to 54.6 days on Bohye. This range of mean fecundity values among varieties correlates with previous studies (Anota & Leuschner, 1983; Smit & Van Huis, 1999; Darko, 2000). Smit & Van Huis (1999), for instance, reported an oviposition period of 110 days on one Kenyan variety "Kalamba Nyerere" while Darko (2000) found 42 days less. Two secondary metabolites boehmeryl acetate and boehmerol, occurring in the periderm of sweet potato stored tuber were identified to act as ovipositional stimulants for *C. formicarius* and the content of these chemicals differs among sweet potato varieties (Wilson *et al.*, 1988; Mao *et al.*, 2001). Wang & Kays (2002) identified three oxygenated monoterpenes especially: nerol, Z-citral, and methyl geranate in the storage roots known as the primary site of oviposition, but not in another part of the plant, and these compounds were shown to be attractants to the female weevils.

On the other hand, the identification of resistance mechanisms in African sweet potato to *C. puncticollis* is not well understood, particularly since it can affect early development of larvae and may also affect oviposition behaviour. Therefore, varieties with higher quantities of hexadecylcaffeic and hexadecyl-p-coumaric acids are less suitable for oviposition and larval development of *C. puncticollis* weevil (Stevenson *et al.*, 2009). Although not tested in the current study, the level of these chemicals may therefore vary among the investigated varieties, and may have contributed to different levels of fecundity in *C. puncticollis*. The highest female fecundity was observed with females that had the highest oviposition period. This suggests that the stimulation of oviposition in the female *C. puncticollis*, may also lengthen the oviposition period and even the post-oviposition period as hypothesized by Darko (2000). Moreover, previous experiments found that, the susceptible varieties upon which the highest numbers of eggs are laid are also in most of the cases the most consumed by the female weevil (Mao *et al.*, 2001; Parr *et al.*, 2016). Thus, increased female longevity on the varieties which were most susceptible for egg laying as observed in this study may also be attributed to the abundance of food obtained by the female weevil.

On all varieties, female *C. puncticollis* lived longer than the corresponding male. This is in agreement with Darko (2000). Difference in longevity between males and females is commonly observed across many insect species, depending on the genetic make-up of insect species and also on the environmental effects especially quality composition of diets (Partridge *et al.*, 2005; Tower & Arbeitman, 2009). For instance, females of the African wild silkmoth, *Gonometa postica* Walker (Lepidoptera: Lasiocampidae) live significantly longer than the males, to offer ample time for egg laying (Fening *et al.*, 2011). Accordingly, the difference between female and male longevity may be attributed to some of the above reasons.

Significant difference in the developmental period, fecundity and survivorship among tested sweet potato varieties was reflected in life table parameters of the *C. puncticollis* including the intrinsic rate of population increase (r_m), net reproductive rate (R_0), mean generation time (G), population doubling time (T) and finite rate of natural increase (λ). Life table parameters of phytophagous insect pest are inherently tied to the host plant characteristics such as nutritional value, secondary chemical compounds and morphology (Norris & Kogan, 1980; Yamaguchi *et al.*, 2006). Reddy & Chi (2015) found that the growth rate of *C. formicarius* differed depending on whether it exploits its major host *Ipomea batatas* or the alternative host *I. triloba*. Das *et al.* (1993) reported Irish potato varietal effect on the life table parameters of *Phthorimaea oprecullela* (Lepidoptera: Gelechiidae).

Life table parameters, particularly r_m are used as indices for host plant susceptibility (Razmjou *et al.*, 2006). Intrinsic rate of population increase is actually a reflective of many biological traits, including survival, fecundity, development and sex ratio and adequately summarizes the physiological qualities of an animal in relation to its capacity to increase; it would be a more appropriate index to evaluate performance of an insect on different host plants (Southwood & Henderson, 2000; Jha *et al.*, 2012). In the current study, the r_m varied significantly from 0.077 ± 0.004 on Bohye to 0.089 ± 0.002 on Okumkom. This suggests different levels of susceptibility of investigated sweet potato varieties to *C. puncticollis* due to different levels of antibiosis mechanisms which affect the development of the insect pests (Dent, 2000; Golizadeh & Razmjou, 2010).

5   Conclusion

This study showed the importance of sweet potato variety in the population build-up of *C. puncticollis*. It is therefore possible to reduce the infestation level by choosing variety in which weevil population grows less rapidly. The lowest intrinsic rate of population increase of *C. puncticollis* on Bohye tubers indicates the relatively lower susceptibility of this variety and can therefore be used in breeding programs as part of IPM of *C. puncticollis*.

Acknowledgements

This work is part of the research undertaken by the first author, towards the award of Master of Philosophy Degree in Entomology under the sponsorship of the German Academic Exchange Service (DAAD). We are also grateful to the Council for Scientific and Industrial Research-Crops Research Institute (CSIR-CRI), Kumasi for providing plant materials used in this study. We appreciate the critical review of the draft manuscript by Fritz Schulthess.

References

Ames, T., Smith, N. E. J. M., Braun, A. R., O'Sullivan, J. N. & Skoglund, L. G. (1997). *Sweetpotato: Major pests, diseases, and nutritional disorders*. International Potato Centre, Lima, Peru.

Anota, T. & Leuschner, K. (1983). *Survival of sweet potato weevil on some promising sweet potato clones*. International Institute of Tropical Agriculture, Ibadan, Nigeria.

Anota, T. & Odebiyi, J. A. (1984). Resistance in sweet potato to *Cylas puncticollis* Boh. (Coleoptera: Curculionidae). *Biologia Africana*, 1, 21–30.

Bouwkamp, J. C. (1985). Production requirement. *In:* Bouwkamp, J. C. (ed.), *Sweet potato products: a natural resource for the tropics*. pp. 9–33, Boca Raton (FL), CRC Press, Florida.

Braendle, C., Davis, G. K., Brisson, J. A. & Stern, D. L. (2006). Wing dimorphism in aphids. *Journal of Heredity*, 97, 192–199.

Chalfant, R. B., Jansson, R. K., Seal, D. R. & Schalk, J. M. (1990). Ecology and management of sweet potato insects. *Annual Review of Entomology*, 35, 157–180.

Darko, Y. S. (2000). *Species composition, biology and management of sweet potato weevils (Cylas species) in Southern Ghana*. Master's thesis, University of Ghana, Ghana, Legon.

Das, G. P., Magallona, E. D., Raman, K. V. & Adalla, C. B. (1993). Growth and development of the potato tuber moth, *Phthorimaea operculella* (Zeller), on resistant and susceptible potato genotypes in storage. *The Philippine Entomologist*, 9, 15–27.

Dent, D. (2000). Host plant resistance. *In:* Dent, D. (ed.), *Insect Pest Management*, (2nd ed.). pp. 123–179, CABI Publishing, Wallingford.

Dixon, A. F. G. (1987). Parthenogenetic reproduction and the rate of increase in aphids. *In:* Minks, A. K. & Harrewijin, P. (eds.), *Aphids: Their Biology, Natural Enemies and Control*. pp. 269–285, Elsevier, Amsterdam.

Ewell, P. T. (2002). *Sweet potato production in Sub-Saharan Africa: Patterns and key issues*. International Potato Center (CIP), Lima, Peru.

FAOSTAT (2016). Production Statistics. FAO, Rome, Italy. Available at: http://faostat.fao.org/

Fening, K. O., Kioko, E. N., Mueke, J. M. & Raina, S. K. (2011). Oviposition preferences of the African wild silkmoth, *Gonometa postica* Walker (Lepidoptera: Lasiocampidae) on different substrates. *Journal of Insect Behavior*, 24, 1–10.

Golizadeh, A. & Razmjou, J. (2010). Life table parameters of *Phthorimaea operculella* (Lepidoptera: Gelechiidae), feeding on of six potato cultivars. *Journal of Economic Entomology*, 103, 966–972.

Goodarzi, M., Fathipour, Y. & Talebi, A. A. (2015). Antibiotic resistance of ranola rultivars affecting demography of *Spodoptera exigua* (Lepidoptera: Noctuidae). *Journal*

of Agricultural Science and Technology, 17, 23–33.

Hulting, F. L., Orr, D. B. & Obrycki, J. J. (1990). A computer program for calculation and statistical comparison of intrinsic rates of increase and associated life-table parameters. *Entomologist*, 73, 601–612.

Jackson, D. M. & Peterson, J. K. (2000). Sublethal effects of resin glycosides from the periderm of sweetpotato storage roots on *Plutella xylostella* (Lepidoptera: Plutellidae). *Journal of Economic Entomolog*, 93 (2), 388–393.

Jha, R. K., Chi, H. & Tang, L. C. (2012). A comparison of artificial diet and hybrid sweet corn for the Rearing of *Helicoverpa armigera* (Hubner) (Lepidoptera: Noctuidae) based on life table characteristics. *Environmental Entomology*, 41, 30–39.

Kabi, S., Ocenga-Latigo, M. W., Smit, N. E. J. M., Stathers, T. E. & Rees, D. (2001). Influence of sweetpotato rooting characteristics on infestation and damage by *Cylas* spp. *African Crop Science Journal*, 1, 165–174.

Magira, P. (2003). *Evaluating sweet potato clones for resistance to the African sweetpotato weevils (Cylas puncticollis) Boheman and Cylas brunneus (Fab.) (Coleoptera: Apionidae)*. Master's thesis, Makerere University, Kampala, Uganda.

Mahmoudi, M., Sahragard, A., Pezhman, H. & Ghadamyari, M. (2015). Demographic analyses of resistance of five varieties of date palm, *Phoenix dactylifera* L. to *Ommatissus lybicus* De Bergevin (Hemiptera: Tropiduchidae). *Journal of Agricultural Science and Technology*, 17, 263–273.

Mao, L., Story, R. N., Hammond, A. M. & Labont, D. R. (2001). Effect of sweet potato genotype, storage time and production site on feeding and oviposition behavior of the sweet potato weevil, *Cylas formicarius* (Coleoptera: Apoinidae). *Florida Entomological Society*, 8, 259–264.

Meyer, J. S., Ingersoll, C. G., McDonald, L. L. & Boyce, M. S. (1986). Estimating uncertainty in population growth rates: jackknife vs. bootstrap techniques. *Ecology*, 67, 1156–1166.

Mirmohammadi, S., Allahyari, H., Nematollahi, M. R. & Saboori, A. (2009). Effect of host plant on biology and life table parameters of *Brevicoryne brassicae* (Hemiptera: Aphididae). *Annals of the Entomological Society of America*, 102, 450–455.

Mohamed, A. E. Z. G. (2005). *Biology, ecology and management of the sweetpotato weevil, Cylas puncticollis Boheman (Coleoptera: Brentidae)*. Ph.D. thesis, University of Khartoum; Soudan, Khartoum.

Moyer, J. W., Jackson, G. V. H. & Frison, E. A. (1989). FAO/IBPGR Technical guidelines for the safe movement of sweet potato germplasm. Food and Agricultural Organization of the United Nation/International Board for Plant Genetic Resources, Rome. 29 pp.

Ndunguru, G., Thomson, M., Waida, R., Rwiza, E. & Westby, A. (1998). Methods for examining the relationship between quality characteristics and economic values of marketed fresh sweet potato. *Tropical Agriculture*, 75, 129–133.

Norris, D. M. & Kogan, M. (1980). Biochemical and morphological basis of resistance. *In:* Maxwell, F. G. & Jennings, P. R. (eds.), *Breeding Plants Resistant to Insects.* pp. 23–61, Wiley, New York.

Nottingham, S. F. & Kays, S. J. (2002). Sweet potato weevil control. *In:* Ames, T. (ed.), *Proceedings of the 1st International Conference on Sweetpotato. Food and Health for the Future.* Vol. 583, pp. 155–161, Acta Horticulturae.

Parr, M. C., Ntonifor, N. N. & Jackai, L. E. (2016). Evaluation of sweet potato cultivars for differences in *Cylas puncticollis* (Curculionidae: Brentidae) damage in South Western Cameroon. *International Journal of Research in Agricultural Sciences*, 3, 1–8.

Partridge, L., Gems, D. & Withers, D. J. (2005). Sex and death: what is the connection? *Cell*, 120, 461–472.

Razmjou, J., Moharramipour, S., Fathipour, Y. & Mirhoseini, S. Z. (2006). Effect of cotton cultivar on performance of *Aphis gossypii* (Homoptera: Aphididae) in Iran. *Journal Economic Entomology*, 99, 1820–1825.

Reddy, G. V. & Chi, H. (2015). Demographic comparison of sweet potato weevil reared on a major host *Ipomoea batatas*, and an alternative host, *I. triofoli*. *Scientific Reports*, 5, 11871. doi:10.1038/srep1187.

Rees, D., Van Oirschot, Q. E. A., Kapinga, R. E., Mtunda, K., Chilosa, D., Mbilinyi, L. B., Rwiza, E. J., Kilima, M., Kiozya, H., Amour, R., Ndondi, T., Chottah, M., Mayona, C. M., Mende, D., Tomlins, K. I., Aked, J. & Carey, E. E. (2003). Extending root shelf-life during marketing by cultivar selection. *In:* Rees, D., Quirien, O. & Kapinga, R. (eds.), *Sweet potato post-harvest assessment. Experiences from East Africa.* pp. 51–56, University of Greenwich, London, UK.

SAS Institute (1997). SAS/STAT Software: changes and enhancements through Release, 6–12. Cary, North Carolina.

Smit, N. E. J. & Van Huis, A. (1999). Biology of the African Sweetpotato Weevil Species *Cylas Puncticollis* (Boheman) and *C. Brunneus* (Fabricius) (Coleoptera: Apionidae). *The Journal of Food Technology in Africa*, 4 (3), 103–107.

Smit, N. E. J. M. (1997). *Integrated pest management for sweet potato in Eastern Africa.* Ph.D. thesis, Wageningen University, Wageningen, The Netherlands.

Smit, N. E. J. M., Downham, M. C. A., Laboke, P. O., Hall, D. R. & Odongo, B. (2001). Mass-trapping male *Cylas* spp. with sex pheromones; a potential IPM component in sweetpotato production in Uganda. *Crop Protection*, 20, 643–651.

Sokal, R. R. & Rohlf, F. J. (1995). *Biometry.* (3rd ed.). W. H. Freeman and Company, San Francisco, CA. 880 pp.

Southwood, T. R. E. & Henderson, P. A. (2000). *Ecological methods.* (3rd ed.). Blackwell, Oxford, United Kingdom.

Stathers, T. E., Rees, D., Kabi, S., Mbilinyi, L., Smith, N., Kiozya, H., Jeremiah, S., Nyango, A. & Jeffries, D. (2003a). Sweet potato infestation by *Cylas* spp. in East Africa: I. Cultivar differences in field infestation and the role of plant factors. *International Journal of Pest Management*, 49, 131–140.

Stathers, T. E., Rees, D., Nyango, A., Kiozya, H. & Mbilinyi, L. (2003b). Sweet potato infestation by *Cylas* spp. in East Africa: II. Investigating the role of tuber characteristics. *International Journal of Pest Management*, 49, 141–146.

Stevenson, P. C., Muyinza, H., Hall, D. R., Porter, E. A., Farman, D. I., Talwana, H. & Mwanga, R. O. M. (2009). Chemical basis for resistance in sweetpotato Ipomoea batatas to the sweetpotato weevil *Cylas puncticollis*. *Pure and Applied Chemistry*, 81 (1), 141–151. doi: 10.1351/PAC-CON-08-02-10.

Talekar, N. S. (1987). Feasibility of the use of resistant cultivars in sweet potato weevil control. *Insect Science and its Application*, 8, 815–817.

Tower, J. & Arbeitman, M. (2009). The genetics of gender and life span. *Journal of Biology*, 8, 38. doi: 10.1186/jbiol141.

Wang, Y. & Kays, S. J. (2002). Sweet potato volatile chemistry in relation to sweet potato weevil (*Cylas formicarius*) behavior. *Journal of the American Society for Horticulture Science*, 127, 656–662.

Wilson, D. D., Severson, R. F., Son, K. C. & Kays, S. J. (1988). Oviposition stimulant in sweet potato periderm for the sweet potato weevil, *Cylas formicarius elegantulus*. *Environmental Entomology*, 17, 691–693.

Yamaguchi, T., Tokimura, K., Hara, Y. & Atari, N. (2006). Mass-production of the Sweet potato weevil *Cylas formicarius* (Fabricius) (Coleoptera: Brentidae) using inexpensive sweet potato cultivars for starch and alcohol production. *Japanese Journal of Applied Entomology and Zoology*, 50, 157–165.

Yu, S. H., Ryoo, M. I., Na, J. H. & Choi, W. I. (2003). Effect of host density on egg dispersion and the sex ratio of progeny of *Bracon hebetor* (Hymenoptera: Braconidae). *Journal of Stored Products Research*, 39 (4), 385–393. doi:10.1016/S0022-474X(02)00032-2.

9

In the search for low-cost year-round feeds: Pen-level growth performance of local and crossbred Ugandan pigs fed forage- or silage-based diets versus commercial diet

Natalie Ann Carter [a,*], Catherine Elizabeth Dewey [a], Delia Grace [b], Ben Lukuyu [c], Eliza Smith [d], Cornelis de Lange [e,**]

[a] *Population Medicine, University of Guelph, Guelph, Ontario, Canada*
[b] *Food Safety and Zoonoses, International Livestock Research Institute, Nairobi, Kenya*
[c] *Integrated Sciences, International Livestock Research Institute, Kampala, Uganda*
[d] *KYEEMA Foundation, Brisbane, Australia*
[e] *Animal Biosciences, University of Guelph, Guelph, Ontario, Canada*

Abstract

Smallholder pig farmers in East Africa report that lack of feed, seasonal feed shortages, quality and cost are key constraints to pig rearing. Commercially prepared pig diets are too expensive and people and pigs compete for food. Smallholder farmers typically feed nutritionally unbalanced diets, resulting in low average daily gain (ADG) and poor farmer profits. Our objective was to compare the ADG of Ugandan pigs fed forage- or silage-based or commercial diets. Ugandan weaner-grower pigs were randomly assigned to forage- or silage-based diets or commercial diet. Pigs were weighed every 3 weeks from 9 to 32 weeks of age. Pen-level ADG and feed conversion were compared across diets using multiple linear regression. The ADG of pigs fed forage- or silage-based diets was lower than those fed commercial diets between 9 and 24 weeks of age ($p < 0.05$). Between 28 and 32 weeks, pigs fed forage-based diets had lower ADG than those on other diets ($p < 0.05$). Least squares mean ADG (g/pig/day) for pigs fed forage- or silage-based diets or commercial diet were 36, and 52, and 294 respectively at 9–15 weeks; 163, 212, 329 at 15–19 weeks; 112, 362, 574 at 20–24 weeks and 694, 994, and 1233 at 28 to 32 weeks of age. It was concluded that forage- and silage-based diets are unsuitable for small, newly weaned pigs. Feeding forage- or silage-based diets to finishing pigs is more suitable. Forage- and silage based diets are year-round low-cost pig-feeding strategies that will improve the growth performance of East African pigs, thereby increasing pig farmer income and food security.

Keywords: average daily gain, East Africa, feed conversion, smallholder, swine

1 Introduction

Lack of feed, poor quality feed and cost of feedstuffs are key constraints to pig rearing for smallholder pig farmers in East Africa (Mutua *et al.*, 2012; Ouma *et al.*, 2015). Pig farmers have limited access to feedstuffs, inadequate stor-age, restricted processing ability and cannot afford commercially prepared pig diets (ibid.). Feedstuffs availability fluctuates seasonally, exacerbating existing food/feed shortages (Kagira *et al.*, 2012; Mutua *et al.*, 2012). Pig farmers consume many of the same products typically fed to pigs (e.g. maize, fruit, sun-dried fish) resulting in food competition (ibid.). A variety of feedstuffs, including kitchen waste, are fed to pigs and pigs typically eat carbohydrate-rich low-or-no-protein nutritionally unbalanced diets (ibid.). Smallholder pig-keeping practices, i.e., free-range management and inadequate and imbalanced feeding contributes to poor

* Corresponding author – natalieacarter001@gmail.com
60 University Private, Ottawa, Ontario, Canada, K1N 6N5

** Deceased

growth performance of pigs on East African smallholder farms (130 g/day) (Kagira et al., 2010; Mutua et al., 2012; Carter et al., 2013), and elsewhere in the tropics (Pheng-savanh et al., 2010; Kambashi et al., 2014) resulting in de-creased farmer profit (Levy et al., 2009). During the dry season, some pigs (33 %) are permitted to scavenge but dur-ing the cropping season, most pigs (65 %) are kept tethered (Kagira et al., 2010). In many cases (31 %) pigs feed mainly on grass while tethered under a tree (Mutua et al., 2012; Mbuthia et al., 2015).

Pig performance in East Africa would be improved through the use of well-balanced cost-effective diets. Ideally, locally available feedstuffs fed either fresh or en-siled would be used to meet the nutrient requirements of pigs (Ly et al., 2012; Kaensombath et al., 2013). Efficient use of these feedstuffs is required for sustainable and prof-itable pig production thereby improving rural livelihoods. Low-cost forage- and silage-based diets containing some zero-cost feedstuffs are needed to improve pig performance and empirical evidence of their efficacy is required. Our hy-potheses were threefold: (1) average daily gain (ADG) of pigs fed forage- or silage-based diets is higher than repor-ted ADG of pigs raised under typical East African small-holder management conditions but lower than pigs fed a commercial diet; (2) feed conversion of pigs fed forage- or silage-based diets is less efficient than pigs fed a commercial diet; and (3) ADG is positively correlated with body weight (BW). The objectives of the study were to determine the ADG and feed efficiency (gain to feed [G : F] ratio) of local (LB) and crossbred (exotic × local; CB) pigs in Uganda fed 1 of 3 diets (forage-based, silage-based, or commercial). This study demonstrated that forage- and silage- based di-ets are low-cost year-round pig-feeding strategies that can improve the growth performance of pigs in East Africa.

2 Materials and methods

2.1 Diet formulation

Feedstuffs included in this study were selected based on nutrient content, cost, relative importance, farmer ease of access, seasonal availability, known safe levels of inclusion due to absence of anti-nutritional factors, and likelihood of securing sufficient volume throughout the trial period. Sweet potato vine and tuber silage was included in response to seasonal feedstuff shortages because silage can be made when vines and tubers are plentiful, and fed when feedstuffs are scarce (Peters, 1997). The 70 % vines, 30 % tubers, 0.05 % salt (as fed basis) ratio reflected optimal pH and nutrient results reported for silage in East Africa (Manoa, 2012).

The nutritional requirements of 8 to 65 kg bodyweight (BW) CB pigs in Uganda were determined using the methods outlined in Carter et al. (2017). Briefly, a static model, based on the dynamic National Research Council of the National Academies (NRC, 2012) nutrient require-ment model for growing-finishing pigs was developed to represent the use of daily intake of digestible energy (DE in kcal/kg of dry matter; DM) for maintenance, body lipid deposition (Ld, gram/day) and body protein deposition (Pd, gram/day), for pigs between 8 to 65 kg BW. Mean ADG (kg/day) was predicted from Ld and Pd using the static model. Nutrient requirements for Ugandan CB pigs (8 to 65 kg BW) were predicted. All other calculations, including nutrient requirement predictions, were taken directly from NRC (2012). Balanced low-cost diets (forage- and silage-based) were formulated for two phases of feeding (newly weaned and finishing pigs (8 to 20 and 20 to 65 kg BW, respectively) using a least-cost diet formulation program based on Skinner et al. (2012) and 26 available feedstuffs (Carter et al., 2015). The composition of diets is presented in Table 1 and the calculated nutrient composition is presen-ted in Table 2.

2.2 Newly weaned and finishing growth study

Animal use was approved by University of Guelph, Guelph, Canada and the International Livestock Research Institute (ILRI), Nairobi, Kenya. In 2014, 45 LB and 45 CB (local × Landrace and/or Large White and/or Cambor-ough) pigs from 16 farms were enrolled in the study. The sample size was calculated based on a 20 gram difference in pen-level ADG using 80 % power and 5 % confidence level. Variance of gain per day was estimated to be 240 grams. Ten pens per diet were required and each pen contained three pigs as replicates. Pig breed was determined by farmer re-call. All pigs were born within 3 days of each other. Pigs were individually weighed after the first feeding of the day at the trial start and every 21 days. Pigs were fed one of three diets ad libitum: (1) forage-based or (2) silage-based or (3) a commercial sow and weaner feed (Ugachick Poultry Breeders, Namulonge Rd, Gayaza, P.O. Box 12337 Kam-pala, Uganda), the only pig ration available for purchase from that company. For each pen, feed of the appropriate diet type was weighed, and the weight (kg) was recorded before feeding.

For the newly weaned growth study – 8-week old pigs (n = 90), blocked by breed, sex, and litter were randomly assigned to 1 of 3 diets (forage-, or silage-based, or com-mercial) and 1 of 2 rooms (5 pens per breed type per diet). Each pen contained 3 pigs of the same sex and breed. Mean BW (kg) across diets and the standard deviation did not dif-

Table 1: *Compositions of diets (g/kg of dry matter) used in a growth study of newly weaned and finishing pigs in Uganda.*

	Forage-based diet*			Silage-based diet*		
	Age (weeks)			Age (weeks)		
	9–15	16–24	28–32	9–15	16–20	21–32
Ingredient (g/kg of dry matter)						
Avocado (*Persea americana*); ripe, with peel, seed removed	155.1	200.0	200.0	0.0	0.0	0.0
Banana leaf (*Musa sapientum*); centre vein removed	7.3	10.0	10.0	0.0	0.0	0.0
Cassava leaf, blade, and axil (*Manihot esculenta*); wilted	0.0	0.0	0.0	50.0	50.0	0.0
Cottonseed meal	0.0	50.0	50.0	50.0	50.0	50.0
Jackfruit (*Artocarpus heterophyllus*); ripe, with peel and seeds	400.0	200.0	200.0	259.6	256.6	254.7
Maize bran	0.0	133.5	239.2	0.0	220.1	266.1
Papaya leaf (*Carica papaya*); wilted	0.0	0.0	0.0	69.2	43.4	45.3
Ground sun-dried fish (*Rastrineobola argentea*)	226.3[†]	77.9[‡]	83.0[‡]	166.3[†]	74.5[‡]	78.1[‡]
Sweet potato vine (*Ipomoea batatas*); wilted	200.0	316.2	204.6	0.0	0.0	0.0
Sweet potato vine and tuber silage (*Ipomoea batatas*)	0.0	0.0	0.0	400.0	300.0	300.0
Limestone	6.0	6.9	7.2	0.0	0.0	0.0
Common table salt	3.8	3.9	4.0	3.5	3.9	3.9
Vitamin and mineral premix[§]	1.5	1.6	2.0	1.4	1.5	1.9

* Non-compliance in diet formulation occurred from 24 through 27 weeks of age. Data not presented.
[†] Pre-ground livestock-grade.
[‡] Whole human-consumption grade ground at study site.
[§] The premix provided the following per kg of complete feed (dry matter): vitamin A 15,000,000 IU; vitamin D_3 2,000,000 IU;
   vitamin E 20,000 IU; vitamin K_3 6000 mg; vitamin B_1 1000 mg; vitamin B_2 5000 mg; nicotinic acid 20,000 mg;
   pantothenic acid 16,000 mg; choline chloride 200,000 mg; biotin 110 mg; folic acid 1500 mg; manganese 40,000 mg;
   iron 150,000 mg; zinc 110,000 mg; copper 40,000 mg; cobalt 280 mg; iodine 1500 mg; selenium 120 mg

Table 2: *Calculated nutrient composition (% of dry matter [DM]) of forage- and silage-based diets used in a growth study of newly-weaned and finishing pigs in Uganda.*

	Forage-based diet*			Silage-based diet*		
	Age (weeks)[†]			Age (weeks)[†]		
	9–15	16–24	28–32	9–15	16–20	21–32
Calculated nutrient composition						
DE[‡] (kcal/kg of DM)	2775	2886	2960	2590	2849	2849
Crude protein	14.0	17.4	16.9	15.9	16.8	16.6
Neutral detergent fibre	24.9	29.5	28.4	23.4	24.0	24.3
Ether extract	5.6	7.5	7.5	2.6	2.5	2.0
Total calcium	0.56	0.56	0.56	0.62	0.60	0.52
Total phosphorous	0.64	0.64	0.65	0.52	0.60	0.59
STTD[§] phosphorous	0.37	0.35	0.35	0.27	0.32	0.30
Total sodium	0.19	0.19	0.20	0.15	0.18	0.18
Total lysine	0.99	1.06	1.06	0.96	1.01	1.00
SID[¶] lysine	0.71	0.74	0.76	0.67	0.73	0.73

* Nutrient composition of commercial diet is unknown due to proprietary confidentiality.
[†] Non-compliance in diet formulation occurred from 24 through 27 weeks of age. Data not presented.
[‡] Digestible energy. Estimated from nutrient composition according to NRC (2012) and based on
   nutrient composition according to pre-trial proximate nutrient analysis (calculated; data not shown) or
   nutrient composition of the diets (analysed).
[§] Standardized total tract digestible phosphorus concentration.
[¶] Standardized ileal digestible lysine concentration.

fer by diet at the start of the trial ($p < 0.05$). During a 7-day acclimation period, all pigs were fed commercial diet, after which they were introduced to their trial diet over a 3-day period, then fed only their trial diet for 11 weeks (9–20 weeks of age), at which point the newly-weaned growth study ended.

For the finishing growth study – Phase 3 (21 to 24-week-old pigs) and Phase 4 (28 to 32-week-old pigs), the 21-week old pigs remained in the same pens with the same pen mates, the same diet formulation was used, and all feeding and sampling methods remained the same. Only diet assignment was changed. Total BW (kg) per pen and pen-level BW tertiles were determined. The two heaviest pens from each tertile were kept on the same diet as during the newly-weaned growth study. All other pens within each tertile were randomly assigned to 1 of the 3 diets. Mean BW (kg) across diets and the standard deviation did not differ. During a 7-day acclimation period pigs assigned to a new diet were fed 40% new diet and 60% old diet for 3 days, then 60% new diet and 40% old diet for 3 days (as-fed) and then 100% new diet. Pigs remaining on the same diet as in the newly-weaned growth study received the same diet during the acclimation period. Non-compliance in diet formulation occurred from 24 through 27 weeks of age. (Data not presented. It is unknown if this had any effect).

2.3 Statistical analysis

Pen-level ADG for each of four phases (9 to 15, 16 to 20, 21 to 24, and 28 to 32-week-old pigs, respectively) were analysed using PROC MIXED of SAS (v.9.4; SAS Institute Inc., Cary, NC). Differences among least square means were assessed using the Tukey's multiple comparisons test for pairwise differences. Pigs were grouped according to diet (commercial, forage-based, silage-based). To control for potential confounding, breed (crossbred or local) and sex (female or barrows) were included as fixed effects. Body weight (kg) at the start of each phase was included as a continuous variable. Interactions between breed and diet as well as starting BW and diet were evaluated to explore any potential effect on ADG. Data are presented as the least square means for the three diets ± standard error. Means were considered different when $p < 0.05$. The same methods as used in the ADG statistical analysis were applied to determine factors associated with gain to feed ratio (kg of gain/kg of feed consumed).

3 Results

Least square means of average daily gain and feed conversion efficiency of newly-weaned and finishing pigs are presented in Table 3. In Phase 1 pigs fed forage- or silage-based diet had lower ADG than pigs fed commercial diet ($p < 0.05$). In Phase 2 pigs fed forage-based diet had lower ADG than pigs fed commercial diet ($p < 0.05$). In Phases 1 and 2 pigs fed forage- and silage-based diet had the lowest feed conversion (gain:feed) ($p < 0.05$). In Phase 3, pigs fed forage-based diet had the lowest ADG, followed by silage-based diet, then commercial diet (0.250 and 0.462 kg/day more respectively for silage-based diet and commercial diet, $p < 0.05$). In Phase 4, pigs fed forage-based diet had the lowest ADG, followed by silage-based diet, then commercial diet (0.300 and 0.539 kg/day more respectively for silage-based diet and commercial diet, $p < 0.05$). Similarly, in Phases 3 and 4, pigs fed forage-based diet had the lowest feed conversion, followed by silage-based diet, then commercial diet ($p < 0.05$). Starting BW was not associated with ADG ($p > 0.05$) because pigs were blocked to have similar mean pen-level BW across diets.

4 Discussion

Across diets ADG was low compared to other pigs in the tropics (Codjo, 2003; Anugwa & Okwori, 2008). Low ADG observed in the newly weaned African growth study may be due to low nutrient density and insufficient DE intake. For newly weaned pigs, the most critical factor limiting growth is daily nutrient intake (Patience et al., 1995). It is unlikely that pigs in this study had sufficient feed intake and fibre digestion capacity to reach their performance potential, as evidenced by low ADG across all three diets. Low ADG may also be due to fairly low mean initial BW (6.8 ± 1.6 kg and 12.1 ± 5.7 kg at 9 and 15 weeks of age respectively). Others report low ADG (97 to 176 g/day) in pigs fed fresh and ensiled forages due to low initial BW (16.9 to 31.4 kg) (Ortega et al., 2012). Pigs' ability to digest dietary fibre is dependent on age, BW, and on-going exposure to fibrous feeds (Noblet & Le Goff, 2001; Wenk, 2001). Feeding fibrous feeds is cost-effective for pigs > 50 kg BW because the animals' ability to digest fibre increases as pigs get older (Machin, 1990). As previous research suggests (Noblet & Le Goff, 2001; Wenk, 2001), pigs may have adapted to fibrous feed through on-going exposure, and their ability to digest dietary fibre may have improved with increased age and BW resulting in higher ADG in Phase 2 than in Phase 1. Higher mean BW in Phase 2 than 1 may have enabled increased feed intake capacity.

Across diets ADG was higher in Phase 3 than in Phases 1 and 2 for the reasons discussed previously. It is important to note that, across diets, pigs in Phase 3 had higher ADG than pigs raised under typical smallholder conditions in East Africa (110 to 130 g/day) (Mutua et al., 2012; Carter

Table 3: *Least square means of average daily gain and feed conversion efficiency of newly-weaned and finishing pigs in Uganda (pen level).*

Item	Diet				Breed			Diet	Breed	Breed × Diet	Start weight
	Forage-based	Silage-based	Com-mercial	SEM	Cross	Local	SEM				
Newly-weaned growth study											
No. of observations (pens)	10	10	10	–	15	15	–	–	–	–	–
Initial bodyweight, kg	7.0	6.4	6.8	–	7.9	5.6	–	–	–	–	–
Average daily gain, g											
Phase 1*	36[a]	52[a]	294[b]	10	125	129	10	< .001	.842	.369	.053
Phase 2	163[a]	212[ab]	329[b]	41.5	253	216	16.5	.017	.175	.148	.099
Gain : Feed (kg body weight gain : kg feed intake)											
Phase 1	.032[a]	.034[a]	.348[b]	.008	.154	.122	.008	< .001	.025	.078	.007
Phase 2	.225[a]	.223[b]	.597[c]	.025	.353	.344	.010	< .001	.577	.265	.059
Finishing growth study											
No. of observations (pens)	12	9	11	–	16	16	–	–	–	–	–
Initial bodyweight, kg	21.4	21.4	24.2	–	25.6	19.1	–	–	–	–	–
Average daily gain, g											
Phase 3	112[a]	362[b]	574[c]	26	402	297	20	.002	.002	.206	.672
Phase 4	694[a]	994[b]	1233[c]	52	1071	876	38	< .001	.002	.136	.241
Gain : Feed (kg body weight gain : kg feed intake)											
Phase 3	.039[a]	.098[b]	.346[c]	.011	.169	.153	.009	< .001	.207	.905	.003
Phase 4	.039[a]	.057[b]	.145[c]	.005	.072	.088	.003	< .001	.003	.081	.007

* Phase 1: Pig age 9–15 weeks; Phase 2: 16–20 weeks; Phase 3: 21–24 weeks; and Phase 4: 28–32 weeks. Non-compliance in diet formulation occurred from 24–27 weeks. Data not presented.

[a,b,c] Means on the same row with different superscripts differ significantly ($p < 0.05$).

et al., 2013). And, although ADG of pigs fed silage-based diet was lower than pigs fed commercial diet, it was higher than that of pigs on smallholder farms in Soroti, Uganda that consumed 400–500 kg of sweet potato for 30 days (Peters, 1997). Similarly Mongcai × Yorkshire (MY) crossbred pigs fed a 40 % sweet potato tuber silage diet, had higher ADG when the starting BW was > 50 kg than at 15–50 kg BW (524 and 423 g/day, respectively), indicating balanced diets including silage can provide the energy and nutrients required for good growth performance (Giang *et al.*, 2004). Ensiling vines and tubers during the wet season converts surplus, highly perishable materials with low marketability, into a much-needed pig feedstuff improving pig growth and farmers' incomes (Peters, 2008; Cargill *et al.*, 2009).

In all growth phases, pigs fed the commercial diet had higher feed efficiency (kg BW gain/kg of feed intake) than pigs fed the forage- and silage-based diets ($p < 0.05$). In this study feed efficiency was rather low for pigs at the BW range tested relative to studies of exotic, commercially raised pigs (NRC, 2012). However, feed conversion ratio (FCR) in this study was similar to FCR reported for LB pigs and CB pigs elsewhere in the tropics (Codjo, 2003; Kanengoni *et al.*, 2004). In this study ADG is low, implying a large proportion of daily energy and nutrient intake is required for maintenance, contributing to reductions in feed efficiency (NRC, 2012). The better feed efficiency in pigs fed commercial diet reflects higher nutrient density in the commercial diet.

This study demonstrated that forage- and silage-based diets are not suitable for small, newly weaned pigs. Feeding commercial diet to newly weaned pigs, and forage-or silage-based diets to finishing pigs is the most suitable solution. Where silage-making resources are available, silage-based diets should be used for growing pigs and when not, forage-based diets can be used. Although in this manuscript we do not report on the cost effectiveness of these diets, in a related pig-level study we determined that it was most cost effective to feed commercial diet until pigs reached sufficient BW, at which time it became cost effective to feed forage- and silage-based diets (10.9 and 11.9 kg BW for forage- and silage-based diets, respectively) (Carter *et al.*, 2017).

We recognize that these forage- and silage-based diets can be improved upon, but they are a better alternative than current smallholder management practices with resulting low ADG (110 g/day) reported elsewhere in eastern Africa (Mutua *et al.*, 2012). The results of this study indicate forage- and silage-based diets are year-round low-cost pig-feeding strategies that will improve the growth performance of East African pigs.

Acknowledgements

Funding from the Smallholder Pig Value Chain Development Project (SPVCD), Irish Aid, the 'Livestock and Fish by and for the Poor' and 'Agriculture for Nutrition and Health' CGIAR Research Programs, International Fund for Agricultural Development (IFAD), European Commission, Ontario Agricultural College, Ontario Veterinary College, and the University of Guelph; the assistance of SPVCD Project team members, Bioversity International Uganda, Masaka District Veterinary Office, Drs. Eliza Smith, Yumi Kirino, Johanna Lindahl, Kristina Rösel, Ian Dohoo, Jane Poole, and Shari van de Pol; Eve Luvumu, Ponsiano Nyombi, John Kiriggwa, Charles Bunnya, Flora Kuteesa Namwanje, John Kato Kalema, Alfred Alifunsi, Jenner Kasibante, Robert Fathke, Karen Richardson, Julia (Cuilan) Zhu, and the participation of the Kyanamukaka - Kabonera Pig Farmers Co-operative Society is greatly appreciated. We are grateful to the editor and reviewer for constructive comments on an earlier version of this paper.

References

Anugwa, F. O. I. & Okwori, A. I. (2008). Performance of growing pigs of different genetic groups fed varying dietary protein levels. *African Journal of Biotechnology*, 7, 2665–2670.

Cargill, C. F., Mahalaya, I. S., Tjintokohadi, I., Syahputra, T. A. & Gray, D. (2009). *Final report: Poverty alleviation and food security through improving the sweetpotato-pig systems in Papua, Indonesia*. Australian Centre for International Agricultural Research (ACIAR), GPO Box 1571, Canberra ACT 2601, Australia.

Carter, N., Dewey, C., Mutua, F. K., de Lange, C. & Grace, D. (2013). Average daily gain of local pigs on rural and peri-urban smallholder farms in two districts of western Kenya. *Tropical Animal Health and Production*, 45, 1533–1538.

Carter, N. A., Dewey, C. E., Grace, D., Lukuyu, B., Smith, E. & de Lange, C. F. M. (2017). Average daily gain and the impact of starting body weight of individual nursery and finisher Ugandan pigs fed a commercial diet, a forage-based diet, or a silage-based diet. *Journal of Swine Health and Production*, 25 (3), 121–128.

Carter, N. A., Dewey, C. E., Lukuyu, B., Grace, D. & de Lange, C. (2015). Nutritional value and seasonal availability of feed ingredients for pigs in Uganda. *Agricultura Tropica et Subtropica*, 48, 91–104.

Codjo, A. B. (2003). Estimation des besoins énergétiques du porc local du Bénin en croissance entre 7 et 22 kg de poids vif. *Tropicultura*, 21 (2), 56–60.

Giang, H. H., Ly, L. V. & Ogle, B. (2004). Digestibility of dried and ensiled sweet potato roots and vines and their effect on the performance and economic efficiency of F1 crossbred fattening pigs. *Livestock Research for Rural Development*, 16 (7), #50. Available at: http://www.lrrd.org/lrrd16/7/gian16050.htm (accessed on: 15 September 2017).

Kaensombath, L., Neil, M. & Lindberg, J. E. (2013). Effect of replacing soybean protein with protein from ensiled stylo (*Stylosanthes guianensis* (Aubl.) Sw. var. *guianensis*) on growth performance, carcass traits and organ weights of exotic (Landrace × Yorkshire) and native (Moo Lath) Lao pigs. *Tropical Animal Health and Production*, 43, 865–871.

Kagira, J. J., Kanyari, P. W. N., Maingi, N., Githigia, S. M., Ng'ang'a, J. C. & Karuga, J. (2010). Characteristics of the smallholder free-range pig production system in western Kenya. *Tropical Animal Health and Production*, 42, 865–873.

Kagira, J. M., Kanyari, P. N., Githigia, S. M., Maingi, N. & Ng'ang'a, J. C. (2012). Risk factors associated with occurrence of nematodes in free range pigs in Busia District, Kenya. *Tropical Animal Health and Production*, 44, 657–664.

Kambashi, B., Picron, P., Boudry, C., Théwis, A., Kiatokoa, H. & Bindelle, J. (2014). Nutritive value of tropical forage plants fed to pigs in the Western provinces of the Democratic Republic of the Congo. *Animal Feed Science and Technology*, 19, 47–56.

Kanengoni, A. T., Dzama, K., Chomonyo, M., Kusina, J. & Maswaure, S. (2004). Growth performance and carcass traits of Large White, Mukota and Large White × Mukota F_1 crosses given graded levels of maize cob meal. *Animal Science*, 78, 61–66.

Levy, M., Dewey, C., Weersink, A. & Mutua, F. K. (2009). Challenges of rural and peri-urban pig butcher businesses in western Kenya. *In:* Proceedings of the 12th Symposium of the International Society for Veterinary Epidemiology and Economics (ISVEE), Durban, South Africa. p. 366, ISVEE.

Ly, N. T. H., Ngoa, L. D., Verstegen, M. W. A. & Hendriks, W. H. (2012). Pig performance increases with the addition of DL-methionine and L-lysine to ensiled cassava leaf protein diets. *Tropical Animal Health and Production*, 44, 165–172.

Machin, D. (1990). Alternative feeds for outdoor pigs. *In:* Stark, B. A., Machin, D. H. & Wilkinson, J. M. (eds.), *Outdoor Pigs Principles and Practices*. Chalcombe Publications, 13 Highwoods Drive, Marlow Bottom, Marlow, Bucks, Sl7 3PU, Great Britain.

Manoa, L. A. (2012). Evaluation of dry matter yields and silage quality of six sweet potato varieties. A thesis submitted in partial fulfillment for the degree of Master of Science in animal nutrition and feed science, Department of Animal Production, Faculty of Agriculture, College of Agriculture and Veterinary Sciences, University of Nairobi.

Mbuthia, J. M., Rewe, T. O. & Kahi, A. K. (2015). Evaluation of pig production practices, constraints and opportunities for improvement in smallholder production systems in Kenya. *Tropical Animal Health and Production*, 47 (2), 369–376.

Mutua, F. K., Dewey, C., Arimi, S., Ogara, W., Levy, M. & Schelling, E. (2012). A description of local pig feeding systems in village smallholder farms of Western Kenya. *Tropical Animal Health and Production*, 44, 1157–1162.

Noblet, J. & Le Goff, G. (2001). Effect of dietary fibre on the energy value of feeds for pigs. *Animal Feed Science and Technology*, 90, 35–52.

NRC (2012). *Nutrient requirements of swine*. National Research Council of the National Academies (NRC), The National Academies Press, Washington.

Ortega, E. A., Van Der Hoek, R., Orozco, R. L., Rodríguez, C., Hoedtke, S., Sarria, P. & Martens, S. (2012). Performance of pigs fed with fresh and ensiled forage of *Vigna unguiculata*, *Lablab purpureus* and *Cajanus cajan*. *In:* Kuoppala, K., Rinne, M. & Vanhatalo, A. (eds.), *Proceedings of XVI International Silage Conference, Hämeenlinna, Finland, 2–4 July 2012*. MTT Agrifood Research Finland, University of Helsinki. Available at: http://urn.fi/URN:ISBN:978-952-487-385-7

Ouma, E., Dione, M., Lule, P., Pezo, D., Marshall, K., Roesel, K., Mayega, L., Kiryabwire, D., Nadiope, G. & Jagwe, J. (2015). Smallholder pig value chain assessment in Uganda: results from producer focus group discussions and key informant interviews. ILRI Project Report, Nairobi Kenya, ILRI. Available at: https://hdl.handle.net/10568/68011

Patience, J. F., Thacker, P. A. & de Lange, C. F. M. (1995). Chapter 7: Feeding the Suckling Pig, and Chapter 8: Feeding the weaned pig. *In:* Swine Nutrition Guide, (2nd ed.). Ch. 7–8, pp. 167–185, Prairie Swine Centre Inc, Saskatoon, Saskatchewan Canada.

Peters, D. (1997). Sweetpotato post-harvest strategies for food security and income generation: the case of Soroti, Uganda. R4D Research for Development Department for International Development UK AID Report. Available at: https://assets.publishing.service.gov.uk/media/57a08dafed915d622c001b1d/R7036c.pdf (accessed on: 15 September 2017).

Peters, D. (2008). Assessment of the potential of sweet potato as livestock feed in East Africa: Rwanda, Uganda and Kenya. A report presented to the International Potato Center (CIP), Nairobi. Available at: http://www.fao.org/fileadmin/templates/agphome/images/iclsd/documents/wk1_c13_claessens.pdf (accessed on: 15 September 2017).

Phengsavanh, P., Ogle, B., Stur, W., Frankow-Lindberg, B. E. & Lindberg, J. E. (2010). Feeding and performance of pigs in smallholder production systems in Northern Lao PDR. *Tropical Animal Health and Production*, 42, 1627–1633.

Skinner, S. A., Weersink, A. & de Lange, C. F. (2012). Impact of dried distillers grains with solubles (DDGS) on ration and fertilizer costs of swine. *Canadian Journal of Agricultural Economics*, 60, 335–356.

Wenk, C. (2001). The role of dietary fibre in the digestive physiology of the pigs. *Animal Feed Science and Technology*, 90, 21–33.

10

Typological characterisation of farms in a smallholder food-cash crop production system in Zimbabwe –opportunities for livelihood sustainability

Nothando Dunjana [a,b,*], Rebecca Zengeni [a], Pardon Muchaonyerwa [a], Menas Wuta [c]

[a] *School of Agricultural, Earth and Environmental Sciences, University of KwaZulu Natal, Private Bag X01, Scottsville 3209, South Africa*
[b] *Marondera University of Agricultural Sciences and Technology, P. O. Box 35, Marondera, Zimbabwe*
[c] *Department of Soil Science and Agricultural Engineering, University of Zimbabwe, P. O. Box MP 167, Mt Pleasant, Harare, Zimbabwe*

Abstract

The diversity of smallholder farms in space, resource endowment, production and consumption decisions are often a hindrance to the design, targeting, implementation and scaling out of agricultural development projects. Understanding farm heterogeneity is crucial in targeting interventions that can potentially contribute to improved crop productivity, food security and livelihood sustainability. The study sought to define and understand farm typology in a resettlement smallholder food-cash crop production area in Zimbabwe. Data was collected from five focus group discussions (FGDs), and 102 household interviews. Principal component analysis (PCA), multiple correspondence analysis (MCA) and cluster analysis were used to analyse quantitative and qualitative data variables and aggregate farms into clusters according to production means, socio-economics and demographics. The three identified farm types were (i) resource-endowed, commercial oriented farms, (ii) medium resourced and (iii) resource constrained farms practising subsistence and income oriented production. Labour was cited as a major challenge, with high labour cost relevant for type I farms, while household size has more bearing for type II and III farms. Ownership of tillage implements and operations varied from mechanised on resource endowed farms, to animal drawn on some medium and resource constrained farms. The farms exhibited variable livelihood strategies and all clusters exhibited market participation, albeit to varying extents. Thus strengthening of market links is imperative. Use of multivariate methods allowed for identification of the most discriminating variables for farm delineation and subsequent clustering of farms forms the basis for further exploring variability across farm types for the targeting of management interventions for livelihood sustainability.

Keywords: diversity, livelihood, maize-tobacco, multivariate analysis, sustainability, targeted interventions, typology

1 Introduction

Smallholder agriculture is a critical component of Africa's agricultural sector responsible for the bulk of food production and income for the developing countries (Muchero, 2008; Wiggins, 2009). Common constraints to productivity in smallholder agriculture include biophysical and socio-economic challenges such as poor soil fertility, rainfall variability and unreliability, and limited access to capital, labour and markets (Ncube *et al.*, 2009; Salami *et al.*, 2010). Although, most challenges faced by smallholder farms are common across spatial and temporal divides, the interaction of biophysical and socio-economic factors results in the heterogeneity of farm types or typologies within and across landscapes. In most regions of sub-Saharan Africa (SSA), soil types and fertility management have culminated in soil heterogeneity within and across farms (Masvaya *et al.*, 2010; Tittonell *et al.*, 2005), while resource endowment, land holdings, production orientation and objectives further compound the stratification of farms into com-

---

* Corresponding author – ntandodunjana@yahoo.com

plex, dynamic and diverse farm typologies (Kuivanen *et al.*, 2016; Mtambanengwe & Mapfumo, 2005; Tittonell *et al.*, 2010). Envisaged impacts of technologies are often not realised in most smallholder farming systems in the developing countries, because they repeatedly fail to match the complexity and diversity of the farming systems (Emtage & Suh, 2005).

An understanding of the diversity across the farms within farming systems is paramount to improved precision in the design of technological interventions (Hilhorst & Muchena, 2000; Tittonell *et al.*, 2010), which could improve applicability, relevance and adoptability of agronomic and technological interventions and recommendations. The concept of farm typologies identifies groups of farms on the basis of similar sets of attributes ranging from social, ownership, operational, production to structural characteristics (Kostrowicki, 1977). Participatory approaches have been used to this end (Mtambanengwe & Mapfumo, 2005; Ncube *et al.*, 2009; Zingore *et al.*, 2007). While the obvious strength of participatory approaches is the involvement of community members in the delineation, individual farm characteristics are likely traded in for general and more encompassing community classification. However, decision making and production orientation of farms is a household decision, hence there is need to aggregate farms according to household characteristics not general community characteristics. Increasingly, multivariate statistical techniques such as principal component analysis (PCA), multiple correspondence analysis (MCA) and cluster analysis have been used in farm typology studies within various farming systems (Bellini & Ramberti, 2009; Kuivanen *et al.*, 2016; Rusinamhodzi *et al.*, 2012; Tittonell *et al.*, 2010). Multivariate analysis techniques applied to household data, systematically reduce data dimensions, farm heterogeneity and give results, that are reproducible in space and time (Kostrowicki, 1977). PCA is more commonly applied to quantitative data, while MCA is adapted to analyse categorical variables (Jolliffe, 2002; Ozden & Mendes, 2005; Savary *et al.*, 1995). Therefore, specific data types can be analysed by specific methods thus presenting a challenge when mixed data is available. Consequently, complementary application of PCA and MCA to mixed data is an attractive option whose effectiveness has been demonstrated in several studies (Ballesteros *et al.*, 2015; Mekkawy *et al.*, 2017; Smith *et al.*, 2002).

The Svosve smallholder farming area, in Zimbabwe, is predominantly a food and cash crop production area. Farms can be generally categorised as smallholder, displaying certain features such as reliance on farm production for livelihoods, limited access to capital inputs as well as depend-

ence on household labour for farm production (Chamberlain, 2007; Ellis, 1988). Maize is grown to meet the household's subsistence, while tobacco is grown as the main cash crop, contributing significantly to smallholder farmers' income (Masvongo, 2013; Shumba & Whingwiri, 2006). Livestock is kept as a source of draft power, organic manures as well as a protein source (Mugwira & Murwira, 1997; Ouma *et al.*, 2003). Although, tobacco production by smallholder farmers is not new, increased participation by smallholder farmers in Africa's leading tobacco producing countries, Malawi, Tanzania and Zimbabwe has been reported (Masvongo, 2013; Prowse, 2013), hence making this farming system a key area of interest upon which smallholders' food and income security are hinged.

Several studies have classified smallholder farms into typologies that seek to aid targeting of interventions (Kuivanen *et al.*, 2016; Tittonell *et al.*, 2010), but maize-tobacco production systems have been largely ignored. This study, thus aims to define and understand the farm types under such systems. The objective of this study was to establish farm typologies in Svosve smallholder farming system and identify farm type-specific constraints and opportunities for the targeting of agricultural interventions and recommendations for improved food security and livelihoods.

2 Materials and methods

2.1 Site description

The study was conducted in Svosve area, in Mashonaland East province of Zimbabwe (18° 21′ 56″ S and 31° 42′ 19″ E). The area is in the agroecological region (AER) IIb receiving an average of 750 mm of rainfall annually between October and April, and has been described as having a "favourable climate for growing crops" (Chimhowu & Woodhouse, 2008). However, seasonal and spatial variations in rainfall distribution are common. Granite-derived sandy soils (Lixisol; FAO, 1998), which are inherently infertile and highly leached, are dominant in the study area.

Smallholder farming is the typical production system mainly characterised by mixed crop-livestock production. Maize (*Zea mays* L.) is grown as the staple crop, while secondary crops include millet (*Pennisetum glaucum* L.), groundnuts (*Arachis hypogaea* L.), bambara nuts (*Vigna subterranea* L.), sweet potato (*Ipomoea batatas* L.) and sugar beans (*Phaseolus vulgaris* L.). Tobacco (*Nicotiana tabacum* L.) is grown as a summer crop in 1 or 2 year rotations with maize across plots within the farm. Horticultural crop production also contributes to both household consumption and income generation. Most horticultural production is conducted on the wetlands (gardens) during the winter season. Livestock consists of cattle, goats, donkeys as well as free ranging chickens. Cattle provide the main source of draught power and organic manure.

The characteristic settlement models consist of the colonial reserves (communal) of the 1900s, the early resettlement farms initiated in the early 1980s until 1997 and resettlement farms instituted under the fast track land reform program (FTLRP) of the 2000s (Moyo, 2006). Most of the farms are made up of individual family homesteads in nucleated villages, individual arable plots with communal grazing, woodlots and water points (Utete, 2003). A few of the farms are self-contained, where the residential, arable and grazing land were allocated in one consolidated farm unit (ibid.). Landholdings in colonial reserves can be as low as 0.5 ha, while in the early resettlement farmland holdings are of variable sizes. Under the FTLRP, the villagised farms (with homesteads concentrated in villages) range from 1 to 20 ha (Matodi, 2012). The self-contained farm types under the FTLRP, range from small, medium to large scale commercial farming with variable farm sizes.

Therefore, for purposes of this study, the villagised farm variants, from the early resettlements and the FTLRP were sampled. Most villages are made up of between 20 to 25 households under the leadership of one administrative headman (*sabhuku*) (Chimhowu & Woodhouse, 2008).

2.2 Data collection

2.2.1 Household survey

A structured questionnaire was administered to a total of 102 farmers in March 2016. Initially, five randomly selected villages were targeted for sampling, but due to inconsistencies of production across seasons, households from villages next to the randomly selected villages were interviewed, so as to reach the sample size. As a result, the interviews extended over a total of eight villages, and interviews per village ranged from 6 to 15. Inconsistency in tobacco production is mostly due to frustrations from low market prices in some seasons. Purposeful random sampling was used in farmer selection, and the criterion for selection was involvement in tobacco production counting back three seasons. Purposeful sampling is useful in the identification of information-rich cases so as to obtain the most knowledge about the phenomenon of interest, while granting the most effective use of limited resources (Cresswell & Clark, 2011; Patton, 2002). Random sampling within the purposefully selected population was used to identify variability within the population of interest (Palinkas *et al.*, 2015). The objective of the study was to understand the diversity of farms that are not only involved in staple crop production, but also engaged in major cash crop production. The exten-

sion officers were crucial in the identification of the farmers. Without judicious record keeping, the risk of wrong information being supplied increases, therefore to improve accuracy of data collected, data from the most recent three seasons, 2012/13, 2013/14 and 2014/15 was collected. The household questionnaire covered aspects on land holding, crops grown, inorganic and organic fertiliser use, agronomic practices, sources of labour, crop residue handling practices, production constraints and crop output. In addition, demographic details and assets ownership data were collected.

2.2.2 Focus groups discussions

Focus group discussions (FGDs) were conducted in five out of eight villages sampled for the household survey. The FGDs were carried out after completion of household interviews so as to minimise public influence and bias in household data collection. The FGDs were facilitated with the assistance of the extension officers. Attendance was open to all farmers, with each FGD having an average attendance of 30 people consisting of both men and women. Farmers were asked to list the crops grown in the village as well as livestock kept and rank them according to importance to the village, with number 1 being the most important. Socioeconomic and bio-physical factors limiting the productivity of farms were identified and prioritised.

2.3 Statistical analysis

Qualitative and quantitative data collected from the household survey and FGDs were processed and analysed using the Statistical Package for Social Sciences (SPSS) version 21 programme, to generate general descriptive trends and frequencies. Variables selected for farm typology characterisation were classified into the following categories; socio-demographic, land holding and use, labour, livestock ownership, access to production inputs, specific capital goods, general capital goods, shelter, transport, information and communication means as used in other similar studies (Chavez *et al.*, 2010; Kuivanen *et al.*, 2016; Tittonell *et al.*, 2010). Quantitative and qualitative variables were subjected to PCA and MCA for dimension reduction. Data exploration included obtaining means, ranges and quartiles for quantitative data and frequencies for qualitative data. To avoid distortions in statistical analysis, outlier detection using boxplots was employed (Hair *et al.*, 2010; Kuivanen *et al.*, 2016).

2.3.1 Principal component analysis (PCA)

PCA was used in the selection and grouping of quantitative variables that influence farm productivity into uncorrelated groups called principal components (PCs). Standardisation of the variables was performed to deal with the complexity of analysing data measured on different measurement scales (Mooi & Sarstedt, 2011). Consequently, z scores for the data were used to run the PCA. Kaiser-Meyer-Olkin (KMO) measure of sampling adequacy and Bartlett test of sphericity were first conducted to test if the data set was adequate for PCA. A KMO value of > 0.5 indicated adequacy of data to be analysed using PCA (SPSS, 2012). Principal components were generated based on a correlation matrix and rotated using the varimax rotation to eliminate multi-collinearity among the PCs. The variance of each PC defined as an eigenvalue is used as the basis for significant PC selection (Lattin et al., 2003; Swan & Sandilands, 1995). Principal components with eigenvalues > 1 were selected as significantly influencing variability in farm types and were selected for further analysis. Within each PC, a variable with factor loading > 0.5 was retained, while those with lower loading factors were discarded. Furthermore, correlated variables within a PC were represented by the variable with the highest loading coefficient (Jagadamma et al., 2008).

2.3.2   Multiple correspondence analysis (MCA)

MCA was used to reduce data dimensions through exploration of relationships among the categorical variables. This analysis uses cross tabulations and gives graphical presentation of the relationships in a low dimensional plane. Correspondence analysis is based on the computation of a data matrix of frequencies (Savary et al., 1995). Two dimensions of data presentation are commonly considered as adequately facilitating data visualisation and interpretation (Gifi, 1996; Savary et al., 1995). Similar, to PCA, eigenvalues are used to determine the significance of a dimension in accounting for variability. The sum of eigenvalues is called inertia and represents the chi-square statistic divided by the total number of observations (Greenacre, 1984). An inertia value > 0.2 and a Cronbach's alpha score, were used as basis for dimension selection and retention (Hair et al., 1998; Johnson & Wichern, 2007).

Discrimination measures plots and joint plots of category points were used to identify category relationships (Costa et al., 2013). Further, correlations transformed variables were explored so as to identify significant ($r < 0.3$) correlations among MCA selected variables with meaningful practical significance (ibid.).

2.3.3   Cluster analysis

Cluster analysis was used to identify homogenous groups of farms based on the PCA and MCA selected variables. The selected variables were subjected to a 2-step cluster analysis. The 2-step clustering procedure combines the principles of hierarchical and non-hierarchical (K means)

methods (Mooi & Sarstedt, 2011). Hierarchical cluster analysis uses an agglomerative clustering algorithm which utilises multi-dimensional distances between entries as the basis for separation and selection of number of clusters. In K means clustering, cluster solutions are optimised until maximum homogeneity within clusters is achieved through reassigning of cases to clusters (Hair et al., 2010). Two-step cluster analysis has the advantage of allowing grouping of cases using a mixture of continuous and categorical variables in a single pass.

3   Results

3.1   General village characterisation

Focus group discussions were used to gain insight into general production trends, constraints and opportunities in the area. The main crops cultivated and listed according to importance included maize, which is cultivated to ensure household food security and occasionally for income, tobacco is the main cash crop, vegetables are cultivated as both household food source and income generation, groundnuts and sweet potatoes are cultivated as secondary crops to lesser extents. Production constraints included limited access to fertiliser inputs due to high prices and transport costs, availability of land preparation implements, variable and unreliable rainfall, labour and poor market prices for produce. Limited access to extension services due to high extension officer: farmer ratio was also cited as a limiting factor. However, farmers said they had access to important information through radio and television media, while some also utilised their cell phones for farming information gathering.

Sources of fertilisers for maize and tobacco production included personal, contracts with private companies and government hand-outs to a lesser extent. Furthermore, use of organic nutrient sources such as cattle manure, composts and leaf litter was constrained by small livestock numbers, bulkiness, lack of transport, diseases associated with use of animal manures as well as lack of information on composting techniques. Cattle manure was the most commonly used organic nutrient resource.

3.2   Household data exploration

Out of the 102 households interviewed, three households were cropping on borrowed land, hence only 99 households who owned land were considered for further analysis. From the data exploration, large variability was obtained on land holdings, which had a large standard deviation. Further exploration of land holding data using box plots indicated positive skeweness, due to outliers in the 90th percentile, from land holdings greater than 40 ha. The outliers were discarded so as to improve the multivariate analysis and its

Table 1: *Quantitative variables used in principal component analysis (n = 95).*

Variable	Minimum	Maximum	Mean	Standard Deviation
Total area of land owned (ha)	1	40	12.27	10.763
Maize area (ha) cultivated across 3 seasons	0.00	7.00	1.32	1.105
Tobacco area cultivated (ha) across 3 seasons	0.00	12.00	1.71	1.964
Number of oxcarts owned	0.00	3.00	0.79	0.634
Number of tractors owned	0.00	3.00	0.15	0.525
Number of ploughs owned	0.00	3.00	1.03	0.721
Number of cattle owned	0.00	60.00	5.65	8.352
Number of goats owned	0.00	12.00	2.44	3.228

generalisation to the population. Ninety five farms were thus retained for statistical analysis. Data descriptives and frequencies for the retained farms are given in Table 1 and 2.

3.3   Quantitative variables selection

A KMO value of 0.8 and a significant (*P* = 0.000) Bartlett's test of sphericity indicated that the variables were related, therefore could be analysed using PCA. Out of the 12 PCs generated, 3 PCs with eigenvalues > 1 which accounted for 72.2 % variability were selected (Fig. 1). The first PC, explained 46.3 % variability in the data set, while the 2nd and 3rd PC explained 16.9 % and 9 % respectively.

The PCs were characterised according to the loading factors within each PC. The variables with significant (> 0.5) factor loading in PC 1 were area planted to maize and tobacco in all 3 seasons under consideration and cattle ownership. Area planted to tobacco during the third season, 2014/15 was the highest loading variable within this PC. The significantly (> 0.5) loading variable in PC 2 was small livestock ownership (goats), while total land holding significantly loaded into PC 3 (Table 3).

3.4   Qualitative variables selection

A two dimensional depiction of the data was achieved through MCA. The two dimensions represented, respectively, Cronbach's alpha 0.74 and 0.63, eigenvalues 3.1 and 2.4 and inertia 0.24 and 0.19. Cumulatively, the two dimensions accounted for 42.5 % variability. Education level of the household head was the clearly discriminating variable in dimension 1, with a discriminating value > 0.5. In addition, age of household head, television and car ownership also contributed to dimension 1. This dimension was a mixture of social, economic and demographic variables. The second dimension was termed the labour dimension, with the significantly discriminating variables being source of labour for maize and tobacco production (Fig. 2). Percentage of land under utilisation had similar and significant loading on both dimensions, but was considered on dimension 2 due to the slightly higher discriminating measure on this

Table 2: *Percentages of categorical variables used in multiple correspondence analysis.*

Variable	Classes	% (n = 95)
Sex of household head	Male	82.1
	Female	17.9
Age of household head (years)	20–29	3.2
	30–39	30.5
	40–49	24.2
	50–59	24.2
	60+	17.9
Highest level of education attained by household head	None	1.1
	Primary school	25.3
	Junior secondary school	16.8
	Senior secondary school	42.1
	Post senior secondary training	14.7
Proportion of land being cultivated (%)	0–10	13.7
	11–30	35.8
	31–50	29.5
	51–80	12.6
	81–100	8.4
Source of labour for maize production	Household males	15.8
	Household females	10.5
	All household members	57.9
	Hired	11.6
	N/A	4.2
Source of labour for tobacco production	Household males	10.5
	Household females	9.5
	All household members	52.6
	Hired	17.9
	N/A	9.5
Access to mineral fertiliser	No	7.4
	Yes	92.6
Access to cattle manure	No	18.9
	Yes	81.1
Main house roof type	Asbestos	94.7
	Thatch	5.3
Main transport mode	Private car	32.6
	Motorbike	3.2
	Public transport	64.2
Ownership of information and communication gadgets	Radio	74.7
	Television	53.7
	Cell phone	95.8

dimension. Sex of household head, access to mineral fertilisers and cattle manure, cell phone and radio ownership, housing type were weakly correlated to the two dimensions with discriminating measures < 0.3, therefore they were of no practical significance and were discarded.

Table 3: *Loading variables in the selected principal components.*

	Principal Component		
	1	2	3
Total area of land owned	0.148	0.012	0.910
Maize area cultivated 2014/15	0.864	0.158	0.094
Tobacco area cultivated 2014/15	0.905	−0.161	0.068
Maize area cultivated 2013/14	0.711	0.353	0.265
Tobacco area cultivated 2013/14	0.890	−0.120	0.128
Maize area cultivated 2012/13	0.719	0.280	0.365
Tobacco area cultivated 2012/13	0.861	−0.120	0.155
Total number of cattle	0.620	0.456	−0.219
Total number of goats	−0.073	0.658	−0.125

Extraction method: Principal Component Analysis.
Rotation method: Varimax with Kaiser normalisation. Rotation converged in 4 iterations.

3.5   Cluster profiles

The silhouette measure of cohesion and separation of clusters was scored as fair. Three farm type clusters were generated from the 2-step cluster analysis of PCA and MCA derived variables.

3.5.1   Type I: Resource-endowed farms with a commercial oriented farming system.

This cluster is the smallest cluster, representing only 6 % of the surveyed farms. The cluster was classified as the resource-endowed commercial oriented farm types. Land holding on these farms is the largest, with an average of 18 ha, of which a significant proportion is utilised for crop production (Table 4). The farms produce maize and tobacco on the largest plots, relative to other clusters. Maize is produced on average on 4 ha, while tobacco is allocated twice this area. Both crops are produced using hired labour (Table 4). A high level of mechanisation is exhibited by possession of high capital assets that include a tractor. In addition, the farms own a car. The highest number of cattle ownership is also found within this cluster. Household heads within type I farms are aged between 40–59 years and possess college level training.

3.5.2   Type II: Medium resourced farms practising mixed subsistence and cash crop production.

Type II farms were classified as the medium resourced farms and represented the largest group (73 %) of the total

Table 4: *Cluster characteristics derived from 2-step clustering.*

	Cluster		
Variable	1	2	3
	(% n = 73.7)	(% n = 6.3)	(% n = 20.0)
	Type II	Type I	Type III
Total land holding (ha)	13.5	18.0	6.5
Mean maize area (ha)	1.3	4.0	0.3
Mean tobacco area (ha)	1.5	8.9	0.6
Total number of cattle	6.0	29.0	1.0
Total number of goats	3.0	0.0	0.0
Categorical variables	% of N		
Age of household head			
20–29	1.4	0.0	10.0
30–39	22.2	0.0	65.0
40–49	25.0	50.0	15.0
50–59	27.8	50.0	5.0
60+	23.6	0.0	0.0
Highest education level reached by household head			
None	2.8	0.0	0.0
Primary school	31.9	0.0	10.0
Junior secondary school	16.7	0.0	15.0
Senior secondary school	34.7	0.0	75.0
College training	13.9	100.0	0.0
Proportion of land being cultivated (%)			
0–10	12.5	0.0	21.1
11–30	36.0	0.0	42.1
31–50	32.0	15.0	21.1
51–80	16.7	20.0	0.0
81–100	2.7	65.0	15.7
Source of labour for maize production			
Household males	19.5	0.0	5.0
Household females	12.5	0.0	5.0
All household members	58.3	25.0	55.0
Hired	9.7	75.0	10.0
N/A	0.0	0.0	25.0
Source of labour for tobacco production			
Household males	13.9	0.0	0.0
Household females	11.1	0.0	5.0
All household members	51.4	25.0	65.0
Hired	16.6	75.0	10.0
N/A	6.9	0.0	15.0
Plough ownership			
Yes	99.0	100.0	25.0
No	1.0	0.0	75.0
Tractor ownership			
Yes	6.9	100.0	0.0
No	93.1	0.0	100.0
Car ownership			
Yes	36.1	100.0	0.0
No	63.9	0.0	100.0
Television ownership			
Yes	58.3	100.0	25.0
No	41.7	0.0	100.0

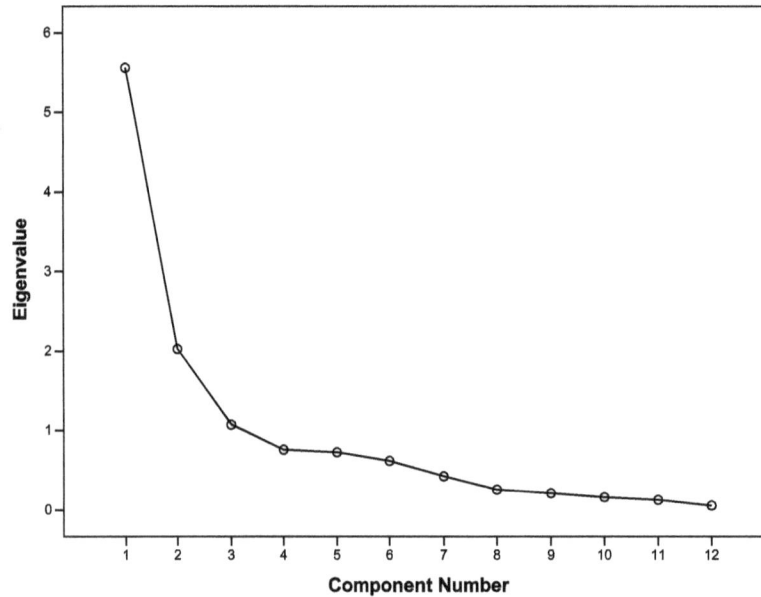

Fig. 1: *Scree plot showing eigenvalues of principal components.*

number of surveyed farms. Mean land holding is 13.5 ha, which is intermediate between the resource endowed and resource constrained farms. A significant number of farms (68%) cropped between 11 to 50% of total land holding against 81–100% cropped under type I farms (Table 4). Approximately equal land area (ca. 1 ha) is allocated to maize and tobacco production. Labour is mainly provided by household members for both maize and tobacco production, which is sometimes complemented with hired labour. Tobacco was not produced on 6.9% of the farms during 2014/15 season. Average number of cattle owned is 6 and 99% own a plough. About 7% own a tractor, 36% own a car and more than half own a television set. This cluster is characterised by almost equal numbers of household heads per age group, except the 20–29 age group which was significantly lower (Table 4).

3.5.3 Type III: Resource constrained farms practising mixed subsistence and cash crop production.

This forms the second largest group and is classified as the resource constrained cluster. This cluster represents 20% of the surveyed farms. The farms are typically the smallest with a mean size of 6.5 ha. Both maize and tobacco are planted on relatively smaller plots, 0.3 and 0.6 ha respectively. Eighty percent of the farms within this cluster cultivate on less than 50% of land owned (Table 4). Although, maize and tobacco production are typical, during 2014/15 season, 25% and 15% of the farms did not produce maize and tobacco, respectively. Family labour is the

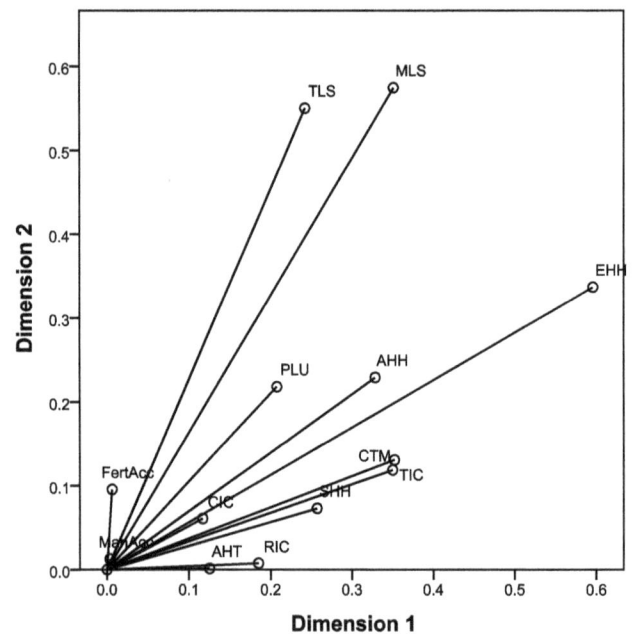

Fig. 2: *Discriminating measures plot of categorical data variables.*

FertAcc, access to fertiliser; ManAcc, access to manure; AHT, main housing type; CIC, Cell phone ownership; RIC, radio ownership; CTM, car ownership; SHH, sex of household head, TIC, television ownership; EHH, household head education level; PLU, percentage of land cultivated; TLS, tobacco labour source and MLS, maize labour source.

main source for crop production. Only 5% own a plough, while average cattle ownership is 1. No high capital assets are owned by this group. The household heads on these are farms are mostly aged between 30–39, and attained senior secondary education (Table 4).

4   Discussion

Findings from FGDs indicated that Svosve smallholder area consists predominantly of a food-cash crop smallholder farming community. Farmers attested to higher market participation as opposed to typical smallholder subsistence farming. This was attributed to the production of tobacco as a cash crop. However, constraints including limited access to fertiliser inputs due to high prices and transport costs, land preparation implements, variable and unreliable rainfall, labour shortage and poor market prices for produce were cited. These constraints are not unique to the Svosve area and have been cited across SSA (Chamberlain, 2007; Ncube et al., 2009; Salami et al., 2010). Limited access to extension services due to a low ratio of extension officer to farms was also cited as a limiting factor. However, alternative sources of information including radio, television and cell phones are being utilised by farmers to access to important information thus indicating a shift towards advanced information and communication technology (ICT) based information systems in the area. This is corroborated by findings from Masuka et al. (2016).

Although farmers collectively cited general challenges in production, it is important to note that the extent to which groups of diverse farms are impacted by the challenges are variable. Therefore, an understanding of farm diversity will allow for identification of specific constraints and impacts across heterogeneous farms. This is important in the tailoring and targeting of interventions for agricultural development. This approach has been adopted by several studies (Kraaijvanger et al., 2016; Mądry et al., 2016; Tittonell et al., 2010). Consequently, household interviews and multivariate analysis methods were used to group farms according to shared salient features that impact farm production orientation so as to facilitate matching of constraints and opportunities accordingly, for maximum impact. The identification of farm types in this study forms a basis for subsequent discussion on constraints and opportunities for agricultural development in the farming system.

The resource endowed type I farms exhibited a high level of commercialisation distinguished from the other farming households by the large areas on which both maize and tobacco are planted. The proportion of land utilisation within this group was higher than for the two other farm types. This was facilitated by the ownership of high mechanisation implements, such as tractors which are utilised for tillage operations thus ensuring optimal utilisation of land. Within

this group, the large maize areas are indicative of the commercial orientation of the farms, as the production exceeds average for household consumption. Although, tobacco has higher tillage and labour demand than maize, it is allocated twice the area for maize possibly because it is commercially more profitable. Timeliness of operations, given the large scale of production is, thus ensured through use of hired labour. Large cattle numbers point to diversification of enterprises, which is a key indicator of well-resourced farms where livestock act as a buffer to shocks, thus decreasing household vulnerability (Kuivanen et al., 2016). Therefore, in terms of constraints, with hired labour being the main source of labour, these farms may be vulnerable to seasonal changes in labour availability and costs which could impact their productivity. On the other hand, mechanisation, draft power and tillage related challenges have little impact on this group. However, intensive crop production and use of highly mechanised systems may exacerbate soil fragility through soil organic matter loss and increase bulk densities due to soil compaction (Chavez et al., 2010); hence adoption of practices such as crop residue retention and reduced tillage practices that result in soil organic matter build-up is imperative. This cluster is characterised by household heads that have attained college training, therefore are generally regarded as educated.

The medium resourced, type II farms, practice food security and income generation oriented production as signified by maize and tobacco production. However, these farms exhibit a low level of mechanisation, relative to the type I farms, mostly relying on manual power for farm operations, as evidenced by 99 % plough ownership and on average 6 head of cattle. Land utilisation is less than that of type I farms, despite these farms being smaller in size. This is indicative of constrained production. Although the farms have access to basic tillage assets, a cattle herd of 6 and a plough are likely inadequate to meet the tillage operation needs of the farms. This challenge was echoed during the FGDs as a major limitation to production. This on the other hand, may present opportunities for soil fertility regeneration through fallowing. There is need for investment towards more mechanised systems, if land is to be fully utilised. In addition, reliance on family labour for production of both maize and tobacco is a major constraint, because its availability is dependent on household size, the age and fitness of household members. Furthermore, both maize and tobacco are summer crops, hence they compete for resources, thus in the absence of hired labour to complement household labour, production and productivity is reduced. Therefore, in the absence of additional resources, staggering of farm operations may offer some relief. This cluster is the most variable in terms of household demographics, where education level and age of household head are almost evenly spread out across the spectrum. Sustainability of livelihoods in a rural setup is related to biophysical conditions, as such management practices that reduce soil degradation and promote soil regeneration (Dunjana *et al.*, 2012; Kuivanen *et al.*, 2016) which should be tailored to efficiently utilise cluster available resources.

Type III farms were characterised as low resource endowment farms, that exhibit typical salient characteristics of resource constrained farms, including least land holdings as reported in other studies, locally and within the SSA region (Kuivanen *et al.*, 2016; Rusinamhodzi *et al.*, 2012; Zingore *et al.*, 2007). Similarly, the relative landholding of this group is the lowest, while the highest percentage of uncropped land (21 % against 0 % and 12.5 % on type I and II farms, respectively) is exhibited by this group of farms. Similar, observations were made by Mtambanengwe & Mapfumo (2005), where in cases where land is not a limiting factor, but other production factors as farm equipment or labour are, uncropped land is common. Maize is planted for subsistence as indicated by the small plots on which it is produced. Although, tobacco is allocated twice the area, cultivated area is relatively lower than that of the medium resourced and resource endowed farms. This could be attributed to limitations on production resources such as inputs and labour for tillage operations. During the 2014/15 season, maize was not cultivated on 25 % of the farms, while 15 % did not cultivate tobacco. This could have been due to a number of reasons and among them, production related constraints, thus forcing the farms to choose only one main crop over the other. This is indicative of households' divergent production orientations. It is common for resource constrained farms to practice subsistence production, however, in this case some farms are clearly prioritising income oriented production over subsistence. Therefore, with regards to food security, such farms are susceptible to both biophysical and market shocks thus making them extremely vulnerable. In this regard, it is vital to provide these farmers with adequate technical back up via extension, improve linkages with contractors to ensure that productivity is improved and livelihoods are sustained.

Dorward (2009) postulated three main livelihood strategies in rural setups, namely, "hanging in" where farmers engage in activities to maintain current livelihood, "stepping up" indicated by expansion of current production activities leading to semi-commercial farming and "stepping out", when farming activities are used to accumulate assets that allow engagement in non-farm activities. In the study area, type I farms fit the criteria of "stepping out". The farms have accumulated capital assets and continue to prosper in crop production as evidenced by high land utilisation. True to the postulate, these farms have stepped out as evidenced

for example, by ownership of passenger vehicles which are used as a form of off-farm income source. Type II farms also closely match the "stepping up" hypothesis as evidenced by consistent 1 ha/year tobacco production, a purely commercial crop. Interventions that seek to address biophysical, economic challenges as well as sustainability in that regard will likely further facilitate "stepping up" for these farms and possibly "stepping out" for others. The "hanging in" farms are described as those that maintain their current livelihood by practising subsistence production. Interestingly, because of the uniqueness of the study area being predominantly tobacco production area, these farms exhibit a propensity for semi-commercial production as well, as indicated by tobacco production on twice the size of plots allocated to maize.

In targeting interventions for improved livelihoods in farming systems, agroecological potential and market opportunities are two dimensions that have been identified as determining opportunities and constraints (IFPRI, 2007). Such is the case within the study area. Due to market availability, all three identified farm types display market participation, albeit to varying extents due to the readily available tobacco market. Consequently, strengthening of market links for inputs and produce within the study area, will likely result in the progression of some farms from one livelihood strategy to the next. Furthermore, within the agroecological potential context, tailored biophysical interventions for sustainable soil use and management will have to be addressed per cluster.

5 Conclusion

Three farm types were identified in the study area, namely resource endowed commercial oriented farms, medium resourced and low resource endowed farms practising subsistence and semi-commercial production. The use of multivariate analysis allowed understanding of farm diversity within the study area and highlighted the varying extents to which the FGD cited constraints affected each farm cluster. For example, while labour constraints are common across the three farm types, issues around the cost and seasonal availability are relevant for type I farms as they depend on hired labour for production, while for types II and III farms, labour is provided by household members hence household size will have a greater bearing. Similarly, draft power challenges are variable across the clusters resulting in varying extents of land utilisation, production and soil degradation, which has a bearing on livelihood sustainability. Consequently, in order to improve livelihoods for each cluster, the challenges and opportunities should be made within the context of the cluster and interventions be recommended accordingly.

Competing interests

The authors declare that they have no conflict of interest.

Funding Acknowledgement

This work was financed through a research grant received from the International Foundation of Science (Grant C/5764-1) and the Organisation for Women in Science for the Developing World (OWSD).

References

Ballesteros, P. W., Lagos, B. T. C. & Ferney, L. H. (2015). Morphological characterization of elite cacao trees (*Theobroma cacao* L.) in Tumaco, Nariño, Colombia. *Revista Colombiana Ciencias Hortícolas*, 9 (2), 313–328. doi:10.17584/rcch.2015v9i2.4187.

Bellini, G. & Ramberti, S. (2009). Is the Italian organic farming model inside rural development? A farm structure survey data analysis. Wye city group on statistics on rural development and agriculture household income, Second Meeting, Italy, Rome, 11–12 June 2009, FAO Head-Quarters. Available at: http://www.fao.org/fileadmin/templates/ess/pages/rural/wye_city_group/2009/Paper_2_b2_Bellini_Ramberti_Organicfarming_RD.doc (accessed on: 24 January 2018).

Chamberlain, J. (2007). Defining smallholder agriculture in Ghana: Who are smallholders, what do they do and how are they linked with markets? GSSP Background Paper No. 6. Washington, DC.

Chavez, M. D., Berentsen, P. B. M. & Oude-Lansink, A. G. J. M. (2010). Creating a typology of tobacco farms according to determinants of diversification in Valle de Lerma (Salta-Argentina). *Spanish Journal of Agricultural Research*, 8 (2), 460–471.

Chimhowu, A. & Woodhouse, P. (2008). Communal Tenure and Rural Poverty: reflections on land transactions in Svosve Communal Area. BWPI Working Paper No. 35. Manchester, U.

Costa, P. S., Santos, N. C., Cunha, P., Cotter, J. & Sousa, N. (2013). The Use of Multiple Correspondence Analysis to Explore Associations between Categories of Qualitative Variables in Healthy Ageing. *Journal of Aging Research*, #302163. doi:10.1155/2013/302163.

Cresswell, J. W. & Clark, V. L. P. (2011). *Designing and conducting mixed method research*. (2nd ed.). Sage Publishing Inc., Thousand Oaks, CA.

Dorward, A. (2009). Integrating contested aspirations, processes and policy: development as hanging in, stepping up and stepping out. *Development Policy Review*, 27, 131–146.

Dunjana, N., Nyamugafata, P., Shumba, A., Nyamangara, J. & Zingore, S. (2012). Effects of cattle manure on selected soil physical properties of smallholder farms on two soils of Murewa, Zimbabwe. *Soil Use and Management*, 28, 221–228.

Ellis, F. (1988). *Peasant Economics: Farm Households and Agrarian Development*. Cambridge University Press, Cambridge.

Emtage, N. & Suh, J. (2005). Variations in socioeconomic characteristics, farming assets and livelihood systems of Leyte rural households. *Annals of Tropical Research*, 27, 35–54.

FAO, ISS & ISRC (1998). *World reference base for soil resources*. World Soil Resource Report. Rome, Italy.

Gifi, A. (1996). *Non-Linear Multivariate Analysis*. John Wiley and Sons, Chichester, UK.

Greenacre, M. J. (1984). *Theory and Applications of Correspondence Analysis*. Academic Press, London.

Hair, J., Black, W., Babin, B. & Anderson, R. (2010). *Multivariate Data Analysis*. Pearson Prentice Hall, New Jersey, USA.

Hair, J. F., Tatham, R. L., Anderson, R. E. & Black, W. (1998). *Multivariate Data Analysis*. (5th ed.). Prentice-Hall, Upper Saddle River, New Jersey, USA.

Hilhorst, T. & Muchena, F. (eds.) (2000). *Nutrients on the move-Soil fertility dynamics in African Farming Systems*. International Institute for Environment and Development, London.

IFPRI (2007). *IFPRI's Africa Strategy: Toward food nutrition security in Africa*. Research and Capacity Building. Washington, DC.

Jagadamma, S., Lal, R., Hoeft, R. G., Nafziger, E. D. & Adee, E. A. (2008). Nitrogen fertilization and cropping system impacts on soil properties and their relationship to crop yield in the central Corn Belt, USA. *Soil Tillage and Research*, 98, 120–129.

Johnson, R. A. & Wichern, D. W. (2007). *Applied Multivariate Correspondence Analysis*. (6th ed.). Prentice-Hall, Upper Saddle River, New Jersey, USA.

Jolliffe, I. T. (2002). *Principal Component Analysis*. (2nd ed.). Springer-Verlag New York Inc., New York, USA.

Kostrowicki, J. (1977). Agricultural typology concept and method. *Agricultural Systems*, 2, 33–45.

Kraaijvanger, R., Almekinders, C. J. M. & Veldkamp,

A. (2016). Identifying crop productivity constraints and opportunities using focus group discussions: A case study with farmers from Tigray. *NJAS - Wageningen Journal of Life Sciences*, 78, 139–151. doi: 10.1016/j.njas.2016.05.007.

Kuivanen, K., Alvarez, S., Michalscheck, M., Adjei-Nsiah, S. & Descheemaeker, K. (2016). Characterising the diversity of smallholder farming systems and their constraints and opportunities for innovation: A case study from the Northern Region, Ghana. *NJAS - Wageningen Journal of Life Sciences*, 78, 153–166. doi: 10.1016/j.njas.2016.04.003.

Lattin, J., Carroll, J. & Green, P. (2003). *Analyzing multivariate data (Duxbury Applied Series)*. Brooks/Cole, CA, USA.

Mądry, W., Roszkowska-Mądra, B., Gozdowski, D. & Hryniewski, R. (2016). Some aspects of the concept, methodology and application of farming system typology. *Electronic Journal of Polish Agricultural Universities*, 19(1), #12.

Masuka, B. P., Matenda, T., Chipomho, J., Mapope, N., Mupeti, S., Tatsvarei, S. & Ngezimana, W. (2016). Mobile phone use by small-scale farmers: a potential to transform production and marketing in Zimbabwe. *South Africa Journal of Agricultural Extension*, 44, 121–135.

Masvaya, E. N., Nyamangara, J., Nyawasha, R. W., Zingore, S., Delve, R. J. & Giller, K. E. (2010). Effect of farmer management strategies on spatial variability of soil fertility and crop nutrient uptake in contrasting agro-ecological zones in Zimbabwe. *Nutrient Cycling in Agroecosystems*, 88, 111–120.

Masvongo, J. (2013). Viability of tobacco production under smallholder farming sector in Mount Darwin District, Zimbabwe. *Journal of Development and Agricultural Economics*, 5, 295–301.

Matodi, P. B. (2012). *Zimbabwe's fast track land reform*. ZED Books, London and New York.

Mekkawy, W., Barman, B. K., Kohinoor, A. H. M. & Benzie, J. A. H. (2017). Characterization of Mono-sex Nile tilapia (*Oreochromis niloticus*) hatcheries in Bangladesh. *Journal of Aquaculture Research and Development*, 8(7), #1000498. doi:10.4172/2155-9546.1000498.

Mooi, E. & Sarstedt, M. (2011). *A concise guide to market research: the process, data, and methods using IBM SPSS Statistics*. Springer, Heilderberg, Germany.

Moyo, S. (2006). The evolution of Zimbabwe's land acquisition. *In:* Rukuni, M., Tawonezvi, P., Eicher, C.,

Munyuki-Hungwe, M. & Matondi, P. (eds.), *Zimbabwe's Agricultural Revolution Revisited*. pp. 148–163, University of Zimbabwe Publications, Harare, Zimbabwe.

Mtambanengwe, F. & Mapfumo, P. (2005). Organic matter management as an underlying cause for soil fertility gradients on smallholder farms in Zimbabwe. *Nutrient Cycling in Agroecosystems*, 73, 227–243.

Muchero, M. (2008). The Development of a SADC Regional Agricultural Policy (RAP). *In:* Agriculture-Led Development for Southern Africa: Strategic Investment Priorities for Halving Hunger and Poverty by 2015. Palm Beach, Gaborone, Botswana.

Mugwira, L. M. & Murwira, H. K. (1997). Use of cattle manure to improve soil fertility in Zimbabwe: Past and current research and future research needs. Networking paper No. 2. Soil fertility for maize based cropping systems in Zimbabwe and Malawi. CIMMYT, Harare, Zimbabwe.

Ncube, B., Twomlow, S. J., Dimes, J. P., Van Wijk, M. T. & Giller, K. E. (2009). Resource flows, crops and soil fertility management in smallholder farming systems in semi-arid Zimbabwe. *Soil Use and Management*, 25, 78–90.

Ouma, E. A., Obare, G. A. & Staal, S. J. (2003). Cattle as assets: Assessment of non-market benefits from cattle in smallholder Kenyan crop-livestock systems. *In:* Proceedings of the 25th International Conference of Agricultural Economists (IAAE). 16–22 August, Durban, South Africa. pp. 328–334, International Association of Agricultural Economists.

Ozden, S. & Mendes, M. (2005). The usage of Multiple Correspondence Analysis in rural migration analysis. *NEW MEDIT - A Mediterranean Journal of Economics, Agriculture and Environment*, 4(4), 36–41.

Palinkas, L. A., Horwitz, S. M., Green, C. A., Wisdom, J. P., Duan, N. & Hoagwood, K. (2015). Purposeful sampling for qualitative data collection and analysis in mixed method implementation research. *Adminstration and Policy in Mental Health*, 42(5), 533–544. doi: 10.1007/s10488-013-0528-y.

Patton, M. Q. (2002). *Qualitative Research and Evaluation Methods*. (3rd ed.). Sage Publishing Inc., Thousand Oaks, CA.

Prowse, M. (2013). A history of tobacco production and marketing in Malawi, 1890–2010. *Journal of Eastern African Studies*, 7(4), 691–712. doi: 10.1080/17531055.2013.805077.

Rusinamhodzi, L., Corbeels, M., Nyamangara, J. & Giller, K. E. (2012). Maize-grain legume intercropping is an attractive option for ecological intensification that reduces climatic risk for smallholder farmers in central Mozambique. *Field Crops Research*, 136,12–22. doi: 10.1016/j.fcr.2012.07.014.

Salami, A., Kamara, A. B. & Brixiova, Z. (2010). Smallholder Agriculture in East Africa: Trends, Constraints and Opportunities. Working Paper Series No. 105. African Development Bank, Tunis, Tunisia.

Savary, S., Madden, L. V., Zadocks, J. C. & Klein-Gebbinck, H. W. (1995). Use of Categorical Information and Correspondence Analysis in Plant Disease Epidemiology. *Advances in Botanical Research*, 21,213–240.

Shumba, E. M. & Whingwiri, E. E. (2006). Commercialisation of smallholder agriculture. *In:* Rukuni, M., Tawonezvi, P., Eicher, C., Munyuki-Hungwe, M. & Matodi, P. (eds.), *Zimbabwe's Agricultural Revolution Revisited*. pp. 577–591, University of Zimbabwe Publications, Harare, Zimbabwe.

Smith, R. R., Moreira, V. L. & Latrille, L. L. (2002). Characterization of dairy productive systems in the Tenth Region of Chile using multivariate analysis. *Agricultura Técnica*, 62 (3),375–395.

SPSS Inc (2012). *SPSS Statistics 21*. IBM, Chicago IL.

Swan, A. & Sandilands, M. (1995). *Introduction to geological data analysis*. Blackwell, London.

Tittonell, P., Muriuki, A., Shepherd, K. D., Mugendi, D., Kaizzi, K. C., Okeyo, J., Verchot, L., Coe, R. & Vanlauwe, B. (2010). The diversity of rural livelihoods and their influence on soil fertility in agricultural systems of East Africa – A typology of smallholder farms. *Agricultural Systems*, 103 (2),83–97. doi: 10.1016/j.agsy.2009.10.001.

Tittonell, P., Vanlauwe, B., Leffelaar, P. A., Shepherd, K. & Giller, K. E. (2005). Exploring diversity in soil fertility management of smallholder farms in western Kenya: II. Within-farm variability in resource allocation, nutrient flows and soil fertility status. *Agriculture, Ecosystems and Environment*, 110,166–184. doi: 10.1016/j.agee.2005.04.003.

Utete, C. (2003). Report of the Presidential Land Review Committee, Volume 1 and 2. Main report to His Excellency, The President of the Republic of Zimbabwe. Harare, Zimbabwe.

Wiggins, S. (2009). Can the smallholder model deliver poverty reduction and food security for a rapidly growing population in Africa? *In:* Proceedings of the FAO Expert Meeting on How to Feed the World in 2050. Food and Agriculture Organization, Rome.

Zingore, S., Murwira, H. K., Delve, R. J. & Giller, K. E. (2007). Soil type, management history and current resource allocation: Three dimensions regulating variability in crop productivity on African smallholder farms. *Field Crops Research*, 101 (3),296–305.

11

Beekeeping adoption: A case study of three smallholder farming communities in Baringo County, Kenya

Renaud Hecklé [a,*], Pete Smith [a], Jennie I. Macdiarmid [b],
Ewan Campbell [a], Pamela Abbott [c]

[a]*Institute of Biological & Environmental Sciences, University of Aberdeen, Aberdeen, UK*
[b]*The Rowett Institute, University of Aberdeen, Aberdeen, UK*
[c]*School of Education, University of Aberdeen, Aberdeen, UK*

Abstract

In Kenya, beekeeping offers benefits which could make it attractive to smallholder farmers as a possible strategy for making their livelihoods more sustainable. However, its potential remains largely unexploited and the lack of new entrants is thought to be one key reason for a decline in beekeeping. This paper reports on a study that examined the factors affecting beekeeping adoption in Baringo County, Kenya with a focus on three smallholder farming communities. Semi-structured interviews were conducted with 90 informants in these communities, including 41 new beekeepers, 21 non-adopters, 13 group leaders, 10 village elders and 5 teenagers. In addition, 28 key stakeholders at national and local levels were approached. The findings show that in high traditional beekeeping areas apprenticeship pathway is predominant, while in low traditional beekeeping areas most of the beekeepers follow the traineeship pathway. The main factors affecting the decision of smallholder farmers to take up beekeeping were access to information, land and beehives, availability of alternative income generating activities, perceptions of beekeeping outcomes and performance, access to market, feelings towards bees, and cultural norms. The importance of these factors varies according to interviewee demographics (gender, age and level of education) and location. The findings suggest that to increase the uptake of beekeeping the following should be considered: (a) increasing awareness and knowledge in all locations but particularly in the low traditional beekeeping areas; (b) improving access to improved harvesting tools (e.g. smokers and protective clothing) and to movable comb or frame hives, especially for young people and women; and (c) supporting local social networks.

Keywords: Africa, bees, honey, livelihoods, qualitative methods, semi-structured interview, sustainable livelihood framework

1   Introduction

Since independence from colonial rule, policies and interventions have been implemented in Kenya to increase the productivity of the agricultural sector, stimulate economic growth and transformation, and reduce poverty. Despite these initiatives, the population still depends mainly on small-scale subsistence farming, with households struggling to make a living (Monitoring African Food and Agricultural Commodities, MAFAP, 2013). Beekeeping, which requires very little resource, land and time, could be incorporated into the livelihood strategy of smallholder farming households in order to provide an additional source of income and to spread risk. In addition, bee products may be used domestically as food or to make traditional healthcare remedies (Lowore *et al.*, 2010).

In Kenya, there are a number of reasons why beekeeping adoption by smallholder farming households could strengthen their livelihoods:

– Honey hunting and beekeeping have been practised since ancient times by a number of ethnic groups in agropastoral systems (Nightingale & Crane, 1983), so indigenous knowledge and skills are locally available;

* Corresponding author
Renaud Hecklé (r01rrh14@abdn.ac.co.uk)

– Kenya has an unexploited potential for beekeeping; only 8 % of the honey potential and 17 % of the beeswax potential are exploited in the country[1];

– On domestic and regional markets for honey and beeswax there is an excess of demand over supply and relatively high local prices (Bees for Development, 2006; Carroll & Kinsella, 2013);

– The Kenyan government and a growing number of development organisations have been supporting beekeepers to improve their practices, to adopt movable comb or frame (MCF) hives[2] and to access markets (Government of Kenya, 2013);

– Cultural restrictions against women keeping bees are declining (Government of Kenya, 2013).

Despite the potentially favourable impact on livelihoods and the underused potential, the number of beehives, and total honey production both declined by a third between 2005 and 2009, although the numbers have slowly increased since then (FAO, 2016).

Various factors may explain the production difficulties, such as (1) natural factors (e.g. degradation of bee habitat), (2) technical factors (e.g. poor beehive quality), (3) human factors (e.g. inadequate practices) and, (4) contextual factors (e.g. access to market) (ibid.; Muli *et al.*, 2015). Another potential factor is the lack of new entrants to replace those moving out of beekeeping (Muriuki, 2010). New entrants may be deterred because traditionally learning beekeeping has involved a long apprenticeship where fathers passed on their knowledge and expertise to their sons (Fisher, 2000; Gichora, 2003). Beekeeping is also not seen as an attractive option by women and young people due to perceived risks (e.g. falling out of trees, and bee stings) (Government of Kenya, 2013). Finally, there is limited access to formal training, credit and land, especially for women and young people (Government of Kenya, 2013; Ahikiriza, 2016).

However, in sub-Saharan Africa generally and specifically in Kenya little research has been carried out on beekeeping decline and very little in-depth qualitative information has been collected. The study aimed to fill this gap by answering three research questions:

– What have the pathways to the adoption of beekeeping been for those that have recently entered (within the last five years)?

– What are the main factors leading smallholder farmers to incorporate beekeeping into their household's livelihood strategy?

– What are the main factors preventing smallholder farmers from taking up beekeeping?

2 Materials and methods

2.1 Research methods

This research used qualitative methods to provide an in depth understanding of peoples' experiences and perceptions. Semi-structured interviews were conducted face-to-face to explore key topics such as motives and capabilities for and constraints on the adoption of beekeeping. General questions were asked first and then followed by more specific or sensitive questions. Follow-up phone interviews were conducted with some informants for two reasons: (1) when some emerging issues were not covered in earlier face-to-face interview and (2) to ask follow-up questions which emerged from analysis of interview. Baringo County was selected as the site for carrying out the research because it is one of the counties with the highest production of honey in Kenya. In addition, based on the most recently available data (Gichora, 2003), beekeeping potential there remains under-exploited, with good bee forage available, traditional skills and knowledge on hand among the communities, and an excess of demand for honey over supply.

Three rural communities were purposively selected in three smallholder farming systems of Baringo County (Fig. 1): (i) the agro-pastoral (AP) system; (ii) irrigated cropping (IR) and (iii) the mixed-farming (MF) system of the highlands.

In AP, people depend largely on pastoralism with indigenous livestock (e.g. meat goats and cattle) and rainfed agriculture with drought-tolerant crops (e.g. finger millet and sorghum) (Kenya Food Security Steering Group (KFSSG), 2015). Bekeeping is a traditional activity and farmers practise it principally using extensive production systems consisting of traditional log hives[3] dispersed in trees. In IR, the Njemps community practises mostly irrigated agriculture with food crops (e.g. maize and beans) and pastoralism with indigenous livestock. They are not traditionally beekeepers and they practise mostly extensive beekeeping. In MF,

---

[1]The potential of honey and beeswax is estimated at over 150,000 and 15,000 metric tonnes per annum, respectively (Muya, 2004) and the levels of production in 2013 were 12,000 and 2,500 metric tonnes per annum, respectively (FAO, 2016).

[2]The main MCF hives promoted in Kenya were the Kenya Top-Bar Hive (KTBH) for the movable comb hives and the Langstroth hive for the movable frame hives.

[3]A log hive consist of a hollowed-out log, split in half to make two troughs, and the two halves are fitted together again to make a cylindrical hive. Two lids with small holes are provided to close both open ends and allow exit and entry of bees. Log hives are quite easy to make and can last about 30 years (Gichora, 2003; Chengo'le *et al.*, 2008). However, management/harvesting of these hives is difficult, as the hives are high up on trees and the combs are fixed. They are also not environmentally friendly because they are made of threatened indigenous wood (Adjare, 1990).

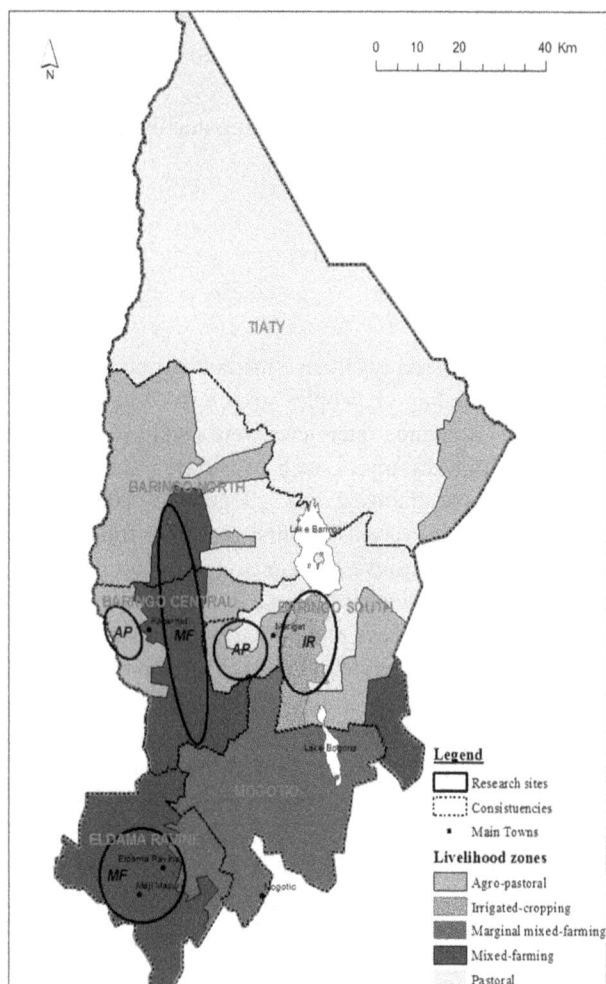

Fig. 1: *Research samplings in Baringo County, Kenya (March 2016 – July 2017).*

small and medium-size farms are predominant, with livestock production (e.g. dairy cattle and sheep) left stationary and rainfed agriculture with a large variety of crops such as food and horticultural crops (e.g. maize, mangoes and coffee). No information existed on beekeeping in MF so far. However, people are likely to be mostly engaged in intensive beekeeping systems (Carroll & Kinsella, 2013) using MCF hives[4], principally Kenyan top bar hives (KTBH), with the hives clustered together in a small piece of land called an apiary.

The honey harvesting seasons differ from one farming system to another (Carroll, 2002). In all systems beekeepers still use traditional methods[5] to harvest their honey. There are two harvesting periods in AP and IR: the main one

between November and January, and the minor one between July and August. These harvesting periods occur after the rainy seasons when there are few other farming activities. In the highlands of Kenya, honey is mostly harvested between August and November, in the same period as the main food crops (e.g. maize and beans).

In each farming system, informants were identified through a purposive sampling. The following community members were interviewed (Table 1): new entrants (NE)[6] with less than 5 years of practise; non-adopters (NA)[6]; group leaders (GL); village elders (VE) and children (CH) aged between 12 and 16 years old. A range of key stakeholders of the beekeeping sector at national ($n = 19$) and local levels ($n = 9$) were also interviewed: from academic centres (AC; $n = 3$), development organisations (NGO; $n = 8$), governmental bodies (GB; $n = 10$), independent consultants or retired people (CN; $n = 2$), and private enterprises (PE; $n = 5$). The informants in the communities were identified and recruited with the help of gatekeepers, and a few additional people were approached on the recommendation of the initial interviewees.

2.2 Data collection and analysis

Interview guides were developed for each type of informant, with a list of topics. The beekeepers were asked directly about their own experience: their motivations and capabilities for adopting beekeeping and what made it easier or more difficult to do so. Similarly, the non-beekeepers in the communities and the key stakeholders were asked to give their perceptions of the adoption of beekeeping, based on the people they knew, to provide another perspective.

The interviews were conducted in English with the key stakeholders and in the local language with community members. Two teams of two local researchers (one man and one woman) with experience of interviewing were recruited and trained. One carried out the interview and the other took detailed notes. Female interviewers were specifically recruited to interview the women.

Before starting the study, ethical approval for the research was obtained through the research governance procedures of the University of Aberdeen. Before each respondent was interviewed, he or she was told the purpose of the study, that their participation was voluntary, and that their privacy and confidentiality were assured. After agreement, they were asked to sign a consent form – in Kiswahili for the communities' members and in English for the key stakeholders. Detailed notes were taken by the local interviewers and the interviews with the key stakeholders were audio-recorded. Finally, transcripts were analysed and data coded by topics (e.g. beekeeping adoption), themes (e.g. reasons for non-adoption) and sub-themes that emerged from the data (e.g.

---

[4]MCF hives are beehives where bees are encouraged to attach their combs to beeswax foundation sheets on movable frames or to construct their combs from the undersides of a series of top-bars. These top-bars then allow individual combs to be lifted from the hive by the beekeeper (Bradbear, 2009).

[5]Beekeepers use a smouldering stick to drive away the bees and do not use protective clothing.

[6]A sample of people who are not keeping bees were matched to the profile of the new entrants from the same specific area or village.

Table 1: *Informants interviewed at community level.*

	AGRO-PASTORAL	IRRIGATED CROPPING	MIXED-FARMING	TOTAL
NEW BEEKEEPERS	**14**	**7**	**20**	**41**
Female	**4**	**5**	**10**	**19**
Youth - 30 or younger	0	1	4	5
Adult - Older than 31	4	4	6	14
Male	**10**	**2**	**10**	**22**
Youth - 30 or younger	5	0	4	9
Adult - Older than 31	5	2	6	13
NON-ADOPTERS	**5**	**4**	**12**	**21**
Female	**2**	**2**	**6**	**10**
Youth - 30 or younger	0	2	3	5
Adult - Oder than 31	2	0	3	5
Male	**3**	**2**	**6**	**11**
Youth - 30 or younger	2	1	4	7
Adult - Older than 31	1	1	2	4
VILLAGE ELDERS	**4**	**2**	**4**	**10**
GROUP LEADERS	**2**	**2**	**9**	**13**
Female	**1**	**1**	**4**	**6**
Youth - 30 or younger	0	0	0	0
Adult - Older than 31	1	1	4	6
Male	**1**	**1**	**5**	**7**
Youth - 30 or younger	0	0	1	1
Adult - Older than 31	1	1	4	6

beekeeping performance). Themes, subthemes and the different types of informants were linked to identify patterns and formulate recommendations.

3 Results

The empirical results are presented under three subheadings corresponding to the three research questions:

3.1 Pathways to beekeeping

Two principal pathways to beekeeping were identified among the communities studied: an *apprenticeship pathway* and a *traineeship pathway* (Fig. 2). These pathways may be differentiated by three main characteristics: the trainer, the knowledge exchanged and the teaching methods.

3.1.1 Apprenticeship pathway

The first pathway is the traditional one: it involves only community members, indigenous knowledge and mostly on the job practical training. A majority of the interviewees that entered beekeeping *via* the apprenticeship pathway were men from ethnic groups that have traditionally kept bees (e.g. Tugen and Keiyo people) and live in AP where beekeeping has been widely practised for many generations. Two sub-pathways may be distinguished depending on the type of trainer (a family member or a skilled beekeeper in the community): the *intra-familial pathway* and the *extra-familial pathway*, with a majority having come *via* the intra-familial pathway.

In the *intra-familial apprenticeship pathway*, knowledge and skills are transferred from older to younger generations. In the past, new-entrants from this pathway became involved in beekeeping by helping a relative (usually a grandfather or father, but in a few cases an uncle or an elder brother) during harvest. They were mostly boys in their early teens (between the ages of 10 and 14 years). Nowadays, young people who follow the intra-familial pathway start their apprenticeship training later, at approximately 18 years of age. A key stakeholder's explanation for this was that children are more likely to be in school than they were in the past[7] and thus not at home to be trained. However, on-the-job training has always been carried out in the evening, usually during weekends. Most beekeepers, nowadays prefer to teach their children after they have finished secondary school as a fall-back in case they are not able to get employment using their educational credentials. For instance, GL3 in AP explained: "*I don't plan to train my children now because of fear that they will drop out of school if they are able to earn money from beekeeping*". Young people who enjoyed helping to harvest continued to help and to learn more about beekeeping. For instance, NGO1, a local development actor, explained: "*Then, I was allowed to climb a tree and assist my father by holding a traditional smoker or holding a container to collect honey*". Apprenticeship training is focused on traditional methods of honey production, in particular harvesting, but

---

[7]The Kenyan government enacted the provision of free and compulsory education at primary level in 2003.

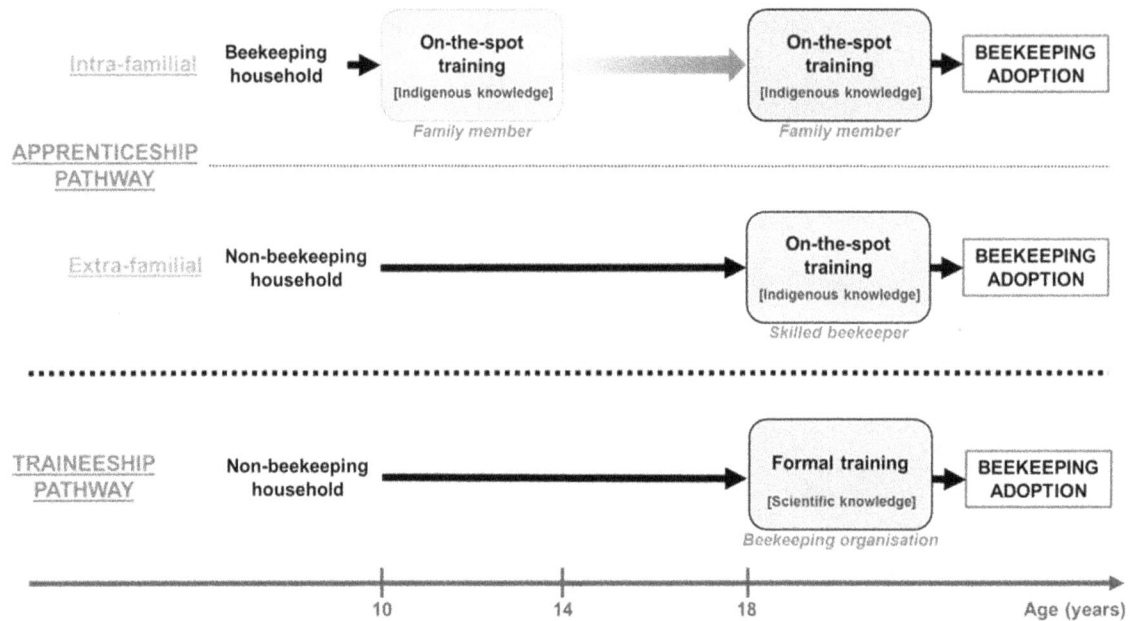

Fig. 2: *Pathways to beekeeping.*

also on how to make and install traditional log hives. Once trained, the young people usually helped their parents with beekeeping activities. They did not set up their own beekeeping enterprise before they had their own household. These new independent beekeepers mostly used inherited or handmade log hives. Some beekeepers started beekeeping with MCF hives (mainly KTBH) after this pathway. However, they were older and had already set up their own enterprise. They bought the hives from personal or family savings or they received them through community-based-organisations (CBOs) (as a donation or loan). A majority of men took up beekeeping on their own after finishing secondary education and before marriage, and were between 18 and 30 years old. Women started later, from 25 onwards, and tended to be married.

In the *extra-familial apprenticeship pathway*, the interviewees mentioned only the training stage before adoption. All of them came from non-beekeeping households and they were taught by skilled beekeepers in the community. They received the same information as the ones from the intra-familial pathway, through similar methods. They started on their own during or just after the training. For instance, NE5, a 35-year-old in IR, said: "*Mr X introduced me to beekeeping. He taught me how to harvest and he invited me to go with him and practice it before I started on my own*". The initial equipment and the age of taking up beekeeping on their own were much the same as for the intra-familial pathway.

3.1.2 Traineeship pathway

In contrast to the apprenticeship pathway, in the second pathway the people learned about beekeeping through formal training provided in an urban area by a formal training organisation (e.g. Baraka College and livestock officers). The training sessions often comprised one or several short seminars and more rarely a period of full-time training. Training was provided mainly on bee health, pollination and business management. Most of the training sessions were provided free, through a partnership between NGOs and livestock officers or a private enterprise. According to many community members and key stakeholders, the main criticism of these training sessions was that they contained too much theory and lacked practice in the field. Most beekeepers from this pathway started beekeeping on their own just after training, using MCF hives and working in CBOs. The beehives were donated or loaned from beekeeping organisations (e.g. development organisations or governmental bodies) or purchased from CBOs' savings. Generally, the producers from this pathway were older than the traditionally trained beekeepers when they took up beekeeping, usually over the age of 30. The respondents who had followed the traineeship pathway were mainly women, members of ethnic groups that were traditionally not beekeepers (e.g. the Njemps people), or people living in MF where a minority of the farmers are beekeepers[8].

---

[8]This aspect was mentioned by most informants as NGO6, a development actor in MF. He said "*About 30 % of the farmers keep bees in the highlands*".

3.1.3 Changes in pathways or beekeeping adoption

Although no recent data are available on the numbers of people that have attended training in Kenya, the majority of the key stakeholders said that the number being trained by formal training organisations is increasing; in particular women are taking it up. Further, many said that the apprenticeship pathway is declining as a path into beekeeping. Finally, a majority of the informants said that the number of beekeepers is increasing in the MF and IR but decreasing in AP[9].

3.2 Factors preventing beekeeping adoption

3.2.1 Lack of knowledge and awareness

According to most informants, the principal obstacle to beekeeping adoption is not growing up in a beekeeping household and not being offered the opportunity to attend a seminar on beekeeping. For instance, NA2, a 29 year-old man in IR said: "*I have never attended any seminar on beekeeping. None of my relatives at home practice beekeeping and that is why I have never been involved in it*". Respondents in MF and IR were more likely not to be aware of the potential of beekeeping as a livelihood option. When asked specifically a large majority said that they were not aware of the possibility of being invited to a seminar. Furthermore, there was no evidence that they had independently sought out the possibility of training.

3.2.2 Priority to other activities

Many informants preferred livelihood activities with higher and regular incomes , education (cf. 3.1.1) or domestic activities to beekeeping. Paid employment (e.g. as officials or in trade) was identified as the preferred and prioritised activity by all types of informants in all locations, whether beekeepers or not.

"*I prefer my job in the County Government because the income is higher*" (NA16, a 28-year-old woman in MF).

Self-employment off-farm (e.g. as trader or casual labourer) activities were also preferred in all farming systems to beekeeping, and in MF and IR on-farm activities (e.g. tending crops and dairy farming) were also mentioned as preferred alternatives to beekeeping.

"*I prefer dairy farming because I get income on a daily basis from my dairy cows*" (NE24, a 39-year-old man in MF).

VE10 in MF also mentioned that there are now more alternative income generating activities than in the past and

young people can prioritise these at the expense of beekeeping. He said: "*In olden days it was only about animal keeping and beekeeping as a source of income. For today, there are so many activities that one [a young person] can undertake to earn a living*".

Several informants, including women, mentioned that women lack time due to their domestic responsibilities, in addition to other farming or income generating activities. Thus, women are less likely to take up beekeeping.

3.2.3 Negative perceptions of beekeeping

The attractiveness of beekeeping was also affected by a number of negative perceptions. A majority of the informants explained that non-adopters perceived it as poorly remunerated and hard manual work which did not fit their image of a modern job.

"*The income earned from beekeeping is low and it would not sustain me and my family*" (NA18, a 32 year-old man in MF).

"*The young people think it is labour-intensive work*" (GB3, a local livestock officer).

"*Nowadays young people see beekeeping as old fashioned*" (NE41, a 36 year-old woman in MF).

Young educated people wanted non-manual jobs with high and regular incomes. Thus, they considered beekeeping as something for uneducated and poor people and as being appropriate for older people.

"*People think it is for poor people who have no other choice*" (GB5 in MF).

"*People go to school and after school young people are expected to find a job and not to keep hives again! This is the parents' expectations. They think beekeeping is a job for old men and not for educated people. If you are a lady, you have to get married or to become a teacher. Life must improve after education. It should be a blessing for the whole family!*" (RI1, a national researcher).

3.2.4 Lack of land and environmental issues

Lack of land was also an obstacle. In AP, many informants no longer had enough trees on their lands (individual or communal lands) to make a log hive or to hang it, whereas in MF, many beekeepers had insufficient space on their own lands to set up an apiary. Many community members in the highlands said that beekeeping is not possible, because the climate is too cold and the nectar is of lower quality. However, GB5 in MF argued that it is not true because: "*[...] there is more bee forage in the highlands and so more potential than in the lowlands*". In IR, climatic shocks like floods were reported as a key reason for people not taking up beekeeping or for abandoning it. For instance, NA5, a 32-year-old man, said that he used to keep bees but he stopped when

---

[9]As the population is growing significantly in Kenya, including in this region, it is more likely that the actual number of beekeepers in AP is increasing but that they are declining as a proportion of the population.

his ten beehives were swept away by floods. Finally, in MF, which is characterised by a high population density, people considered beekeeping not compatible with farming for two main reasons: competition for land with other agricultural production, and risk of neighbours complaining about being stung by bees.

"They can stop [keeping bees] in places when landholdings are small and you have conflicts with your neighbour. The problem is the distance between the beehives and the neighbours. Sometimes people are very biased against bees" (GB1, a national livestock officer).

"Farmers know by experience that pyrethrum kills bees. Pesticides are also a problem. There is a conflict between crop farming and beekeeping" (GB5 in MF).

3.2.5   Socio-cultural restrictions

Socio-cultural norms were reported as a key factor preventing women from becoming beekeepers using traditional systems. Women are not expected to keep bees because climbing trees to harvest log hives is seen as too difficult and risky for them. Furthermore, GB3, a local livestock officer, explained: *"Women could not take part because the men are removing their clothes and their gloves and they climb and they harvest mainly at night. Only men can do that"*, while AC1, a national academic, pointed out that there is a superstition that: *"If a women opens a beehive, the bees will abscond"*. In the Njemps community, women were not even allowed to manage MCF beehives unless working with men. For instance, NE15, a woman in IR, said: *"We installed our first five beehives [KTBH] but later the village elder forced us to remove them because of the Njemps norms. Women are not allowed to keep beehives unless assisted by men in our society"*. Another informant mentioned that a man expected not to take up beekeeping if a forefather died while practising it. He said: *"The father of my father fell from the tree during harvesting and died. So I am not supposed to practise beekeeping"* (GB4 in AP). However, a number of key stakeholders informed us that these traditional beliefs and restrictions are fading away.

3.2.6   Fear of bees

Fear of bees influences people's feelings towards beekeeping. Many key stakeholders and community members, in particular in MF and IR, mentioned that non-adoption was due to the perception that bees are dangerous. For instance NA13, a 25-year-old man in MF, explained: *"I decided not to keep bees because they are dangerous to human beings and animals"*. A number of beekeepers said that they were afraid of bees when they first became beekeepers. NE14, a 29-year-old woman in IR, said: *"I feared bee stings when I started but later came to realize that bees do not sting unless they are disturbed"*.

3.2.7   Limited access to beehives

Difficulty in obtaining beehives is another factor preventing people, especially young people, from taking up beekeeping. A number of beekeepers from the traineeship pathway did not take up beekeeping because they found it difficult to get MCF hives especially purchasing Langstroth hive[10]. As formal training is focused on MCF hives, if they cannot get a MCF hive, they do not have the knowledge and skills to manage or make a log hive.

The few informants who followed the apprenticeship pathway and experienced difficulties in obtaining a hive cited lack of knowledge and/or wood, tools or time to make one.

3.2.8   Lack of knowledge on honey marketing

A few beekeepers mentioned the marketing of the honey as a barrier to taking up beekeeping, in particular, knowing how and where to market it. For instance, NE11, a 31-year-old man in AP, said: *"marketing was a big problem when I started beekeeping because I didn't know how and where to sell my honey"*. This issue was very rarely mentioned by the key stakeholders. In contrast, beekeepers generally said that they could easily sell raw or semi-refined honey to a range of customers (e.g. intermediaries and neighbours). So, a lack of knowledge about marketing honey does not seem to be a key barrier.

3.2.9   Beekeeping threats

A number both of beekeepers and of non-adopters said that empty beehives and pests (e.g. honey badgers (*Mellivora capensis*)) discourage some people from taking up beekeeping, or in some cases, lead them to abandon it, as these threats lead to low productivity or having nothing at all to harvest. Empty beehives at harvest time is a common issue either because the bees did not colonise the beehive or because they absconded before harvesting for a variety of reasons (e.g. lack of bee forage, smoke from nearby charcoal production sites). We found that only a few informants in the communities knew how to catch a swarm to colonise a beehive.

"I attempted to practise beekeeping but abandoned it because the bees do not move into the beehives" (NA8, a 28 year-old woman in MF).

"I heard of a few cases were modern beehives (MCF hives) have been abandoned. This is due to bees not having colonised the beehives and the owners not knowing how to trap bees into them" (NE37, a 32-year-old man in MF).

---

[10]A KTBH and a Langstroth hive cost respectively two and three times as much as a log hive. MCF hives are never made by the beekeepers themselves because they are complex and require expensive woodworking tools.

3.3 Factors positively influencing beekeeping adoption

3.3.1 The outcomes of beekeeping

The principal factor leading people to take up beekeeping was the income generated from selling honey, especially for those who took it up before the age of 30. It was generally seen as a way of supplementing income rather than as the only or main livelihood activity. For many, especially young people and those living in AP and IR, it was an important way to supplement income from subsistence farming when no other alternatives were available. For example, NE16, a 25-year-old man in AP, said: "*It is difficult to raise enough money to pay school fees. So, I thought of beekeeping as an alternative source of income for me and my family*", while CH5, a 15-year-old girl in AP said: "*I want one day to become a beekeeper and to practise it as a hobby or part time job that can bring me an income while I am in my office*". The priority is given to employment with high and regular income but this does not necessarily mean that beekeeping cannot be taken up as well.

Although the majority of beekeepers considered it a profitable activity, it was generally an additional activity and rarely the principal source of income. However, for some it was their only source of income. For example, NE37, a 32-year-old man who started beekeeping at the age of 28, told us: "*I stayed at home for 2 years after my degree without any job opportunities. This prompted me to start beekeeping as an alternative source of income*".

Market opportunities such as the presence of customers and an attractive price were also important inducements to venturing into beekeeping. These factors were mostly reported by group leaders and key stakeholders.

"*Many are keeping bees because of the high demand for honey and the available market*" (GL9 in MF).

"*People in the village know who has beehives and they will ask you for honey. Even if you want to give up, people will encourage you to continue*" (GB1, a national livestock officer).

"*Beekeeping is a most profitable activity because the price of honey is usually very high compared to the labour it takes*" (NE4, a 48 year-old woman in AP).

While income is the greatest motivation for starting beekeeping, for some informants honey and beeswax were also important, for their own domestic uses. Many beekeepers said that they usually sold most of the honey they produced and kept only a small amount for own use. Respondents told us that they use honey as a food, as a medicine for themselves and their animals, for brewing honey beer, as dowry payments and as gifts for visitors. According to the key stakeholders, traditional uses of honey have become less common, in particular beer brewing and payment of dowry.

Some women were also attracted to start beekeeping on their own in order to have more control over their life, especially by controlling honey uses and incomes from beekeeping. For instance, NGO2, a national development officer, explained: "*Women are interested in beekeeping because [...] they can have cash. [...] Because men are careless about decisions on the cash they get from their business. Women love that power of decision*".

3.3.2 Available assets

The second category of factors which play an important role in beekeeping adoption is the availability of assets. Social networks are important, as skilled beekeepers are a source of aspiration and motivation. As illustrated below, children were influenced by their parents, men by an elder and women by seeing others in the community doing it.

"*I went harvesting with my father and this motivated me to become a beekeeper*" (NE10, a 30-year-old man in AP).

"*I took up beekeeping because an old man in my community encouraged me to take it up*" (NE5, a 35-year-old man in IR).

"*I decided to start because everybody was doing it in the village, so I decided to do it too*" (NE4, a 48-year-old woman in AP).

Beekeeping knowledge and skills (human capital) are also essential for starting beekeeping. Besides the invitation to participate to one or another type of training, GB1, a national livestock officer, mentioned that learning how to avoid bee stings is important. She said: "*People become interested in beekeeping when they are made aware that it is possible to protect themselves from bee stings*".

Availability of physical capital such as beehives, harvesting tools (e.g. smokers and protective clothing) or land, also seems to be a great help when taking up beekeeping. Many beekeepers informed us that they were able to start beekeeping because they inherited one or more beehives and/or materials to make log hives were locally available.

"*If they [the young people] venture into beekeeping, it is because they inherited beehives*" (GL9 in MF).

"*The availability of materials to make beehives also helped me to decide on the activity*" (NE33, a 26-year-old man in MF).

Many producers from the traineeship pathway took up beekeeping thanks to donations (e.g. equipment and tools) or loans (money and equipment) provided by development organisations or livestock officers. For example, NE18, a 32-year-old woman in TH, said: "*Our group [SHG] was motivated to keep bees by Hand in Hand East Africa, which gave us seven free Langstroth hives*".

According to a few producers, beekeeping was preferred to other income-generating activities because of the low inputs required. For instance, NE13, a 35-year-old woman in

IR, said: *"Our group [SHG] was once invited to Marigat by the Kenya Red Cross society [...] we were taught beekeeping and poultry farming. When we went back to the village we decided to do beekeeping as it required less capital investment compared to poultry farming"*.

Financial capital was also an important factor. For instance, VE7 in MF explained: *"At this age [late twenties], they have sufficient capital to enable them to purchase one or two beehives to start their enterprise with"*.

Finally, natural capital was clearly reported as also being important. Key stakeholders cited the favourable ecological conditions (e.g. presence of bee forage) required to take up beekeeping. In addition, a majority of the key stakeholders and the informants in MF talked about the benefits of beekeeping for the natural environment such as forest conservation and pollination. It is unlikely that these aspects were direct incentives to start beekeeping. However, as these aspects were promoted during seminars, these would probably have had a positive influence on the perception of beekeeping and in turn beekeeping adoption among the trainees.

3.3.3 Mitigation risks

A number of community members talked about beekeeping as a strategy for mitigating risks and stabilising household incomes. According to them, beekeeping is better suited to the climatic conditions than other types of agricultural production. For instance, NE10, a 30-year-old man in AP, said: *"Beekeeping is less vulnerable to climate conditions than crops or livestock"* and VE1 in AP explained: *"I prefer to keep bees because it is not like cattle and goats; bees are not adversely affected by change in rain-fall patterns, though they are somewhat affected by lack of water, but they do not die in drought like other animals"*. In a few cases, beekeeping adoption was presented by the informants as a strategy to spread climatic risks associated with other forms of agricultural production, or to cope after a climatic shock such as drought. For instance, NE9, a 35 year-old man in AP, is an interesting example. *"I started beekeeping two years ago when the climatic conditions were so harsh that it resulted in the failure of mangoes and vegetable crops and left me without any income. At that time I had not secured a job at X. I was really affected and my brother advised me to diversify my source of income such that when one sector was not doing well, the other sector was doing well"*.

3.3.4 Beekeeping performance

Perceived performance was another key factor influencing beekeeping adoption. Many informants explained that producers spend little time on beekeeping because bee colonies colonise beehives naturally and it is usually not necessary to manage log or MCF hives to produce honey. The principal activities were to install and harvest beehives. For instance, NE5, a 35-year-old man with four log hives in IR, explained: *"Beekeeping doesn't need a lot of time after the installation of the hives. I can leave them until harvesting time comes, unlike other activities [irrigated crops and livestock] which need daily supervision"*. Some new entrants also mentioned that little strength or physical activity is required. NE36, a 58-year-old man in MF, even mentioned that he could practice it in his old age. He said: *"Beekeeping requires less effort for my old age"*. Finally, many informants mentioned that they started beekeeping because they perceived it as a profitable activity to start or maintain with low start-up and running costs.

> *"The reason why I started beekeeping is because of the profit my father was getting. Because he was able to provide everything for my family through beekeeping"* (NE22, 36-year-old man in AP).
> *"The main reason why we started beekeeping is that it is cheaper to start than other forms of agricultural production"* (GL1, a group leader in AP).

However, these factors were rarely reported by the beekeepers in MF.

3.3.5 Attractiveness of MCF hives

Many key stakeholders and a few beekeepers reported the attractiveness of MCF hives. Firstly, MCF hives were a way for young people to practice beekeeping easily and safely. GB3, a local livestock officer, summarised this aspect as follows: *"The young people want something easy to do. [...] So, they are interested in beekeeping thanks to the modern technologies"*.

Secondly, using MCF hives was a way to modernise the image of beekeeper for some producers in IR and MF. For instance, GB4 explained: *"They [the Njemps people] have started recently when the new technologies came in. The good point is that you are not considered as a poor person"*.

Finally, MCF hives improved accessibility for women. As explained by PE4, a national entrepreneur selling frame beehives, they do not need to climb trees to manage their beehives. He said: *"They are mostly in modern beehives since it is accessible to them. No need for a lot of energy"*.

3.3.6 Social norms

Finally, social pressures were identified as a factor affecting the adoption of beekeeping in two situations. Firstly, some children were pressured to keep bees by their parents. Concerning this, PE5 said: *"Beekeeping was like an obligation. I had to continue what my parents did"*. Secondly, a few informants with traditional religious beliefs pointed to the importance of producing honey to participate in traditional ceremonies: *"During traditional ceremonies like weddings, honey is a compulsory ingredient. It is used to make honey beer that is taken by the old men before the initiation of any wedding activity"* (VE3 in IR).

4 Discussion

This study confirms that AP is a high traditional bee-keeping area because most of the farmers are beekeepers and follow the *apprenticeship pathway*. Furthermore, bee-keeping adoption was particularly favoured by the access-ibility of log hives and social pressure. In MF and IR, a few farmers were beekeepers and the *traineeship pathway* was important. The main barriers[11] were limited access to formal training and technologies and negative feelings to-wards bees, especially among women and young people. Thus MF is a low traditional beekeeping area, as IR. In all farming systems, formal knowledge has been gaining in importance with the development of training opportuni-ties offered to beekeepers and a growing number of people follow the traineeship pathway. However, there is no evi-dence that more formal knowledge has been incorporated into apprenticeship training. Furthermore, the loss of indi-genous knowledge did not appear prominently in this study as assumed by Muli & Frazier (2011). The three small-holder farming communities present other differences dis-cussed below depending on socio-cultural norms, climatic conditions and socio-economic context.

The main factors mitigating for or against the adoption of beekeeping which emerged in this study are:

(1) Men are more likely to take up beekeeping, especially in the apprenticeship pathway, because women face a num-ber of barriers' including socio-cultural restrictions and the lack of time available to keep bees in addition to their other activities (e.g. farming and household responsibilities). Fur-thermore, many informants, women and men, said that it is difficult for them (women) to harvest honey from the log hives which are hung high in trees. Therefore, they face additional costs compared to men, because they have to hire men to harvest the honey or use expensive MCF hives[12] (Baraka Agricultural College, 2016). This finding is in line with that of Ahikiriza (2016) in Uganda. However, the cur-rent research did not identify, as she did, any difficulty for women in accessing technologies and training. In contrast, many key stakeholders said that women have been receiving growing support to access formal training and technologies. On the other hand, this study showed that traditional atti-tudes do change and women are attracted by this activity as it enables them to generate an income and take more control over their lives. Women's access to beekeeping is likely to improve in the coming decades, especially via the trainee-ship pathway.

(2) Access to beekeeping tends to be more difficult for young people. The first reason is that most young people

from non-beekeeping households have quite negative per-ceptions of beekeeping. The second reason is that this ac-tivity is not an attractive option for young educated people, even for those living in high traditional beekeeping areas. They prefer non-manual and more secure salaried employ-ment. Therefore, this study confirmed Muriuki's (2010) as-sumption that young people are reluctant to take up bee-keeping because negative attitudes. It also suggested that one possible reason for the decline in new entrants to bee-keeping in Kenya is the growing importance of education, which makes beekeeping less popular among young people. In other words, there is a negative relationship between level of education and involvement in beekeeping, as it is the case with other forms of agriculture in Africa (Asciutti *et al.*, 2016). On the other hand, education does not necessarily mean that people cannot keep bees. Indeed, access to know-ledge and awareness on beekeeping benefits were found to support beekeeping adoption, even among educated people. The latter can take it up to supplement their other liveli-hood activities. Furthermore, some young people in AP end up in beekeeping because there is no alternative employ-ment. The final reason preventing young people from mov-ing into beekeeping is that MCF hives and improved har-vesting tools are attractive to young people, but they can-not easily access these items. As Ahikiriza (2016) pointed out, improved equipment and tools are more expensive than traditional ones, and young people have limited resources and difficulty in accessing capital unless they are involved in CBOs.

(3) We also noted that the factors influencing adoption vary depending on the climatic and socio-economic con-texts. In MF and IR, on-farm activities were preferred al-ternatives to beekeeping. Indeed, agriculture in general is favoured in these locations thanks to colder and damper cli-mate or presence of irrigation schemes. In contrast, bee-keeping was considered as an important income-generating activity in AP. Smallholder household incomes are indeed more vulnerable to climatic risks affecting farming incomes.

(4) Finally, knowledge on honey marketing was con-sidered by a few beekeepers as a constraint, whereas the key stakeholders and the group leaders perceived selling the honey as a positive factor for beekeeping adoption. Thus, this finding may confirmed the argument of Lowore *et al.* (2010) that beekeepers do not necessarily benefit from a high local demand. A proactive behaviour is needed to seize market opportunities and CBOs are better positioned to do so than individual beekeepers, especially in remote areas.

In order to support beekeeping adoption, efforts should be focussed on continuing and intensifying initiatives that increase awareness and knowledge, particularly in low tra-ditional beekeeping areas. Early awareness at school may be effective in raising motivation and reducing inequalities or negative perceptions and feelings based on false informa-

---

[11] These factors were also mentioned as key constraints for beekeeping development in Kenya, especially in the highlands (Carroll, 2002; Muya, 2004; Muriuki, 2010).

[12] A KTBH and a Langstroth hive cost respectively two and three times much as a log hive.

tion. In addition, better access to improved harvesting tools (e.g. manufactured smokers and protective clothing) and MCF hives could help to limit fear of bees and facilitate beekeeping access, especially for young (educated) people and women. However, three disadvantages of MCF hives, especially Langstroth hives, have been noted: high costs, lack of skills, and low occupation rates. These issues, also raised by Bradbear (2002), would have to be addressed by key organisations supporting beekeepers in Kenya to ensure sustainable benefits and the continuation of this activity. Finally, we suggest that key organisations, particularly formal training organisations, and skilled beekeepers willing to share their experience, along with the CBOs, be supported to improve access to knowledge and technologies.

Acknowledgements

This study was partly supported by an Elphinstone Scholarship from the University of Aberdeen and two travel grants from the Sir Maitland Mackie Scholarship fund and the Charles Sutherland Scholarship fund were awarded to the first author. We wish to acknowledge help from our research assistants, the respondents and those who supported us in the identification of informants. J.I.M. acknowledge funding from the Scottish Government's Rural and Environment Science Analytical Services Strategic Research Programme.

References

Adjare, S. O. (1990). Beekeeping in Africa. FAO Agricultural Services Bulletin 68/6. Food and Agriculture Organization of the United Nations, Rome, Italy.

Ahikiriza, E. (2016). *Beekeeping as an Alternative Source of Livelihood in Uganda.* Master's thesis, Ghent University.

Asciutti, E., Pont, A. & Sumberg, J. (2016). Young people and agriculture in Africa: A review of research evidence and EU documentation. IDS Research Report 82. Institute of Development Studies (IDS), Brighton, UK.

Baraka Agricultural College (2016). College prospectus 2016. Molo, Kenia.

Bees for Development (2006). African Honey Trade Forum. Bees for Development Journal, 81.

Bradbear, N. (2002). Beekeeping and sustainable livelihoods. *In:* Bradbear, N., Fisher, E. & Jackson, H. (eds.), *Strengthening Livelihoods, Exploring the Role of Beekeeping in Development.* pp. 11–22, Bees for Development, Monmouth, UK.

Bradbear, N. (2009). *Bees and their role in forest livelihoods: A guide to the services provided by bees and the sustainable harvesting, processing and marketing of their products.* Non-Wood Forest Products 19, FAO, Rome, Italy.

Carroll, T. (2002). *A study of the beekeeping sector in Kenya: June 2001 – January 2002.* Baraka Beekeeping Development Unit / Self Help Development International (SHDI), Molo, Kenya.

Carroll, T. & Kinsella, J. (2013). Livelihood Improvement and Smallholder Beekeeping in Kenya: The Unrealised Potential. *Development in Practice,* 23, 332–345.

Chengo'le, J. M., Duyu, J. J., Musila, F. & Chesang, S. K. (2008). Honey production in dry hot areas. Kenya Agricultural Research Institute, Nairobi, Kenya.

FAO (2016). Statistics Division. (accessed on: 7 February 2017).

Fisher, E. (2000). Forest livelihoods: Beekeeping as men's work in western Tanzania. *In:* Creighton, C. & Omari, C. K. (eds.), *Gender, Family and Work in Tanzania.* pp. 138–176, Ashgate Publishing Ltd, Aldershot, United Kingdom.

Gichora, M. (2003). *Towards Realization of Kenya's Full Beekeeping Potential: A Case Study of Baringo District.* Ph.D. thesis, University of Göttingen, Germany.

Government of Kenya (2013). Sessional paper no7 of 2013: National beekeeping policy.

KFSSG & Baringo County Steering Group (2015). Baringo County 2014 Short rains food security assessment report. Kenya Food Security Steering Group, Baringo County Steering Group, Kenya.

Lowore, J., Bradbear, N., Ndyabarema, R. & Okello, B. (2010). Market access for beekeepers. Bees for Development.

MAFAP (2013). Review of food and agricultural policies in the Kenya 2005–2011. MAFAP Country Report Series, Monitoring African Food and Agricultural Policies, FAO, Rome, Italy.

Muli, E. & Frazier, M. (2011). Beekeeping: Indigenous Knowledge Lost and Found. Department of Entomology, The Pennsylvania State University. Available at: http://ento.psu.edu/news/2011/beekeeping-indigenous-knowledge-lost-and-found (accessed on: 7 February 2017).

Muli, E., Kilonzo, J. W. & Ngang'a, J. K. (2015). Adoption of Frame Hives: Challenges Facing Beekeepers in Kenya. *Global Science Research Journals,* 3, 251–257.

Muriuki, J. M. (2010). Beekeeping Technology Adoption and its Effect on Resource Productivity in Southern Kenya Rangelands. BSc. Thesis, University of Nairobi.

Muya, B. I. (2004). An analysis report on Kenya's apiculture sub-sector. BENFLO consultants, Nairobi, Kenya.

Nightingale, J. & Crane, E. (1983). A lifetime's recollections of Kenya tribal beekeeping. International Bee Research Association, UK.

12

Smallholder goat production in the Namaacha and Moamba districts of southern Mozambique

Gracinda A. Mataveia [a,c,*], Carmen M. L. P. Garrine [b], Alberto Pondja [b],
Abubeker Hassen [c], Carina Visser [c]

[a] Department of Clinics, Faculty of Veterinary, University of Eduardo Mondlane, Maputo, Mozambique
[b] Department of Animal Production, Faculty of Veterinary at University of Eduardo Mondlane, Maputo, Mozambique
[c] Department of Animal and Wildlife Sciences, Faculty of Natural and Agricultural Sciences, University of Pretoria, South Africa

Abstract

Goat rearing is one of the most common livestock farming activities in Mozambique and has the potential to play a powerful role in improving the livelihoods of resource-poor farmers. This study was conducted to investigate the status of goat husbandry practices in rural areas of southern Mozambique. Data were collected from a total of 45 smallholder goat keepers in three different villages through questionnaires complemented by interviews. Most households were dependent on crop production and livestock as their main source of income. Goats were reared under extensive systems where free grazing and tethering were the common feeding management practices with limited supplementation during the dry season. The flock sizes per household were predominantly small (13 ± 2.4) with uncontrolled breeding of goats. The goats were reared mainly as a source of meat for home consumption and a means of reserve cash income. All household members were involved in goat production but women and children had a minor role in terms of decision making. The main constrains limiting goat production were diseases, lack of veterinary services, limited size of grazing land and scarcity of feed resources. Intervention programs focused on improving the husbandry practices and veterinary assistance should be initiated to improve goat production and thereby improve the income and livelihood of the resource-poor farmers in Mozambique. This paper presents a summary of the results of a baseline study in the Namaacha and Moamba districts of Mozambique.

Keywords: communal, extensive, goat, husbandry, small-scale

1 Introduction

Goats are kept in a wide range of agro-ecological zones and management systems, and are mainly owned by smallholder farmers in developing countries (Casey & Webb, 2010), where they contribute to improved livelihoods for many resource-poor communities (De Vries, 2008; FAO, 2012; Hossain *et al.*, 2015; Ouchene-Khelifi *et al.*, 2015). Their role and relative importance varies noticeably across regions and cultural groups. In addition to providing meat and milk for household consumption, goats are one of the easiest and most readily accessible sources of income available to meet the immediate social and financial needs of rural farmers (Boogaard *et al.*, 2012; Boogaard & Moyo,

2015). Furthermore, they are used for cultural purposes such as traditional ceremonies and birthday festivities (Kosgey *et al.*, 2008; Rumosa Gwaze *et al.*, 2009; Oluwatayo & Oluwatayo 2012; Boogaard *et al.*, 2012). Goats are mostly owned by smallholder farmers and have comparative advantages over other livestock species in the traditional farming systems due to their rapid turnover, adaptability to harsh environmental conditions and the efficient use of available feeding resources (Braker *et al.*, 2002). Goat production worldwide grew steadily in the last decade, particularly in the developing world, with Africa contributing approximately 36.2 % to the global goat population (FAOSTAT, 2014).

In Mozambique, goat population is estimated at about 5 million head, of which almost 95 % are kept by rural smallholder farmers and less than 1 % is farmed commercially (INE, 2014). Mozambique has two indigenous goat breeds,

* Corresponding author – gracindaamataveia@gmail.com,
gmataveia@yahoo.com

namely the Landim breed which is spread across the country, and the Pafuri breed which is mostly located in the semi-arid area of Pafuri in South-West Mozambique (Garrine *et al.*, 2010). Goats are commonly raised under a mixed crop-livestock management system, where they subsist on grazing on natural veld and shrubs or marginal lands, and sometimes on crop residues (Devendra & McLeroy 1982). However, the prevalence of a long dry season and droughts in the country poses major challenges to most goat keepers as it leads to shortages of forage and water. In addition, the reduction of grazing land for ruminants associated with increasing human population size and its subsequent degradation, uncontrolled fires and an absence of pasture management exacerbate the shortage of fodder for goats (Timberlake & Jordão, 1985). This problem leads to underfeeding of goats and consequently loss of body condition, reduced productivity, increased susceptibility to diseases, and high mortality rates (Kanani *et al.*, 2006).

In Mozambique past efforts aimed at improving goat production are limited. The development of goat improvement programs would be more effective if information regarding the prevailing goat farming systems in the country were available. In order to design appropriate strategies aimed to improve goat production and to explore the potential contribution of goats to food security in resource-poor areas, there is a need to evaluate the existing goat production system and its role in these rural communities. This study was therefore conducted with the objective of generating baseline information with regard to the current goat husbandry practices in the rural areas of southern Mozambique (Maputo province). It aims to characterize the existing rural goat production system with regards to socio-economic factors, general management and limits encountered.

2 Materials and methods

2.1 Study setting

The study was conducted in two districts (Namaacha and Moamba) of the Maputo province in southern Mozambique. These districts were selected because of their importance in goat production and their proximity to the Extension Centre of the Eduardo Mondlane University. The Namaacha district covers an area of 2,196 km^2 and is characterised by a tropical humid climate with an average annual rainfall of 751 mm. However, the district has experienced a substantial decrease in rainfall over the last years, having received an annual rainfall of 260 mm in 2015 and 471 mm in 2016. Most parts of the district are classified as semi-arid, with visible land degradation due to poor management caused by overgrazing (MAE, 2005). The Moamba district, covering

an area of 4,598 km^2, is characterised by a subtropical dry climate, with an annual rainfall ranging between 580 and 590 mm. In both districts, the average annual temperature varies between 23 °C and 24 °C, with maximum highs of 36 °C. The rainy season is from October to April and the dry season is from May to September. According to Timberlake & Jordão (1985) and Morgado (2007), the vegetation consists mainly of grasses (*Andropogon gayanus*, *Cynodon dactylon*, *Eragrostis superba*, *Panicum maximum*, *Setaria holstii*, *Themeda triandra*, *Urochloa mosambicensis*), and shrubs and trees (*Acacia nigrescens*, *Acacia nilotica*, *Dichrostachys cinerea*, *Sclerocarya birrea*).

2.2 Sampling and data collection

Prior to the study, goat keepers from both districts were approached to evaluate their willingness to participate in the study. Three villages (Michangulene and Mahelane from Namaacha district, and Moamba-sede from Moamba) were chosen and fifteen goat keepers were randomly selected from each village to participate in the study, resulting in a total of 45 goat keepers. Information regarding household demographics and goat management practices (e.g. feeding, health, reproduction and constraints) was collected through questionnaires. Participants ranked certain parameters such as major sources of income and reason for keeping goat on a scale of 1 to 3, with 1 being the most important and 3 the least important. The questionnaires were complemented by directed observations to collect additional qualitative data. The interviews were performed by the principal investigator and a trained enumerator. In order to ensure that all questions were clear to the interviewees, the questionnaire was pre-tested before the survey and was translated into the local language where necessary. Before the commencement of the study, consent was obtained from the villages' leaders and from each individual respondent.

2.3 Data analysis

Data were captured in EpiData Entry Client version 4.0 (Lauritsen & Bruus, 2005) and exported to SPSS version 20.0 (IBM Corp, 2011) for analysis. Data were analysed using descriptive statistics, wherein means and standard deviations were obtained for quantitative data and frequency and percentages were obtained for categorical data. The source of income, purpose of rearing goats, reasons for choice of buck, and marketing/culling of goats were subjected to a rank analysis according to the perceived grade provided by the goat keepers. Indices were calculated using the following formula: Index = sum of [3 for rank 1 + 2 for rank 2 + 1 for rank 3] given for an individual use divided by the sum of [3 for rank 1 + 2 for rank 2 + 1 for rank 3] summed over all uses.

Table 1: *Socio-economic characteristics of goat keeping households in the three study villages.*

Parameters	Villages			Total
	Michangulene	Mahelane	Moamba	
Land holding (%)				
Own	86.7	100	100	95.6
Lease	6.7	0.0	0.0	2.2
Other	6.7	0.0	0.0	2.2
Sex of household head (%)				
Male	66.7	53.3	93.3	71.1
Female	33.3	46.7	6.7	28.9
Age group of household head (%)				
≤ 30	0.0	13.3	6.7	6.7
31–45	26.7	0.0	33.3	20.0
46–60	33.3	73.3	20.0	42.2
> 60	40.0	13.3	40.0	31.1
Marital status of household head (%)				
Married	73.3	80.0	66.7	73.3
Single	20.0	13.3	13.3	15.6
Widower / Widow	6.7	6.7	20.0	11.1
Level of education of the household head (%)				
Primary	60.0	40.0	33.3	44.4
Secondary	6.7	6.7	20.0	11.1
Tertiary	0.0	13.3	13.3	8.9
None	33.3	40.0	33.3	35.6
Household size (mean ± sd)				
Male	1.5 ± 1.1	1.6 ± 0.8	2.2 ± 1.6	1.8 ± 1.2
Female	1.8 ± 0.9	1.9 ± 0.9	1.9 ± 1.8	1.8 ± 1.3
Children (< 15 years)	2.1 ± 1.5	4.1 ± 3.0	4.7 ± 12.9	3.6 ± 7.7
Total	5.3 ± 2.6	7.6 ± 3.6	8.8 ± 13.5	7.2 ± 8.2

3   Results

3.1   Socio economic characteristics of households

Socio economic characteristics of the households included in the study are presented in Table 1. The majority of the respondents (95.6%) surveyed in all three villages owned their land. Of the households surveyed, most (71.1%) were headed by males. However, there was a substantial number of female-headed households in Michangulene (33.3%) and Mahelane (46.7%). Most household heads (73.3%) were over 45 years old, and had attained some level of formal education (64.4%). Results on additional household characteristics (marital status and household size) are also presented in Table 1.

Generally, the household members shared roles and responsibilities regarding goat husbandry activities. Overall, Table 2 shows that in Moamba and Mahelane villages mainly men are responsible for goat husbandry, while in Michangulene village, activities are more evenly spread over men and women. However, in latter village women are largely responsible for some activities, such as breading, purchasing and selling, although this percentage is clear lower in the other two villages. Table 2 also shows that children are to a high extent responsible for the herding/feeding of the goats in Michangulene and Mahelane villages.

The majority of surveyed households in the study villages were dependent on mixed crop and livestock production as their main source of income. Crop production such as maize, beans and cassava was ranked as the primary source of income in Michangulene and Mahelane, while in Moamba, livestock was ranked highest (Table 3). In addition to crops, other sources of income regarded as important in Mahelane and Michangulene were earning a salary as farm workers and livestock, while informal business related activities also played an important role as source of income in Moamba village.

Table 2: *Extent of household members' participation (%) in various goat husbandry activities in the three study villages.*

Activity	Michangulene			Mahelane			Moamba		
	Men	*Women*	*Children*	*Men*	*Women*	*Children*	*Men*	*Women*	*Children*
Herding/Feeding	14.3	42.9	42.9	6.2	18.8	75.0	83.3	16.7	0.0
Breeding decisions	35.7	64.3	0.0	66.7	33.3	0.0	93.3	6.7	0.0
Slaughtering	57.1	42.9	0.0	71.4	28.6	0.0	93.3	6.7	0.0
Selling	47.1	52.9	0.0	70.0	30.0	0.0	93.3	6.7	0.0
Purchasing	38.5	61.5	0.0	71.4	28.6	0.0	92.9	7.1	0.0
Animal health care	54.5	45.5	0.0	66.7	33.3	0.0	90.0	10.0	0.0

Table 3: *Ranking of source of income in households in the three study villages.*

Source of income	Rank (Index)		
	Michangulene	*Mahelane*	*Moamba*
Salary	2nd (0.12)	2nd (0.33)	4th (0.05)
Crops	1st (0.56)	1st (0.40)	2nd (0.38)
Livestock	3rd (0.32)	3rd (0.24)	1st (0.44)
Business	4th (0.00)	4th (0.03)	3rd (0.14)

The average livestock holding per household was higher in Moamba village (14.0 ± 3.55 TLU) than the other two villages (Table 4). Among livestock type, average cattle holding was also higher in Moamba (11.1 ± 3.03 TLU). In terms of number of heads, goats were the major livestock species kept by the households in Michangulene followed by chickens and pigs, while in Mahelane and Moamba chickens were kept in higher numbers, followed by goats and cattle. Irrespective of village, goats were kept in larger numbers (13.04 ± 2.41 head) when compared to cattle (6.76 ± 1.88 head) and pigs (1.20 ± 0.44 head). With regard to the number of goats kept by village, Moamba had larger flock sizes (23.0 ± 6.39 head) compared to Michangulene (8.0 ± 1.2 head) and Mahelane (8.13 ± 1.48 head).

3.2 Reason for keeping goats

The reason for rearing goats was evaluated based on the rank attributed to each specific purpose by the goat keepers. Generally, most goat keepers primarily used goats as a source of meat for home consumption and cash income from sales (Table 5). In Mahelane and Michangulene villages, the use of goats for social ceremonies and for investments/insurance, respectively, were indicated as other important reasons for rearing goats.

3.3 Important traits for goat keepers

Goat keepers indicated their preferences in terms of phenotypic traits and the ranking thereof is presented in Table 6.

In general, all traits were considered important, however, body size, growth rate, disease and drought tolerance were considered the most important traits for male goats, while prolificacy and fertility traits were ranked very high for female goats. Irrespective of the sex of the goat, the traits that were considered as being the foremost important were body size in Michangule, while growth rate and quality of meat were ranked at the top in Mahelane, and grow rate and body size in Moamba.

3.4 Production system

Goats were raised under extensive conditions where they were allowed to graze either freely on communal grazing areas, herded or tethered. In Michangulene and Mahelane villages, children were at large responsible for herding the goats to grazing areas during the day, while in Moamba village goats mostly grazed unsupervised during the day and confined at night. Tethering was also a common practice in the Michangulene and Mahelane villages (50–93.3 %). Although supplementary feeding was not common, some goat keepers (7.1–53.3 %) in the Michangulene and Mahelane villages provided crop residues and leaves from fodder trees, such as *Leucaena leucocephala* and *Moringa oleifera* mainly during the dry season.

The major sources of water for goats were boreholes in the Michangulene and Mahelane villages, and a river in the Moamba village. These water sources provided water for the goats throughout the year and were usually located near to the households in the case of boreholes, while the river was distant from the households.

Most goat keepers (60–98 %) housed their goats in own kraals throughout the year. The kraals were used to keep the goats safe during the nights, while they were either left to browse or tethered during the day. The kraals were mostly traditional, made of untreated wood and with earth floors. Approximately half of the kraals in the Namaacha villages (Michangulene and Mahelane) had an iron sheet roof to protect the animals from the rain, while in Moamba the kraals were mostly open.

Table 4: *Herd size (Mean ± SE) per household in the three study villages.*

Livestock type	Number of heads			TLU		
	Michangulene	Mahelane	Moamba	Michangulene	Mahelane	Moamba
Cattle	0.3 ± 0.21	4.2 ± 2.33	15.8 ± 4.32	0.2 ± 0.14	2.9 ± 1.64	11.1 ± 3.03
Goats	8.0 ± 1.20	8.1 ± 1.48	23.0 ± 6.39	0.8 ± 0.12	0.8 ± 0.15	2.3 ± 0.64
Sheep	0.0 ± 0.00	0.3 ± 0.26	3.7 ± 2.80	0.0 ± 0.00	0.0 ± 0.03	0.4 ± 0.28
Chicken	5.3 ± 0.82	12.2 ± 3.17	24.9 ± 7.40	0.1 ± 0.01	0.1 ± 0.03	0.2 ± 0.07
Pigs	1.5 ± 0.82	1.9 ± 1.01	0.2 ± 0.20	0.3 ± 0.16	0.4 ± 0.20	0.0 ± 0.04
Other [†]	1.67 ± 0.62	7.93 ± 6.54	1.20 ± 0.52	–	–	–
Total herd size				1.3 ± 0.25	4.3 ± 1.93	14.0 ± 3.55

[†] Includes ducks, rabbits and donkeys
TLU=Tropical Livestock Unit

Table 5: *Purpose of keeping goats as ranked by goat keepers in the three study villages.*

Purpose	Rank (Index)		
	Michangulene	Mahelane	Moamba
Meat	2nd (0.20)	1st (0.29)	1st (0.26)
Skin	6th (0.01)	7th (0.00)	6th (0.00)
Cash from Sales	1st (0.21)	3rd (0.13)	1st (0.26)
Ceremonies	5th (0.08)	2nd (0.21)	2nd (0.17)
Breeding	3rd (0.17)	4th (0.11)	3rd (0.16)
Insurance/emergency	2nd (0.20)	6th (0.06)	5th (0.04)
Cultural rites	6th (0.01)	5th (0.10)	4th (0.07)
Investment	4th (0.12)	5th (0.10)	5th (0.04)

Table 6: *Preferred traits as ranked by goat keepers in the three study villages.*

Reason	Rank (Index)					
	Michangulene		Mahelane		Moamba	
	Bucks	Does	Bucks	Does	Bucks	Does
Growth rate	2nd (0.10)	2nd (0.10)	1st (0.10)	1st (0.10)	1st (0.10)	1st (0.10)
Body size	1st (0.11)	1st (0.11)	2nd (0.09)	2nd (0.09)	1st (0.10)	1st (0.10)
Meat quality	4th (0.07)	3rd (0.09)	1st (0.10)	1st (0.10)	1st (0.10)	2nd (0.09)
Prolificacy	3rd (0.09)	1st (0.11)	2nd (0.09)	1st (0.10)	2nd (0.09)	2nd (0.09)
Disease tolerance	1st (0.11)	2nd (0.10)	3rd (0.08)	3rd (0.08)	2nd (0.09)	2nd (0.09)
Drought tolerance	2nd (0.10)	3rd (0.09)	3rd (0.08)	3rd (0.08)	2nd (0.09)	2nd (0.09)
Heat tolerance	2nd (0.10)	4th (0.08)	2nd (0.09)	2nd (0.09)	2nd (0.09)	2nd (0.09)
Temperament	2nd (0.10)	2nd (0.10)	1st (0.10)	1st (0.10)	2nd (0.09)	2nd (0.09)
Body shape	2nd (0.10)	2nd (0.10)	2nd (0.09)	3rd (0.08)	4th (0.07)	3rd (0.08)
Colour	5th (0.03)	5th (0.02)	2nd (0.09)	3rd (0.08)	3rd (0.08)	3rd (0.08)
Fertility	1st (0.11)	2nd (0.10)	1st (0.10)	1st (0.10)	1st (0.10)	2nd (0.09)

Goat flocks consisted of local breeds where kids and weaners formed the major part of the flock structure. The main source of goats for the majority of goat keepers (46.7–93.3 %) was purchasing from other goat keepers across the study villages. Within their own flocks, most goat keepers (53.3–63.6 %) used their own breeding bucks for natural breeding. However, in Mahelane village, goat keepers were largely dependent on the use of communal breeding bucks (60 %).

Although breeding was uncontrolled, the choice of bucks for mating was based mainly on their body size (39.0–100 %), while other selection criteria such as body shape and performance (11.0–33.0 %) were also important. Buck performance and body shape were ranked second in Mahelane (11.0 and 29 %, respectively) and Moamba (33.0 and 19.0 %, respectively). Other traits, such as colour and availability, were generally perceived as being less important.

Bucks were used for breeding from as young as 6 months (46.7–86.7 %) and their breeding life lasted typically between 2 to 4 years. The majority of goat keepers (46.7–64.3 %) reported ages at first kidding to be between 12 and 18 months. However, a substantial portion of respondents (28.6–50.0 %) also reported early kidding ages of between 6 and 12 months. The kidding interval was commonly between 6 and 8 months, but sometimes it lasted as long as 12 months. Natural weaning was the sole practice of weaning and many goat keepers did not allow kids to wean before 4 months of age.

In general, culling was a common practice among goat keepers (20–73.3 %) in the villages. Old age and temperament were the main reasons for removing male goats from the flocks, while poor fertility and old age were the main reason for culling of females. All culled animals were marketed either to consumers or to other goat keepers. Apart from culling, selling of goats was also a common practice (50.0–68.8 %). Male goats constituted the major proportion of goats sold (20–73.3 %) as compared to females (0–13.3 %). Goats were sold mainly to cover household needs, such as food, school fees, medicines and traditional ceremonies.

3.5 Constraints to goat production

The households generally considered health issues as the most important constraint for goat production, where diarrhoea was stated as a main concern, followed by respiratory problems and ectoparasites. Theft of goats, limited grazing areas particularly in Michangulene and Mahelane, as well as shortage in quantity and good quality pastures in the dry season, and insufficient veterinary/extension assistance were the other constraints reported (Fig. 1).

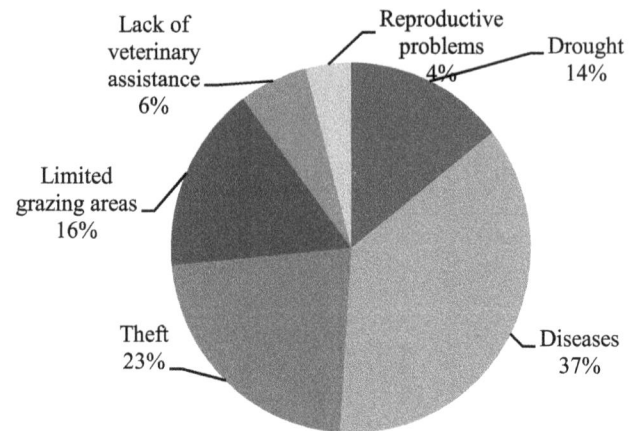

Fig. 1: *Constraints for goat production in the three study villages.*

4   Discussion

Communal, indigenous goats are mostly kept by rural communities and play a crucial role in food security and improving of livelihoods (Hossain *et al.*, 2015; Ouchene-Khelifi *et al.*, 2015). Most research on goats has been performed in controlled research conditions, and is usually not applicable to the rural conditions in which animals are kept. Rumosa Gwaze *et al.* (2009) stressed that surveys to collect baseline data using questionnaires, discussions and direct observations from goat keepers, are essential for goat development in Southern Africa.

Demographic characteristics of this study indicated that while males are still dominating as heads of households, women do have reasonably good participation in the goat production systems. However, goat husbandry is largely a male activity in Moamba village, probably due to traditional habits prevailing in that area whereby men own and are responsible for livestock, while women are relegated to crop production and domestic duties. In contrast, in Michangulene and Mahelane villages the involvement of women in goat activities is more pronounced, likely because men are engaged in other activities, either as farm workers, or other occupations locally or at nearby industrial plants in Maputo province. Furthermore, the university centre located at Michangulene village has been developing gender-based livestock programs which might have contributed to involvement of more women in goat production. According to Guèye (2009), development programmes aimed to enhance the role of rural female farmers in agro-production systems have a potential to empower women over time. On the other hand, the absence of children in goat activities observed in Moamba village might reflect the husbandry system used in that area, whereby cattle and goats are not herded to the grazing and watering points, whereas the increasing human population in Michangulene and Mahelane may have contributed to an increased level of goat's thefts, resulting on the need of a supervised grazing or tethering.

Although most households owned some land in the study areas, it has mainly been used for crop farming. While in Moamba village, grazing land is not yet a problem, the increasing need of land for habitation in Michangulene and Mahelane villages, has been reducing the areas formerly used for grazing. Previous studies have pointed that the land available for agriculture activities has been negatively affected by the increasing human population (Berihu *et al.*, 2015; Kalema *et al.*, 2015). Goats are only allowed to graze in fallow lands or shared grazing areas within the communities. These findings are in line with those reported in other studies targeting small ruminants (Simela & Merkel, 2008; Kosgey *et al.*, 2008; Oluwatayo & Oluwatayo 2012). Limited grazing land pose an important challenge to smallholder goat farming since it negatively affects the quantity as well as the quality of available fodder and consequently the reproduction efficiency and health of goats, and therefore the role these animals play in the livelihoods of the rural poor.

The larger TLU and goat flock size observed in Moamba village may probably reflect the availability of grazing areas and other conditions when compared to Michangulene and Mahelane villages. However, overall average flock size (13.04 ± 2.41 head) found in the present study corresponds to the large range previously reported for rural goat herd sizes. Average goat flock sizes in Southern Africa varies from 9.7 in Mozambique (van Niekerk & Pimentel, 2004) and 12.0 in Zimbabwe (Assan & Sibanda, 2014) to 16.0 and 25.3 in South Africa as reported by Mahanjana & Cronjé (2000) and Mdladla *et al.* (2017), respectively. Flock sizes in communal areas are generally limited by little available grazing land available, low reproduction efficiency and high prevalence of parasites and diseases (Rumosa Gwaze *et al.*, 2009).

Rural smallholder farmers in Mozambique depend on mixed crop-livestock farming for their subsistence. Goats are raised primarily as a source of meat for home consumption and to use as cash reserve, and the finding of this study is in agreement with observations from previous studies conducted in other African countries (Collins-Lusweti, 2000; Kosgey *et al.*, 2008; Rumosa Gwaze *et al.*, 2009; Semakula *et al.*, 2010; Oluwatayo & Oluwatayo 2012; Hassan & Tesfaye, 2014). Variation in the importance of livestock as a source of income is normal, as it depends on the production environment as well as the proximity to markets (Monau *et al.*, 2017). In villages closer to urban areas, goat keepers are more prone to have other means of income such as informal employment. The use of goats in social ceremonies was ranked second in some areas (Moamba and Mahelane), which emphasizes the socio-cultural importance of goats in rural areas of Mozambique. The importance of selection criteria is vital in goat breeding and has been reported to vary according to production systems in the tropics (Kosgey & Okeyo, 2007). In this study, body size and growth rate as well as disease and drought tolerance were considered the most important traits for male goats. In addition to these traits, prolificacy was considered as a major trait in female goats. Body size and growth rate are valued since they are linked to improved weight gains and hence to increased income and meat. Similarly, disease and drought tolerance were emphasised by farmers due to their influence on flock production. This corresponds to the traits used for selection in West African goats (Dossa *et al.*, 2015), which also ranged from health status and body conformation to tolerance and drought and disease resistance. In Botswana, Monau *et al.* (2017) also reported that body conformation and body size were the two most important characteristics for selection of Tswana goats. Lack of tolerance to droughts and diseases predisposes animals to loss of body condition, and therefore results in reduced productivity (Kanani *et al.* 2006). When selecting male breeding animals, farmers put most emphasis on body size, being an indicator of meat production. These findings are consistent with previous reports from Ethiopia (Tadesse *et al.*, 2014), Uganda (Byaruhanga *et al.*, 2015) and West Africa (Dossa *et al.*, 2015).

This study shows that farmers value animals that have shown an ability to survive and thrive under stressful environmental conditions. Prolificacy and fertility are also valued traits for female goats since they influence the growth and productivity of the flock. It is clear from this study that goat farmers rely on multiple selection criteria to ensure adaptability to the local environment and increase goat production.

This study showed that most of the animals grazed on poor-quality natural veld in communal land. This practice is common in extensive smallholder systems and is used in many resource-poor areas of the developing countries where cattle, goats and sheep depend on natural vegetation as their primary source of feed (Kusiluka & Kambarage, 1996; Salem & Smith, 2008; Kumar *et al.*, 2010; Byaruhanga *et al.*, 2015). Tethering was a common management practice and was used throughout the year, particularly in Michangulene and Mahelane. This practice was used to prevent stock theft and destruction of crops during the cropping season, while it also limit the animals to a specific area with sufficient vegetation. While tethering is a common practice in goat keeping in many parts of Africa (Banda *et al.*, 1993; Lovelace *et al.*, 1993; Webb & Mamabolo, 2004; Boogaard *et al.*, 2012), it can have an adverse effect on goat production. It generally leads to restricted feeding and therefore results in inadequate nutrition (Salem & Smith, 2008; Byaruhanga *et al.*, 2015), particularly if supplementary feeding is not provided or the alternative available is of low quality.

Supplements were provided for goats mainly during the dry season when feeding resources were scarce. Supplementation consisted mainly of crop residues and leaves from fodder tree species such as *Leucaena leucocephala* and *Moringa oleifera*. This practice was largely observed in villages where tethering was common, suggesting that the practice of tethering forced farmers to provide supplementary fodder to meet feeding and nutritional needs of the animals. However, supplementation was not practiced at all in the Moamba village, where farmers indicated that they were not aware of the nutritional qualities of the fodder trees such as *Leucaena leucocephala* and *Moringa oleifera*. This indicates the importance of extension and knowledge transfer regarding such alternative fodder resources. The fodder trees can easily be grown in this study area, and will relieve grazing pressure during the prolonged droughts. The use of supplementary feed sources, such as maize grain and flours has been reported as a common practice in other studies conducted in resource poor areas of Asia and Africa (Collins-Lusweti, 2000; Kumar *et al.*, 2010; Boogaard *et al.*, 2012; Tadesse *et al.*, 2014; Byaruhanga *et al.*, 2015) as a way of meeting maintenance requirements and sustaining body condition and flock productivity during the dry seasons.

Goats usually graze communal fields that are unfenced, and this makes them vulnerable to predators and thieves. Protection from stock theft during the night is the main reason for providing housing. The housing was basic and did not allow systematic separation of animals based on their physiological status. This preventative measure seems to work, as theft accounted for 23 % of stock losses, which is lower than the 52 % reported by Collins-Lusweti (2000) in South African village goats and 40 % reported by Monau *et al.* (2017) for Tswana goats in Botswana.

In the present study purchasing was the main way of acquiring goats, corroborating findings from previous studies conducted in some African countries (Assan & Sibanda, 2014; Byaruhanga *et al.*, 2015; Dossa *et al.*, 2015). Goats were also acquired via government programs in the Michangulene village. Most farmers used their own breeding buck(s) for natural breeding although farmers in Mahelane village relied on a communal breeding buck. Regardless of the source of the male, uncontrolled breeding took place. Breeding bucks were used for mating from as young as 6 to 12 months. As soon as males reached puberty, they were free to mate as all animals graze together. The lack of structured breeding systems and appropriate infrastructure, such as paddocks, as well as limited knowledge regarding

herd management facilitate does and bucks run together all year round (Rumosa Gwaze *et al.*, 2009). A lack of controlled breeding results in inbreeding and no fixed kidding seasons (Monau *et al.*, 2017) compounding the poor management as kids are born throughout the year.

A male is usually kept within a production system for between 2 to 4 years, after which they were slaughtered for meat or sold. The age at first kidding of 12 to 18 months reported by the majority of farmers, was similar to that reported for the Mashona breed (16–18 months) in Zimbabwe (Ndlovu & Royer, 1988) and the Nguni breed (16–18 months) in South Africa (Webb & Mamabolo, 2004). Earlier ages between 6 and 12 months were also reported in this study, which can be expected in traditional management systems where bucks run continuously with does (Chukwaka *et al.*, 2010). Kidding intervals of 6–8 months for goats reported across the study villages were in line with results reported by Webb & Mamabolo (2004) for Nguni goats in South Africa. The longer kidding intervals reported in the Michangulene and Mahelane villages, corresponds with that reported by McKinnon & Rocha (1985), Wilson (1989) and Rumosa Gwaze *et al.* (2009). A large variation in kidding intervals are associated with traditional management systems where random mating and continuous mating throughout the year is common (Chukwaka *et al.*, 2010).

Droughts, theft and diseases are commonly reported as major constraints to rural goat farming (Collins-Lusweti, 2000; Monau *et al.*, 2017). Health problems were frequent during the rainy season in which diarrhoea was most prevalent. The occurrence of diarrhoea can be attributed to grazing on regrowth of natural vegetation with high moisture content and nutritive value, after periods of scarcity and poor quality vegetation during the dry season (Payne, 1990). Respiratory disorders and ticks were also frequent, indicating poor or lack health management and limited or non-existent veterinary assistance in the study areas. Similar findings were described in other studies (Devendra & McLeroy, 1982; Kusiluka & Kambarage, 1986, Nsereko *et al.*, 2015; Onzima *et al.*, 2017), who reported gastrointestinal, infection diseases on extensive systems with limited veterinary assistance.

Culling of goats was a common practice among goat farmers across the study villages. Old age and temperament were the main reasons for culling male goats from the flock. This is not in agreement with previous studies conducted in Ethiopia (Demissie *et al.*, 2014; Seid *et al.*, 2015) and West Africa (Dossa *et al.*, 2015), where health problems were the main reason for farmers to cull goats, irrespective of their sex. However, in the present study, poor fertility and old age were the main causes for culling female goats, which is in agreement to the findings reported in Kenya (Bett *et al.*, 2009).

This study has been limited by the low number of respondents inquired, due to several challenges for data collection in the rural villages. These include a lack of understanding of the benefits of a survey for goat keepers, the limited number of possible participants and their unavailability during the cropping season as they prioritize farm activities. Furthermore, smallholders do not have any phenotypic records on the productive and reproductive parameters of their animals, making it very difficult to provide data to inquirers. Similar numbers of households per village were used in surveys of goat production in South Africa by Collins-Lusweti (2000) and Mdladla *et al.* (2017). As baseline data for goat production is virtually non-existent in the rural regions of Mozambique, these findings will contribute to future research and assist in baseline knowledge.

5   Conclusions

Goat production plays an important role in the livelihoods of rural Mozambican farmers. It is comprised of indigenous goats reared under extensive system, browsing natural pasture throughout the year. Although the goats are hardy and well adapted to local conditions, their production is limited by poor nutrition, a lack of management and a high prevalence of diseases and parasites. Therefore, there is a need for appropriate intervention strategies to improve goat production, through education of farmers on good husbandry practices, such as better breeding and feeding practices as well as disease control strategies. Also, the baseline information provided in this study will contribute in the development of coordinated and comprehensive goat production improvement programs and ultimately improve goat productivity and the livelihood rural farmers.

Acknowledgements

The authors are gratefully to the Project Sida/SAREC 2 - UEM and *Fundo Nacional de Investigação – Projecto No 164 - Inv/FNI* for their financial support. The authors are thankful to goat farmers for their participation in the present study. Also we acknowledge the *Serviços Distritais de Actividades Económicas (SDAE)* in Moamba and Namaacha for their valuable contribution for the realization of this study.

References

Assan, N. & Sibanda, M. (2014). Goat production in the smallholder section in the Matobo district in semi arid areas of Zimbabwe. *Agricultural Advances*, 3 (8), 218–228.

Banda, J. W., Ayoade, J. A., Karua, S. K. & Kamwanja, L. A. (1993). The local Malawi goat. *In:* Chupin, D., Daldin, J., Roland, N. & Gumprecht, T. (eds.), *Ticks in a changing world*. World Animal Revist. FAO.

Berihu, M., Berhane, G. & Gebrechiristos, S. (2015). Feeding and Management Practices of Free Range Goat Production in Tahtay Koraro District Northern Ethiopia. *American Journal of Social and Management Sciences*, 6 (2), 40–47.

Bett, R. C., Kosgey, I. S., Kahi, A. K. & Peters, K. J. (2009). Analysis of production objectives and breeding practices of dairy goats in Kenya. *Tropical Animal Health Production*, 41, 307–320. doi:10.1007/s11250-008-9191-9.

Boogaard, B. & Moyo, S. (2015). The multi-functionality of goats in rural Mozambique: Contributions to food security and household risk mitigation. ILRI Research Report 37, International Livestock Research Institute (ILRI), Nairobi, Kenya. 30 pp., Available at: https://hdl.handle.net/10568/67395

Boogaard, B. K., Hendrickx, S. C. J. & Swaans, K. (2012). Characterization of smallholder goat production and marketing systems in Inhassoro District, Mozambique: Results of a baseline study. ILRI Research Brief 1. ILRI, Nairobi, Kenya. Available at: https://cgspace.cgiar.org/handle/10568/21698

Braker, M. J. E., Udo, H. M. & Webb, E. C. (2002). Impacts of intervention objectives in goat production within subsistence farming. *South African Journal of Animal Science*, 32 (3), 185–191.

Byaruhanga, C., Oluka, J. & Olinga, S. (2015). Socioeconomic Aspects of Goat Production in a Rural Agropastoral System of Uganda. *Universal Journal of Agricultural Research*, 3, 203–210.

Casey, N. H. & Webb, E. C. (2010). Managing goat production for meat quality. *Small Ruminant Research*, 89, 218–224. doi:10.1016/j.smallrumres.2009.12.047.

Chukwuka, O. K., Okoli, I. C., Okeudo, N. J., Opara, M. N., Herbert, U., Ogbuewu, I. P. & Ekenyem, B. U. (2010). Reproductive Potentials of West African Dwarf Sheep and Goat: A Review. *Research Journal of Veterinary Sciences*, 3, 86–100. doi:10.3923/rjvs.2010.86.100.

Collins-Lusweti, E. (2000). The performance of the Nguni, Afrikander and Bonsmara cattle breeds in developing areas of Southern Africa. *South African Animal Science*, 30, 28–29.

De Vries, J. (2008). Goats for the poor: some keys to successful promotion of goat production among the poor. *Small Ruminants Research*, 77, 221–224.

Demissie, C., Zeleke, M. & Mengistie, T. (2014). Husbandry practices of Western highland goats in Enebse Sar Midir district, East Gojjam Zone, Ethiopia. *Livestock Research for Rural Development*, 26, #137.

Devendra, C. & McLeroy, G. B. (1982). *Goat and Sheep Production in the tropics*. Intermediate tropical agriculture series. Longman, London, UK. 271 pp.

Dossa, L. H., Sangaré, M., Buerkert, A. & Schlecht, E. (2015). Production objectives and breeding practices of urban goat and sheep keepers in West Africa: Regional analysis and implications for the development of supportive breeding programs. *SpringerPlus*, 4, 281. doi: 10.1186/s40064-015-1075-7.

FAO (2012). *Livestock sector development for poverty reduction: an economic and policy perspective – Livestock's many virtues*. by J. Otte, A. Costales, J. Dijkman, U. Pica-Ciamarra, T. Robinson, V. Ahuja, C. Ly & D. Roland-Holst. FAO, Rome.

FAOSTAT (2014). Food and Agriculture Organization Statistics (FAOSTAT), Food and Agriculture 466 Organization Statistics. FAO, Rome, Italy. Available at: http://faostat3.fao.org/home/E (accessed on: 12 Februay 2018).

Garrine, C. M. L. P., Kotze, A., Heleen, E. & Grobler, J. P. (2010). Genetic characterization of the indigenous Landin and Pafuri goat breed from Mozambique. *African Journal of Agricultural research*, 5 (22), 3130–3137.

Guèye, E. F. (2009). Gender issues in family poultry production systems in low income Food deficit countries Gender issues in family poultry production systems in low income food deficit countries. *American Journal of Alternative Agriculture*, 18, 185–195.

Hassen, A. S. & Tesfaye, Y. (2014). Sheep and goat production objectives in pastoral and agro-pastoral production systems in Chifra district of Afar, Ethiopia. *Tropical Animal Health and Production*, 46, 1467–1474.

Hossain, M. S., Akhtar, A., Hossain, M. H., Choudhury, M. P. & Islam, F. (2015). Goat husbandry practices in Southern region of Bangladesh. *Journal of Bioscience and Agriculture Research*, 5 (2), 59–64. doi: 10.18801/jbar.050215.55.

IBM Corp. (2011). IBM SPSS Statistics for Windows, Version 20.0. IBM Corp., Armonk, NY.

INE (2014). Instituto Nacional de Estatística de Moçambique (INE). Censo Agro-Pecuário. 71–78.

Kalema, V. N., Witkowski, E. T. F., Erasmus, B. F. N. & Mwavu, E. N. (2015). The Impacts of changes in Land use on Woodlands in an Equatorial African Savanna. *Land Degration & Development*, 26, 632–641.

Kanani, J., Lukefahr, D. S. & Stanko, R. L. (2006). Evaluation of tropical forage legumes. *Medicago sativa, Dolichos lablab, Leucaena leucocephala* and *Desmanthus bicornutus* for growing goats. *Small Ruminant Research*, 65, 1–7. doi:10.1016/j.smallrumres.2005.04.028.

Kosgey, I. S. & Okeyo, A. M. (2007). Genetic improvement of small ruminants in low-input, smallholder production systems: Technical and infrastructural issues. *Small Ruminant Research*, 70, 76–88. doi: 10.1016/j.smallrumres.2007.01.007.

Kosgey, I. S., Rowlands, G. J., van Arendonk, J. A. M. & Baker, R. L. (2008). Small ruminant production in smallholder and pastoral/extensive farming systems in Kenya. *Small Ruminant Research*, 77, 11–24. doi: 10.1016/j.smallrumres.2008.02.005.

Kumar, S. C. A., Rama Rao, K. & Venkateswarlu, B. (2010). Role of Goats in Livelihood Security of Rural Poor in the Less Favoured Environments. *Indian Journal of Agriculture Economics*, 65 (4), 761–781.

Kusiluka, L. & Kambarage, D. (1996). Diseases of small ruminants – A Handbook. Common Diseases of Sheep and Goats in Sub-Saharan Africa. VETAID, Centre for Tropical Veterinary Medicine Easter Bush Roslin, Scotland. 2–5

Lauritsen, J. M. & Bruus, M. (2005). EpiData (Version 4.0). A comprehensive tool for validated entry and documentation of data. The EpiData Association, Odense, Denmark.

Lovelace, C. E. A., Lungu, J. C. N., Masebe, P. O. C. S., Sakala, B., Nyirenda, I., Sikazwe, G. & Mizinga, K. M. (1993). Reproductive performance of Zambian goats under drought conditions. *In:* Lovelace, C. E. A., Masebe, P. O. C. S., Sakala, B., Nyirenda, I., Sikazwe, G., Mizinga, K. M. & Lungu, J. C. N. (eds.), *Improving the Productivity of Indigenous African Livestock. IAEA-TECDOC-708*. International Atomic Energy Agency (IAEA), Vienna, Austria.

MAE (2005). Perfís dos Distritos de Namaacha, Magude (Maputo) e Distrito de Angónia (Tete). 1-2. Ministério de Administração Estatal, Série Perfís Distritais de Moçambique.

Mahanjana, A. M. & Cronjé, P. B. (2000). Factors affecting goat production in a communal farming system in the Eastern Cape region of South Africa. *South African Journal of Animal Science*, 30 (2), 149–154.

McKinnon, D. & Rocha, A. (1985). Reproduction, mortality and growth of indigenous sheep and goats in Mozambique. *In:* Wilson, R. T. & Bourzat, D. (eds.), *Small ruminants in African agriculture: Proceedings of a conference held at ILCA, Addis Ababa, Ethiopia, 30 September – 4 October 1985.* pp. 154–162, ILCA, Ethiopia.

Mdladla, K., Dzomba, E. F. & Muchadeyi, F. C. (2017). Characterization of the village goat production systems in the rural communities of the Eastern Cape, KwaZulu-Natal, Limpopo and North West Provinces of South Africa. *Tropical Animal Health Production*, 49, 515–527.

Monau, P. I., Visser, C., Nsoso, S. J. & Van Marle-Köster, E. (2017). A survey analysis of indigenous goat production in communal farming systems of Botswana. *Tropical Animal Health Production*, 49, 1265–1271.

Morgado, P. F. (2007). A pecuária no Sul de Moçambique As províncias do sul: Inhambane, Gaza e Maputo. 221-222. Maputo, Moçambique.

Ndlovu, L. & Royer, V. (1988). A comparative study of goat productivity in three regions of Zimbabwe. *In:* Harrison, J. (ed.), *Goat Development Workshop held in Bikita, Masvingo, Zimbabwe. 11–13 January, 1998.* French Embassy, Harare, Zimbabwe.

Nsereko, G., Emudong, P., Mulindwa, H. & Okwee-Acai, J. (2015). Prevalence of common gastro-intestinal nematode infections in commercial goat farms in Central Uganda. *Uganda Journal Agriculture Science*, 16, 99–106.

Oluwatayo, I. B. & Oluwatayo, T. B. (2012). Small Ruminants as a Source of Financial Security: A Case Study of Woman in Rural Southwest Nigeria. IMTFI Working Paper 2.

Onzima, R. B., Gizaw, S., Kugonza, D. R., van Arendonk, J. A. M. & Kanis, E. (2017). Production system and participatory identification of breeding objective traits for indigenous goat breeds of Uganda. *Small Ruminants Research*, 163, 51–59.

Ouchene-Khelifi, N., Ouchene, N., Maftah, A., Da Silva, A. B. & Lafri, M. (2015). Assessing admixture by multivariante analyses of phenotypic differentiation in the Algerian goat livestock. *Tropical Animal Health Production*, 47, 1343–1350.

Payne, W. J. A. (1990). *An introduction to animal husbandry in the tropics.* (4th ed.). ELBS, Singapore. 881 pp.

Rumosa Gwaze, F., Chimonyo, M. & Dzama, K. (2009). Communal goat production in Southern Africa: a review. *Tropical Animal Health Production*, 41, 1157–1168.

Salem, B. H. & Smith, T. (2008). Feeding strategies to increase small ruminant production in dry environments. *Small Ruminant Research*, 77, 174–194. doi: 10.1016/j.smallrumres.2008.03.008.

Seid, A., Kebede, K. & Effa, K. (2015). Breeding Objective, Selection Criteria and Breeding Practice of Indigenous Goats in Western Ethiopia: Implications for Sustainable Genetic Improvement. *Greener Journal of Agricultural Sciences*, 5, 167–176. doi: 10.15580/GJAS.2015.5.072715105.

Semakula, J., Mutetikka, D., Kugonza, R. D. & Mpairwe, D. (2010). Smallholder Goat Breeding Systems in Humid, Sub-Humid and Semi Arid Agro-Ecological Zones of Uganda. *Global Veterinaria*, 4 (3), 283–291.

Simela, L. & Merkel, R. (2008). The contribution of chevon from Africa to global meat production. *Journal Meat Science*, 80, 101–109.

Tadesse, D., Urge, M., Animut, G. & Mekasha, Y. (2014). Perceptions of households on purpose of keeping, trait preference, and production constraints for selected goat types in Ethiopia. *Tropical Animal Health Production*, 46 (2), 363–370. doi: 10.1007/s11250-013-0497-x.

Timberlake, J. & Jordão, C. (1985). Inventory of feed resources for small scale livestock production in Mozambique. *In:* Kategile, J. A., Said, A. N. & Dzowela, B. H. (eds.), *Animal feed resources for small-scale livestock producers - Proceedings of the second PANESA workshop, held in Nairobi, Kenya, 11–15 November 1985.* International Development Research Centre.

van Niekerk, W. A. & Pimentel, P. L. (2004). Goat production in the smallholder section in the Boane district in Southern Mozambique. *South African Journal of Animal Science*, 34, 123–125.

Webb, E. C. & Mamabolo, M. J. (2004). Production and reproduction characteristics of South African indigenous goats in communal farming systems. *South African Journal of Animal Science*, 34, 236–239.

Wilson, R. T. (1989). Reproductive performance of African indigenous small ruminants under various management systems: a review. *Animal Reproduction Science*, 20, 265–286.

13

Impact of agricultural activities on pesticide residues in soil of edible bamboo shoot plantations

Yali Wang [a], Hongshi Yu [b], Wei Gao [c], Liqun Bai [b,*], Jiafu Hu [a,b,*]

[a]Collaborative Innovation Center of Green Pesticide, Zhejiang A & F University, Lin'an, Zhejiang Province, 311300, P.R. China
[b]Zhejiang Provincial Key Laboratory of Chemical Utilization of Forestry Biomass, Zhejiang A & F University, Lin'an, Zhejiang Province, 311300,P.R. China
[c]State Forest Protection Station, Shengyang, 110034, P.R. China

Abstract

Edible bamboo shoot is one of the most important vegetables in Asian countries. Intensive agricultural management measures can cause many negative influences, such as soil acidification and excessive pesticide residues. In the present study, more than 300 soil samples were collected from edible bamboo shoot plantations in six areas throughout Zhejiang province, China, to investigate the soil pesticide pollution and its change after different agricultural activities. Thirteen organic chemicals were detected; nine less than that detected during a similar study executed in 2003–2004. All the detected residues were far below the Chinese national environmental standards for agricultural soils. The pesticide residues in bamboo plantations showed a decline over the past decade. Organic materials used for mulching and plantation's background of being formerly a paddy field are two important factors increasing the pesticide residues. Conversely, lime application to acidified soil and mulching with uncontaminated new mountain soil could decrease the residues significantly. Our results indicated that the current agricultural activities are efficient in reducing pesticide residues in the soil of bamboo shoot plantations and should be further promoted.

Keywords: agricultural activity, bamboo plantation, remediation of soil, soil pollution

1 Introduction

Bamboo shoots are new culms that come out of the ground of many bamboo species, such as *Bambusa vulgaris* Schrad. and *Phyllostachys edulis* (Carrière) J. Houz. Due to its rich nutrients, crisp texture and delicious taste, bamboo shoots have been used as vegetables for thousands of years in numerous Asian countries. In China, *Phyllostachys praecox* f. *prevernalis* Chen et Yao (Lei bamboo) is one of the most popular species for bamboo shoot production (Jiang *et al.*, 2002). Since the 1990s, Lei bamboo has been transplanted and cultivated in large areas and many paddy fields have been transformed to bamboo shoot cultivation in several provinces, and particularly in Zhejiang because of the

increase in market demand (Fang *et al.* 1994; Zhou *et al.*, 1998).

To gain higher profits, intensive agricultural management measures have been adopted for bamboo shoot production, like weeding, chemical fertiliser application, as well as pesticide usage for pest control. In addition, bamboo plantations are mulched with a thick layer of straw, bran or some other organic material before the shoot harvest in winter time (Huang *et al.*, 2007). Through mulching, the bamboo shoot yield could be brought forward with two months, in order to coincide with the Chinese Spring Festival and obtain highest prices. On the other hand, however, these intensive management measures have caused many problems, including soil acidification, declining shoot yield, pest infestation, organic pollution of the soil, and excessive pesticide residues in the bamboo shoots (Guo *et al.*, 2011; Gui *et al.*, 2013).

* Corresponding authors
Liqun Bai (bailiqun78@163.com); Jiafu Hu (hujiafu2000@163.com)

In order to achieve a sustainable production and to guarantee the quality of bamboo shoots, several farmer cooperative societies were established, and professional experts were hired to provide technical support to the farmers. By these means, several novel agricultural activities have been extensively carried out in the past decade, such as using low-toxicity and easily degradable pesticides, applying lime to increase pH in acidified soil, and mulching the bamboo plantation with new mountain soil (Zhuang *et al.*, 2014).

In the present paper, a field survey on organic pollutants, including organochlorine pesticides (OCPs), organic phosphorus pesticides (OPPs) and pyrethroids (PYs) was conducted during 2015–2016 in bamboo plantations from 6 major Lei bamboo shoot production areas in Zhejiang province. The pollution was analysed with respect to agricultural activities and also compared to historical data collected in the same areas during 2003–2004. Pesticide residues in bamboo shoots were not further analysed since these were all lower than the detection lines and not detectable from the samples collected in 2015–2016. The aims of this work were to examine the current state of organic pollutants in bamboo plantation soils, to elucidate their spatiotemporal dynamic characteristics, and to assess the efficacy of different agricultural activities for decreasing soil pesticide residues.

2 Materials and methods

2.1 Soil sampling

A total of 349 surface soil samples were collected from the six major bamboo shoot production counties in Zhejiang province (Fig. 1). All the bamboo plantations selected for soil sampling were once investigated in 2003–2004. Bamboo plots located in hilly uplands are usually sloping fields. In contrast, those fields that were originally paddy farmlands are relatively small, flat fields and are distributed separately on terraces or platforms as a result of land fragmentation.

The sampling plots were defined according to their former use [former paddy field (FPF) or hilly upland (HU)] and the different soil management measures previously conducted: non-mulched (NOMM) or organic material mulched (OMM), lime applied (LA), non-lime applied (NLA), new mountain soil mulched (NSM) plots and non-NSM. Soil samples taken from the surface (0–20 cm in depth) at five random points in each plot were thoroughly mixed to form a composite sample. One such sample was collected from each plot. The number of soil samples collected from the different types of plots is listed in Table 1.

Fig. 1: *Location of the six sampling areas in Zhejiang Province (shown divided into prefectures).*

2.2 Sample preparation, extraction and clean-up

The soil moisture content was determined by weighing subsamples of the soils before and after drying for ca. 20 h at 105 °C. The samples were air-dried at 30 °C and sieved through a 1 mm polyethylene sieve. The prepared soil samples were sealed in polyethylene bags and stored at −20 °C for further analysis.

Pesticide residues in all samples were extracted, cleaned up and analysed using the national standard methods (AQSIQ, 2003a, b), which were also used in 2003–2004 for data collection. For OPPs, the extraction and clean-up procedures were as follows: 20 g of the air-dried soil sample was extracted for 8 h with 100 ml acetone. After 4 h oscillation, the mixture was passed through a Buchner funnel with vacuum suction filtration packed with one layer of filter aid agent (celite 545) and several layers of filter paper. The eluate was added with 10 ml coagulation liquid (20 g NH_4Cl + 40 ml 85 % H_3PO_4 at constant volume of 2000 ml with pH 4.5–5.0) and 1 g filter aid agent (celite 545), oscillated 1 min, stood for 3 min and filtered. The eluate was then extracted 3 times with 50 ml dichloromethane and passed through a funnel packed with 1 g anhydrous sodium sulphate and 1 g celite 545. The solvent extracts were collected in a 250 ml flat bottom flask, added with 0.5 ml ethyl acetate and concentrated to ca. 3 ml with a rotating evaporator. The

Table 1: *Number of soil samples collected from different types of plots.*

Site	Location	OMM Plantation					NOMM Plantation	
		FPF			HU			
		LA	NLA		LA	NLA	FPF	HU
			NSM	non-NSM				
1	Deqing	20	4	6	7	8	10	6
2	Linan	20	5	7	7	6	12	7
3	Fuyang	20	3	8	8	6	8	6
4	Longyou	20	3	3	6	9	9	6
5	Jinyun	21	6	7	2	9	7	6
6	Qingyuan	17	5	6	2	8	7	6
Total		118	26	37	32	46	53	37

Note: *OMM*, organic material mulched; *NOMM*, non-mulched; *FPF*, former paddy field; *HU*, hilly upland; *LA*, lime applied; *NLA*, non-lime applied; *NSM*, new mountain soil mulched.

residue was dried with a nitrogen flux and then dissolved in 5 ml acetone for further chromatographic analysis.

For OCPs and PYs, a 20 g soil sample was mixed with 4 g diatomite and 100 ml acetone/petroleum ether (1 : 1, v/v) for 12 h. Then residues were extracted for 4 h at 80 °C in a soxhlet extractor. After being cooled down, the solvent were added with 100 ml sodium sulfate solution, oscillated for 1 min and stood for several min until getting different layers. The acetone solution in the lower layer was discarded and the supernatant solution was eluted through a Florisil column with 100 ml petroleum ether/ethyl acetate (95 : 5, v/v). The eluate was collected, condensed with a rotating evaporator, dissolved in petroleum ether and finally reduced to 1 ml with a nitrogen flux.

2.3 Chromatographic analysis

All standard chemicals, including HCHs, DDX, quintozene, chlorothalonil and dicofol were supplied by J&K Chemica, Beijing, China and used as references. For OPPs, the samples were analysed using an Agilent Gas Chromatograph 7890A equipped with a flame photometric detector (FPD) and a DB-1701 column (30 m × 0.32 mm i.d. × 0.25 μm film thickness). The samples were injected by auto-sampling in the splitless mode with a venting time of 1 min. The oven temperature was programmed to increase from 100 to 200 °C at a rate of 10 °C min^{-1}, then to 250 °C at 25 °C min^{-1} and maintained for 10 min. Nitrogen was used as the carrier gas (2 ml min^{-1}) and make-up gas (25 ml min^{-1}). The injector and detector temperatures were 200 and 220 °C, respectively. For OCPs and PYs, samples were analysed using

an Agilent Gas Chromatograph 6890N and 5973-MSD. A HP-5 column 30 m × 0.32 mm i.d. × 0.25 μm film thickness was used. The samples were injected by auto-sampling in the splitless mode with a venting time of 2 min. The temperature program was as follows: initial temperature held at 50 °C for 2 min, increased to 150 °C at a rate of 10 °C min^{-1}, then to 240 °C at 3 °C min^{-1} and maintained for 28 min. Nitrogen was used as the carrier gas (1 ml min^{-1}) and make-up gas (37.25 ml min^{-1}). The injector and detector temperatures were 240 and 280 °C, respectively.

2.4 Quality control

The recovery rates of pollutants with samples fortified by pesticides (0.032–4.000 μg in weight for different ones) ranged from 88.95 % (methamidophos) to more than 100 % (γ-HCH, chlorothalonil, chlorpyrifos, parathion-methyl and parathion). The detection limits for the soil samples ranged from 0.05 (α-HCH) to 3.0 (malathion) ng g^{-1} (Table 2). A procedural blank was run with every set of twenty samples to check for contamination. Three replicates for every soil sample were performed to reduce error levels in the analysis.

2.5 Statistical analysis

Average concentrations of pesticide residues were calculated by geometric mean. Concentrations from the same type of plantation were tested by normal distribution. All data were in line with normal distribution and homogeneity of variance, and were analysed using single-factor analysis of variance (ONE-WAY ANOVA) (SAS OnlineDoc®, Version 8.01, Statistical Analysis System Institute, Cary, NC).

Table 2: *Organic pollutants in soils collected from bamboo shoot plantations.*

		Concentration (ng g^{-1})			Occurrence rate	Detecting limit	Recovery
		Lowest	Highest	Average†	(%)	(ng g^{-1})	(%)
OCPs‡	α-HCH	0.17 (0.31)	0.24 (0.61)	0.20 (0.38)	16.62 (33.03)	0.05	96.01
	β-HCH	0.10 (0.19)	9.66 (24.79)	1.49 (2.44)	35.24 (77.46)	0.08	99.40
	γ-HCH	0.55 (0.59)	14.22 (37.61)	2.35 (5.01)	32.38 (93.51)	0.07	100.40
	δ-HCH	1.17 (3.11)	2.19 (4.22)	1.27 (3.35)	6.88 (11.46)	0.18	96.22
	$\sum$ HCH	0.22 (0.58)	17.13 (35.57)	2.43 (6.03)	46.70 (100)		
	p,p'-DDT	0.52 (0.78)	1.82 (3.73)	0.91 (1.43)	11.75 (37.81)	0.49	93.71
	p,p'-DDE	0.36 (0.73)	3.13 (5.66)	1.85 (2.46)	25.47 (36.55)	0.17	98.20
	p,p'-DDD	N.D. (0.51)	N.D. (2.31)	N.D. (0.92)	N.D. (43.68)	0.48	99.65
	o,p'-DDT	N.D. (0.24)	N.D. (6.12)	N.D. (1.91)	N.D. (37.81)	0.19	92.87
	$\sum$ DDT	0.24 (0.36)	4.33 (8.82)	1.88 (2.55)	31.23 (83.10)		
	Quintozene	0.30 (0.30)	0.48 (0.66)	0.29 (0.53)	16.33 (79.58)	0.20	99.60
	Chlorothalonil	0.62 (1.55)	27.67 (72.25)	8.83 (20.20)	24.93 (67.23)	0.30	103.41
	Dicofol	0.96 (1.62)	32.62 (118.32)	2.35 (17.16)	17.19 (72.54)	0.80	101.00
	$\sum$ OCP	1.01 (2.48)	61.25 (151.35)	11.16 (33.23)			
OPPs§	Chlorpyrifos	2.8 (2.8)	200.00 (421.22)	23.13 (74.14)	47.85 (83.32)	2.00	103.41
	Methamidophos	N.D. (13.22)	N.D. (37.57)	N.D. (23.23)	N.D. (64.78)	1.00	88.95
	Dichlorvos	N.D. (4.12)	N.D. (68.90)	N.D. (36.77)	N.D. (4.67)	1.00	94.88
	Malathion	N.D. (3.61)	N.D. (37.22)	N.D. (4.66)	N.D. (7.12)	3.00	98.79
	Methidathion	N.D. (2.33)	N.D. (11.15)	N.D. (3.24)	N.D. (4.56)	2.00	97.89
	Parathion-methyl	N.D. (1.21)	N.D. (26.65)	N.D. (4.38)	N.D. (97.12)	1.00	108.43
	Parathion	N.D. (1.53)	N.D. (3.69)	N.D. (2.32)	N.D. (36.61)	1.00	102.41
	Dimethoate	1.46 (3.17)	9.85 (26.62)	4.31 (14.45)	4.87 (9.12)	1.00	97.39
	$\sum$ OPP	1.52 (2.11)	203.65 (518.12)	27.29 (82.13)			
PYs¶	Cypermethrin	1.27 (1.33)	5.22 (4.21)	1.76 (2.32)	16.33 (20.13)	1.00	98.89
	Fenvalerate	2.46 (5.84)	132.36 (312.25)	16.61 (123.38)	18.91 (71.16)	2.00	97.39
	Deltamethrin	N.D. (2.01)	N.D. (5.17)	N.D. (2.41)	N.D. (3.73)	2.00	95.88
	$\sum$ PY	2.26 (2.96)	134.56 (312.25)	16.92 (147.55)			

Note: data in parentheses are historical data collected during 2003–2004 (unpublished). These original data were obtained from Forestry Bureau of Zhejiang Province. N.D. means not detectable. Detecting limit and recovery refer to data collected during 2015–2016.
† Average means the geometric mean. ‡ OCP represents organochlorine pesticides, § OPP organic phosphorus pesticides and ¶ PY pyrethroid.

3 Results

3.1 Concentration of pesticide residues

The results are summarized in Table 2. A total of 22 compounds were investigated. Residues of nine OCPs, including four HCHs, two DDX and three others; two OPPs and two PYs were detected in the surface soil samples. Both the concentration and occurrence rate of all these residues were obviously lower than those found during 2003–2004. Moreover, nine residues (p,p'-DDD, o,p'-DDT, deltamethrin, dichlorvos, malathion, methamidophos, methidathion, parathion and parathion-methyl), which had been detected 10 years earlier, were not detectable in our study. Our data showed a dramatic decrease in the pesticide residues during the last 10 years.

From Table 2, it can be seen that DDX and HCHs were still detectable although these pesticides have been banned for more than 30 years in China (since 1983), indicating their persistent-polluting characteristics in soil. The detected values are all below the Chinese national environmental standards for agricultural soil: < 0.05 mg kg^{-1} for $\sum$ DDT and < 0.01 mg kg^{-1} for $\sum$ HCH (Environmental Quality Standard for Soils of China, GB15618-2008).

3.2 Effects of mulching with organic material

The samples collected on plantations mulched with organic material (OMM) were compared with those from non-mulched (NOMM) plantations. However, plantations which were seriously deteriorated due to long-term intensive management and mulched with new mountain soil were not

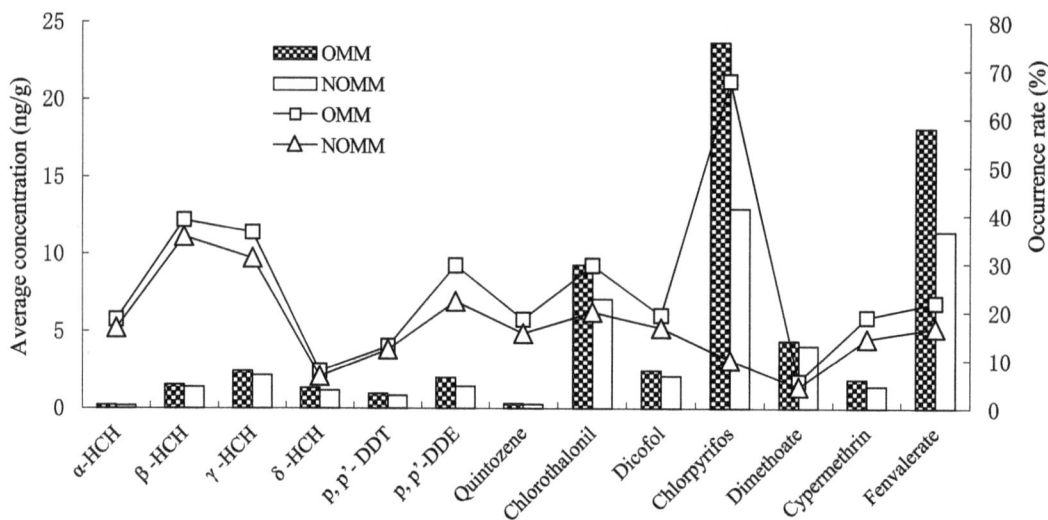

Fig. 2: *The mean concentrations and occurrence rates of residues from different bamboo shoot production plots classified according to mulching. (OMM: organic material mulched plots; NOMM: non-mulched plots).*

included in the analysis because we used the surface soil for residue analysis and the effect of new mulched soil on residues was significant. So the samples were divided into two groups based on organic material mulching, one for the OMM ($n = 233$) and the other for the NOMM ($n = 90$) (see Table 1). The results of the residue analysis are presented in Fig. 2.

From Fig. 2, it can be seen that the concentration of pesticide residues found on OMM plantations were nearly all higher than those from NOMM plantations; the differences in chlorothalonil ($p = 0.042$), chlorpyrifos ($p = 0.0001$), and fenvalerate ($p = 0.006$) were significant. A similar picture was found for the occurrence rates of pesticide residues.

3.3   Effects of lime application

In order to control soil acidification, lime is extensively used by farmers in their plantations, which have been mulched with straw, bran or other organic materials for several years. We divided the samples from OMM plantations into two groups based on lime application, one for the lime applied (LA) ($n = 150$) and the other for non-lime applied (NLA) ($n = 83$) (see Table 1). Again, the samples from plantations which were mulched with new mountain soil were not included. The results are presented in Fig. 3.

From Fig. 3, we can see that the concentrations of all types of residues in LA plantations were significantly lower than those from NLA plantations ($p = 0.037$ for $\sum$ OCP, 0.006 for $\sum$ OPP and 0.028 for $\sum$ PY). The soil pH values were 3.89 ± 0.17 and 6.73 ± 0.14 in NLA and LA plantations respectively and significant differences were detected ($F = 27.62$, df $= 1$, $p = 0.0186$). Our results show that lime

can increase the soil pH value and accelerate the decomposition of residues.

3.4   Effects of land-use history

Before 1990s, nearly all bamboo fields were planted in hilly upland and most lowland areas were used to grow rice in Zhejiang province. From the beginning of 1990s, some of the paddy farmlands have been changed and used for bamboo cultivation because of the better economic benefits. Currently, more than 60 % of all Lei bamboo shoot plantations have been planted on former paddy fields (FPF). The soil samples collected from FPF plantations ($n = 208$) were compared with those from hilly uplands (HU) ($n = 115$). Again, the samples from plantations which were mulched with new mountain soil were not included. The results are presented in Fig. 4.

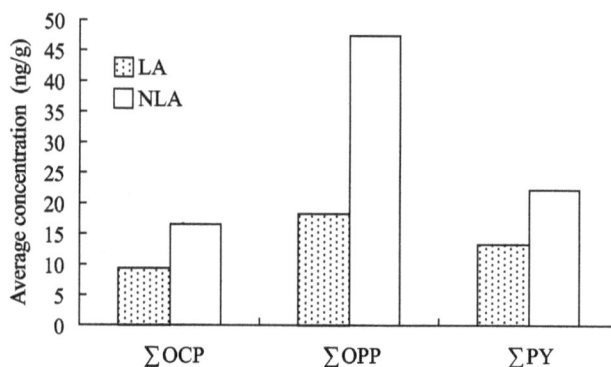

Fig. 3: *The average concentrations of different types of residues from bamboo shoot production plots classified by lime application (LA: residues from lime applied plots; NLA: residues from non-lime applied plots; OCP: organochlorine pesticides; OPP: organic phosphorus pesticides; PY: pyrethroid).*

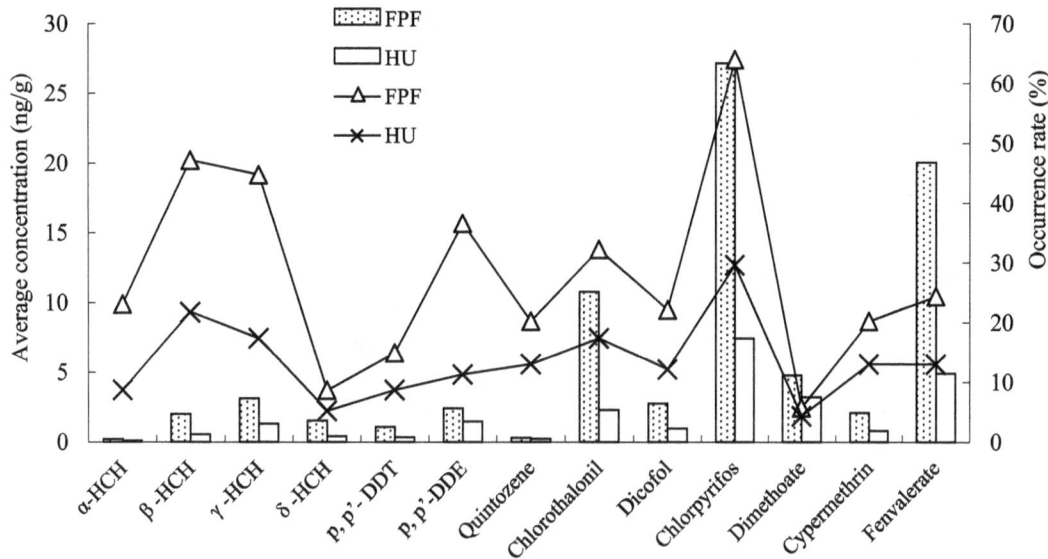

Fig. 4: *The mean concentrations and occurrence rates of residues from bamboo shoot production plots classified by different land-use history (FPF: former paddy field; HU: hilly upland).*

From Fig. 4, it was clear that both the average concentration and occurrence rate of pesticide residues were higher in FPF plantations than those in HU plantations. For eleven residues, significant differences were observed between the plantations with different land use histories (α-HCH: $p = 0.048$; β-HCH: $p = 0.016$; γ-HCH: $p = 0.035$; δ-HCH: $p = 0.022$; p,p'-DDT: $p = 0.030$; p,p'-DDE: $p = 0.043$; chlorothalonil: $p = 0.0001$; chlorpyrifos: $p = 0.0006$; cypermethrin: $p = 0.036$; dicofol: $p = 0.028$; fenvalerate: $p = 0.0003$).

3.5 Effects of mulching with new mountain soil

After several years of mulching with organic materials, the yield of bamboo shoot declined seriously because the soil properties were changed with excessive organic-matter input (Huang *et al.*, 2007; Gui *et al.*, 2013). In such cases, uncontaminated soils from mountain areas were used to mulch the seriously deteriorated plantations with a layer in thickness of about 15 cm. After the reestablishment of uncontaminated mountain soil system, yield of bamboo shoots can return to the normal level. In this study, no pesticide residue was detected for the samples collected from plantations mulched with new mountain soil. Although the pesticide residues in this soil are not really degraded and just buried in the underlying soil, the bamboo root tends to grow upward into the newly added surface soil, reducing the final concentration of residues in the edible bamboo shoot.

4 Discussion

Thirteen types of pesticide residues were detected in the surface soil from bamboo plantations in six major bamboo shoot producing areas of Zhejiang province. During the 2003–2004 study (unpublished data), nine further residues were detected. Both the concentrations and occurrence rates of the detected residues were reduced. These results indicate that pesticide residues in bamboo plantations have decreased over the past 10 years. Feng *et al.* (2003) reported that HCH would decrease faster than DDT in soil-water system of lake, but our results showed that both the concentrations and occurrence rates of HCH residues were higher than those of DDT in bamboo plantations, despite the fact they were banned at the same time in 1983 in China. This seems to indicate that HCH was not completely stopped in the first few years after the ban in Zhejiang province. For DDX, p,p'-DDT and p,p'-DDE were detected, but p,p'-DDD and o,p'-DDT were not detectable in our samples. The possible reason might be that under the conditions found in bamboo plantations, p,p'-DDT and p,p'-DDE are more difficult to decompose and more persistent than p,p'-DDD and o,p'-DDT, a case which has been found by others (Aguilar, 1984; Gong *et al.*, 2004). In China, methamidophos, parathion-methyl, and parathion have been banned since 2008. This might be the reason why the residues of these three pesticides were not detected in our study. In contrast to DDX and HCHs, other pesticides are still widely used in China as these decompose easier. For example, chlorpyrifos is intensively used for controlling the rice borer pest, *Tryporyza incertulas*. Straw and bran with residues of chlorpyrifos are used as organic materials to mulch bamboo plantations and this might have led to chlorpyrifos being the highest residue in our study.

The kind of organic material used for mulching the bamboo plantation was an important factor in increasing the residues in the soil. For all of the detected residues, both the concentration and occurrence rate were higher in the samples from OMM plantations than those from NOMM plantations. Significant differences were found with chlorothalonil, chlorpyrifos and fenvalerate. The possible reason might be that mulching with organic materials helped subterranean pest insects to overwinter, causing infestation in the next year, like *Melanotus cribricollis* (Shu *et al.*, 2012). The concomitant pest control measures increased the usage of pesticides, leading to a higher level of residues. In addition, organic materials used for mulching might contain some pesticides per se and so increase the residues in soil as shown above for chlorpyrifos. Furthermore, long-term mulching with organic materials could change the soil properties and decrease its pH value (Naramabuye & Haynes, 2006; Gui *et al.*, 2013). An acidic environment might slow down the degradation rate of some pesticides, such as fenamiphos, chlorpyrifos, DDT and its metabolites (Singh *et al.*, 2003a, b; Gong *et al.*, 2004).

Compared to hilly-upland bamboo plantations, residues were significantly increased in soil samples from plantations that had formerly been paddy fields, implying that the historical background of a plantation is an important factor affecting the level of residues. The residues left from the original paddy field might contribute to the differences found in our investigation, especially for persistent organic pollutants like HCHs, DDT and its metabolites. In another study, it was found that the population of subterranean pest insects in FPF plantations is much higher than that in HU plantations (Shu *et al.*, 2012). This causes the application of more chemicals for pest control, increasing the difference in residue concentrations. Also the higher runoff in sloping HU plantations could have contributed to the residue differences found. In contrast, lime application in the acidified soil was found to reduce the residues efficiently because organic pesticides tend to decompose faster with an increase in pH. For example, the half-life of imazapyr has been negatively correlated with the pH of soil: with an increase of alkaline, the degradation rate of imazapyr is faster (Wang *et al.*, 2004). Similar results have also been found in other studies (Bobé *et al.*, 1998; Singh *et al.*, 2003a, b; Gunasekara *et al.*, 2007; Gianelli *et al.*, 2014; Martin-Reina *et al.*, 2017). On the other hand, however, exogenous lime might also affect the balance of soil microbial flora, which could adversely reduce the function of microbes in decomposing pesticides. Therefore, further research is necessary

to be conducted on the reasonable application of lime in Lei bamboo shoot plantations. Moreover, no residues were detected in surface soil samples collected from plantations mulched with uncontaminated mountain soil, implying that such an application on plantations deteriorated through excessive mulching is an efficient measure, although pesticide residues are just buried in the deeper soil layers.

Our findings inferred that the present-day agricultural activities suggested by professional experts, such as application of lime and mulching with uncontaminated mountain soil, as well as using chlorpyrifos, chlorothalonil and malathion instead of methamidophos, parathion-methyl and parathion as pesticides, were efficient in reducing the pesticide residues in soil of bamboo plantations. These agricultural activities could be applied extensively or used as reference for other bamboo shoot producing areas.

Competing interests

The authors declare that they have no conflict of interest.

Author contributions

JH designed the study, and drafted the manuscript. YW carried out the soil sampling, sample preparation and chromatographic analysis. HY performed in the statistical analysis. WG participated in the sample collection, preparation and chromatographic analysis. LB participated in the design of the study and helped to draft the manuscript. All authors read and approved the final manuscript.

Acknowledgements

We are greatly indebted to Huiming Wu for his technical assistance in the chromatographic analysis. We thank Zhenghui Dong for his kind help in data calculation and Teresa Gatesman for her English language modification. We would also like to thank the anonymous referees for providing constructive comments and help in improving the contents of this paper. This work was financially supported by the Major Project of Zhejiang Province Cooperating with the Chinese Academy in Forestry (2013SY12).

References

Aguillar, A. (1984). Relationship of DDE/DDT in marine mammals to the chronology of DDT input to the ecosystem. *Canadian Journal of Fisheries and Aquatic Sciences*, 41 (6), 840–844. doi:10.1139/f84-100.

AQSIQ (2003a). *GB/T 14550-2003, Method of gas chromatographic for determination of BHC and DDT soil.* General Administration of Quality Supervision, Inspection and Quarantine of China (AQSIQ), Standards Press of China, Beijing. (in Chinese)

AQSIQ (2003b). *GB/T 14552-2003, Method of gas chromatographic for determination of organophosphorus pesticides in water and soil.* General Administration of Quality Supervision, Inspection and Quarantine of China (AQSIQ), Standards Press of China, Beijing. (in Chinese)

Bobé, A., Meallier, P., Cooper, J.-F. & Coste, C. M. (1998). Kinetics and mechanisms of abiotic degradation of fipronil (hydrolysis and photolysis). *Journal of Agricultural and Food Chemistry*, 46(7), 2834–2839. doi: 10.1021/jf970874d.

Fang, W., He, J. C., Lu, X. K. & Chen, J. H. (1994). Cultivation techniques of early shooting and high yielding for Lei bamboo sprout. *Journal of Zhejiang Forestry College*, 11, 121–128. (in Chinese)

Feng, K., Yu, B. Y., Ge, D. M., Wong, M. H., Wang, X. C. & Cao, Z. H. (2003). Organo-chlorine pesticide (DDT and HCH) residues in the Taihu Lake Region and its movement in soil–water system. I. Field survey of DDT and HCH residues in ecosystem of the region. *Chemosphere*, 50(6), 683–687. doi:10.1016/S0045-6535(02)00204-7.

Gianelli, V. R., Bedmar, F. & Costa, J. L. (2014). Persistence and sorption of imazapyr in three Argentinean soils. *Environmental Toxicology and Chemistry*, 33(1), 29–34. doi:10.1002/etc.2400.

Gong, Z. M., Tao, S., Xu, F. L., Dawson, R., Liu, W. X., Cui, Y. H., Cao, J., Wang, X. J., Shen, W. R., Zhang, W. J., Qing, B. P. & Sun, R. (2004). Level and distribution of DDT in surface soils from Tianjin, China. *Chemosphere*, 54(8), 1247–1253. doi: 10.1016/j.chemosphere.2003.10.021.

Gui, R. Y., Sun, X. & Zhuang, S. Y. (2013). Soil acidification in *Phyllostachys praecox* f. *preveynalis* cultivation with intensive management. *Communications in Soil Science and Plant Analysis*, 44, 3235–3245.

Gunasekara, A. S., Truong, T., Goh, K. S., Spurlock, F. & Tjeerdema, R. S. (2007). Environmental fate and toxicology of fipronil. *Journal of Pesticide Science*, 32(3), 189–199. doi:10.1584/jpestics.R07-02.

Guo, Z. W., Chen, S. L. & Xiao, J. H. (2011). Organochlorine pesticide residues in bamboo Shoot. *Journal of the Science of Food and Agriculture*, 91(3), 593–596. doi: 10.1002/jsfa.4233.

Huang, M. Z., Chen, J. H., Wang, L. Z., Wu, J. X., Hu, Z. P. & Zhu, X. (2007). The reconstruction of Lei bamboo in

degradation. *Applied Techniques in Forestry*, 11, 12–13. (in Chinese)

Jiang, P. K., Zhou, G. M. & Xu, Q. F. (2002). Effect of intensive cultivation on carbon pool of soil in Phyllostachys praecox stands. *Scientia Silvae Sinicae*, 38, 6–11. doi:10.11707/j.1001-7488.20020602. (in Chinese)

Martin-Reina, J., Duarte, J. A., Cerrillos, L., Bautista, J. D. & Moreno, I. (2017). Insecticide reproductive toxicity profile: organophosphate, carbamate and pyrethroids. *Journal of Toxins*, 4(1), 1–7. doi: 10.13188/2328-1723.1000019.

Naramabuye, F. X. & Haynes, R. J. (2006). Effect of organic amendments on soil pH and Al solubility and use of laboratory indices to predict their liming effect. *Soil Science*, 171(10), 754–763. doi: 10.1097/01.ss.0000228366.17459.19.

Shu, J. P., Teng, Y., Chen, W. Q., Shi, J., Liu, J., Xu, T. S. & Wang, H. J. (2012). Control techniques of *Melanotus cribricollis* (Coleoptera: Elateridae). *Forest Research*, 25, 620–625. (in Chinese)

Singh, B. K., Walker, A., Morgan, J. A. W. & Wright, D. J. (2003a). Effect of soil pH on the biodegradation of chlorpyrifos and isolation of a chlorpyrifos-degrading bacterium. *Applied and Environmental Microbiology*, 69(9), 5198–5206. doi:10.1128/AEM.69.9.5198-5206.2003.

Singh, B. K., Walker, A., Morgan, J. A. W. & Wright, D. J. (2003b). Role of soil pH in the development of enhanced biodegradation of fenamiphos. *Applied and Environmental Microbiology*, 69(12), 7035–7043. doi: 10.1128/AEM.69.12.7035-7043.2003.

Wang, X. D., Zhou, H. B., Wang, H. L. & Fan, D. F. (2004). Studies on degradation dynamics of imazapyr and its influencing factors in different kinds of soil. *Chinese Journal of Pesticide Science*, 6, 53–57. (in Chinese)

Zhou, G. M., Jin, A. W., Zhen, B. S. & Fang, W. H. (1998). Structure analysis of Lei bamboo stand with protective cultivation. *Journal of Zhejiang Forestry College*, 15, 111–115. (in Chinese)

Zhuang, S. Y., Ji, H. B., Cheng, L., Li, G. D. & Gui, R. Y. (2014). Effect of liming on loss of nitrogen and phosphorus in soil of *Phyllostachys praecox* cv. *prevernalis* stand. *Journal of Zhejiang Forestry Science & Technology*, 34, 68–71. (in Chinese)

14

Intensification of rain-fed groundnut production in North Kordofan State, Sudan

Elgailani Adam Abdalla [a], Jens B. Aune [b,*], Abdelrahman K. Osman [a], Aldaw M. Idris [a]

[a] ElObeid Research Station, ElObeid, Sudan
[b] Department of International Environment and Development Studies, Norwegian University of Life Sciences, Aas, Norway

Abstract

The main objective of this study was to evaluate intensification pathways for groundnut production in the marginal rain-fed environment of North Kordofan State, Sudan. The effect of intensification on yields was assessed in three different experiments. In the first experiment, the treatments were organised according to increasing level of intensification from the traditional production package to the improved production package (the ladder experiment). The complete improved package in the ladder approach consisted of increased density, new variety, seed priming, micro-dosing (0.6 g NPK per pocket) and mulching. Three levels of mulching and two levels of intensification constituted the second experiment (mulching experiment), while the third was an on-farm experiment involving 20 farmers testing two levels of intensification (on-farm experiment). The average yield increases were 75, 61, and 32 %, from the ladder, mulch and on-farm experiments, respectively. Results from the ladder experiment showed that farmers' gross margin increased by 83 % compared to traditional practices. Resource limited farmers can increase yield by 18 % and gain 25 % additional cash incomes by only adopting increased plant density. As farmers differ in their wealth status, they can choose low-cost, low-risk components of the technological package whereas farmers with more access to resources can achieve high cash incomes by adopting the complete improved production package. There was no clear effect of mulching on yields in these experiments.

Keywords: plant density, variety, seed priming, micro-dosing, mulching, profitability

1 Introduction

Crop production in the sandy rain-fed areas of North Kordofan State in Sudan is mainly practised by small-holder, resource-limited farmers who constitute 75–80 % of the population. The environment is characterised by low and erratic seasonal rainfalls, short growing periods, and poor soil conditions with very low extractable phosphorus (4 mg / 100 g), low carbon (0.056 %) and low nitrogen (0.03 %) due to continuous cultivation. Kordofan State is generally classified as one of the most vulnerable areas of Sudan due to frequent droughts (NAPA, 2007). Average yield for groundnut is estimated at 184 kg ha^{-1} (NKS-

MARD, 2014). Farmers expand cultivated areas to increase production in order to attain food security and improve their livelihoods. Agriculture is practiced without improved technologies such as the use of fertilisers, irrigation, water conservation or improved high yielding and disease resistant varieties. In West Africa, crop and soil management research has shown that the introduction of high yielding and disease resistant improved crop varieties accompanied by appropriate crop management practices can more than double sorghum and pearl millet grain yields while reducing soil and organic carbon depletion (Bagayoko *et al.*, 1996; Coulibaly *et al.*, 2000). In Sudan, Elobeid Research Station has recommended improved high yielding and disease resistant crop varieties and crop husbandry practices for increasing the yield of the major crops grown on sandy soils

* Corresponding author – jens.aune@nmbu.no

under marginal rain-fed conditions. Yield of groundnut crop was increased by 20 % due to improved varieties (Abdalla & AlAhmedi, 1997). Elements of the production package were individually evaluated in groundnut cultivation using the 'minus one' trial technique. The study showed that farmers should plant early in order to obtain high yields. However, since that study, new technologies such as seed priming and NPK fertiliser micro-dosing have been recommended (Osman & Elamin, 2000). Therefore, there was a need to test and evaluate the complete production package as well as the effect of each component individually. The approach used in this study was the ladder approach in which the treatments are ordered according to increasing levels of intensification. It was expected that yields increase as intensification levels increase, but so will the cost and associated risks. Farmers can choose different levels of intensification depending on the available resources. As farmers get more resources and confidence in the technologies, they can "climb" the intensification ladder and reap more benefits. The ladder approach has previously identified low-cost, low-risk technologies for food crops in West Africa that could be adopted by resource limited and wealthy farmers (Aune & Bationo, 2008). Resource-efficient technologies can make it more appealing for farmers to intensify production rather than to clear land for cultivation. The study assessed if increasing levels of intensification can increase yields and gross margins in groundnut production under the marginal rain-fed environment of North Kordofan State, Sudan.

2 Materials and methods

Studies were conducted during the cropping seasons of 2013/2014 and 2014/2015 on sandy soil at Elobeid Research Station farm (30°14'13.230" E; 13°12'36.801" N) in Elobeid City, North Kordofan State, Sudan. As shown in Table 1, the treatments were organised according to increasing levels of intensification. Treatment 1 represents farmers' own production package, while treatment 6 represents a complete improved production package. In treatments 2–5, one component of farmers' own production package was replaced by a component of the improved package in ascending order.

Farmers' current production package consisted of a popular variety (Barberton), conventional spacing (40 cm × 40 cm) and single seed per hole corresponding to 62,500 seeds ha^{-1}, no priming, no micro-dosing and no mulching, while the complete improved production package consisted of one improved variety (Gubeish), increased density (60 cm × 20 cm) and two seeds per hole corresponding to 166,666 seeds ha^{-1}, seed priming, NPK micro-dosing (0.6 g / hole) and sorghum straw mulching (0.5 ton ha^{-1}). The Gubeish and the Barberton varieties mature in 85 and 100 days, respectively and the yield is about 20 % higher for Gubeish compared to Barberton. Seed priming consists of soaking the seeds in water for 8 hours, followed by air-drying at room temperature for one hour before being sown. The treatments were laid out in a randomized complete block design (RCBD) with four replications.

Simultaneously, a second experiment was carried out to test the effect of mulching on groundnut under rain-fed conditions. This experiment was a 3 × 2 factorial, consisting of three mulching levels (zero, 0.5 ton ha^{-1}, and 1 ton ha^{-1}) and two production packages (farmer practice and improved package). The improved production package included seed priming, micro-dosing of 0.6 g NPK fertiliser per planting pocket and the improved variety (Gubeish), while the farmers' package was without these applications. Treatments were laid out in a RCBD with four replications and a plot size of 6 rows, each 5 m long. The fertiliser used was NPK 15-15-15, applied with the seeds at planting as a micro-dose of 0.6 g per hole. Mulching was applied before planting by spreading sorghum straw on the soil. The data collected included vigour scores based on visual observation after two and four weeks from seeding 100-seed weight, shelling percentage, pod and hay yields.

The on-farm trials included 20 farmers each in three villages where the improved package tested consisted of an improved variety, NPK micro-dosing of 0.6 g per pocket and mulching. The plot size was 10 m × 10 m for each treatment.

Table 1: *Components of the production package using the 'plus trial' technique.*

Treatments	Elements of the package				
1: Farmers' own package	Conventional density	Local variety	No seed priming	No micro-dosing	No mulching
2	Increased density	Local variety	No seed priming	No micro-dosing	No mulching
3	Increased density	Improved variety	No seed priming	No micro-dosing	No mulching
4	Increased density	Improved variety	Seed priming	No Micro dosing	No mulching
5	Increased density	Improved variety	Seed priming	Micro dosing	No mulching
6: Complete improved package	Increased density	Improved variety	Seed priming	Micro dosing	Mulching

Table 2: *Effect of improved production package on yield (kg ha^{-1}) and yield components of groundnut and gross margin in Sudanese Pounds† (SDG ha^{-1}) combined over 2013/2014–2014/2015 cropping seasons.*

Treatment	Stand 1000 ha^{-1}	Pod yield	Hay yield	Gross margin SDG (Pod + hay)‡	100 seed weight (g)	Vigour	Shelling %
Farmers' package (FP)	42 ^{c}	707 ^{c}	958 ^{b}	3962 ^{c}	34 ^{bc}	4 ^{a}	64 ^{b}
Increased density (ID)	71 ^{b}	836 ^{bc}	1132 ^{b}	4967 ^{bc}	32 ^{c}	3 ^{b}	67 ^{ab}
ID+Improved variety (IV)	78 ^{ab}	942 ^{bc}	1201 ^{b}	5733 ^{abc}	35 ^{b}	3 ^{b}	67 ^{ab}
ID+IV+priming (P)	92 ^{a}	1035 ^{ab}	1542 ^{a}	6577 ^{ab}	36 ^{ab}	2 ^{c}	69 ^{a}
ID+IV+P+Micro-dose (M)	83 ^{ab}	1080 ^{ab}	1563 ^{a}	6565 ^{ab}	34 ^{bc}	2 ^{c}	70 ^{a}
ID+IV+P+M+Mulching (M)	66 ^{b}	1236 ^{a}	1708 ^{a}	7258 ^{a}	38 ^{a}	2 ^{c}	69 ^{a}
SE	6 **	90 **	62 **	607 **	0.8 **	0.2 **	1 **
CV %	23	26	13	29	7	29	5

Figures with different letters are significantly different at 0.05 probability level. Vigour score from 1 to 5 with 1 as highest score at harvest.

† 1 SDG = 0.045 Euro (03/2018)

‡ Prices: Prices were obtained from local market and were 6.7 SDG kg^{-1} and 0.8 SDG kg^{-1} for groundnut and ground hay respectively. Total variable cost were 1,450 SDG ha^{-1} for farmers practice.

The data were analysed using MSTATC software (MSTAT, 1993). The gross margin was calculated by deducting the variable costs (land cleaning, seed dressing, seed price, fertiliser cost, labour cost, cost of bags, transportation cost and land rental costs) from the total cash income (pods and hay). The production costs were taken from the annual agricultural survey report (North Kordofan State Ministry of Agriculture and Rural Development, 2014). The fertiliser cost was added in the micro-fertiliser treatment based on the amount of fertiliser applied.

3 Results

3.1 Ladder experiment

Over the two seasons in the ladder experiment, groundnut seedling vigour, crop establishment, pod yield, hay yield, 100 seed weight, and shelling percentage were enhanced each time an improved element was added to the production package (Table 2). Yield increments of 18, 33, 46, 53, and 75 % were observed due to the ascending effects of plant density, improved variety, seed priming, NPK fertiliser micro-dosing and mulching, respectively. The highest pod yield (1,236 kg ha^{-1}) obtained from the complete package was significantly higher than farmers' package with a pod yield of 707 kg ha^{-1}. The treatment including micro-dosing was significantly higher than farmers practice. Increased plant density alone increased yield by 18 %, while the improved variety increased yield by an additional 13 %. The highest hay yield and shelling percentage were also recorded under the complete improved technological package. Shelling percentage, which is indicative for seed filling, was highest in the treatment including fertiliser micro-dosing. The yield increase with increasing level of intensification is a result of improved stand, superior seedling vigour and

improved grain filling. The gross margin increased as new technologies were added and the highest return was attained from the complete package (Table 4). The gross margin increased by 83 % from the lowest to the highest intensification level, representing an additional revenue of 3,296 SDG ha^{-1} ($p < 0.01$).

3.2 Mulching experiment

The experiment with three mulch levels and two production packages (improved and farmers' package) showed that the improved package performed better than farmers' package, while there was no effect of mulching. Over the two cropping seasons, the improved production package enhanced seedling vigour, pod yield, hay yield, 100-seed weight and stand at harvest compared to farmers' package (Table 3). Pod and hay yields were improved by 61 and 50 %, respectively. Mulching as an individual component did not significantly improve yield and yield components, and its effect was only significant on plant vigour which was significantly better at the highest mulching treatment. The gross margin was improved by 80 % with the introduction of the improved technological package; this improvement was not, however, due to mulching.

3.3 On-farm trials

Results from the on-farm trials showed a significant improvement in pod and hay yields when farmers adopted the improved production package with 0.5 ton ha^{-1} sorghum straw mulching (Table 4). Pod yield increased by 32 % and hay yield by 19 % due to the improved production package. Gross margin increased by 30 % as the result of the improved package, with a net gain of 2,778 SDG ha^{-1} (125 Euro).

Table 3: *Effect of mulching and improved package on yield and yield components of groundnut (mulching experiment), combined over 2013/2014 and 2014/2015 cropping seasons.*

Treatment	Farmers' package	Improved package	Mean
Pod yield (kg ha^{-1})			
Zero mulching	708 ^{bc}	1070 ^{a}	889
Mulching (0.5 ton)	644 ^{c}	932 ^{ab}	788
Mulching (1 ton)	624 ^{c}	1168 ^{a}	896
Mean	658	1057	
SE	51 **		
Hay yield (kg ha^{-1})			
Zero mulching	771 ^{b}	1276 ^{a}	1024
Mulching (0.5 ton)	903 ^{b}	1390 ^{a}	1146
Mulching (1 ton)	952 ^{b}	1274 ^{a}	1113
Mean	875	1313	
SE	63 **		
Gross margin (SDG ha^{-1})			
Zero mulching	3,817 ^{bc}	6,182 ^{a}	4,999
Mulching (0.5 ton)	3,098 ^{c}	4,947 ^{ab}	4,023
Mulching (1 ton)	2,600 ^{c}	6,035 ^{a}	4,318
Mean	3,172	5,722	
SE	359 **		

Figures with different letters are significantly different at 0.05 probability level.
Vigour score from 1 to 5 with 1 as highest score at harvest.

Table 4: *Pod yield, hay yield and gross margin (GM) in SDG ha^{-1} from on-farm trials.*

Treatment	Season 2013/2014		Season 2014/2015		Combined over locations and seasons		
	Pod yield (kg ha^{-1})	Hay yield (kg ha^{-1})	Pod yield (kg ha^{-1})	Hay yield (kg ha^{-1})	Pod yield (kg ha^{-1})	Hay yield (kg ha^{-1})	GM (Pod + hay yield)
Farmers' package + seed priming	1,428 ^{b}	885 ^{b}	1,224 ^{b}	3,937	1,275 ^{b}	3,174	9,159 ^{b}
Improved package + mulching (0.5 ton ha^{-1})	1,916 ^{a}	1,470 ^{a}	1,611 ^{a}	4,546	1,687 ^{a}	3,777	11,937 ^{a}
SE	105**	175**	22**	297	19**	223	235**
CV	9	21	12	54	11	57	20

Figures with different letters are significantly different at 0.05 probability level. Vigour score from 1 to 5 with 1 as highest score at harvest.

4 Discussion

Groundnut production in the sandy, rain-fed areas of North Kordofan State is characterised by very low yields (185 kg ha^{-1}) (NKSMARD 2014). For this reason, farmers cultivate large areas as a strategy to sustain, and increase cash income. North Kordofan is classified as the most vulnerable area to climate change in Sudan and there is a need to develop and promote the use of improved technological packages that can sustain groundnut production

and productivity. The yield increasing effect of combining seed priming and fertiliser micro-dosing as compared to farmers practice has previously been shown in groundnut, cowpea, pearl millet and sorghum in the drylands of Africa (Osman & Aune, 2011; Aune & Osman, 2011; Maman *et al.*, 2000). In this study, groundnut pod yield was significantly increased due to the introduction of different components of an improved technological package. The yield increase observed in the improved package was 75 % in the experiment with increasing levels of inputs, 61 % in the experiment with three mulch levels, and 32 % in the on-farm experiment. These yield records show that there is a consistent yield increase with the adoption of improved groundnut packages. By increasing yields it is also likely that the need for clearing new land for cultivation could be reduced. The tendency of the farmers to avoid land clearing as they adopt more intensive management technologies has been shown in West Africa (Aune & Bationo, 2008; McCalla, 1999). Thus, the transformation of subsistence-oriented farming to small-scale commercial farming units can be achieved by adopting improved technologies. Wealthier farmers could adopt the full package while resource-limited farmers can make use of the least costly technologies in the intensification ladder. Results from the ladder study showed that the gross margin can be increased up to 83 % (3,296 SDG ha^{-1}). This can be seen as a result of combining all the yield enhancing technologies.

Mineral fertiliser in Sudan has been mainly used in the irrigated sector, but there is now a national recommendation to use micro-dosing in dryland crops as well. However, availability of mineral fertiliser is still a problem. Resource-limited farmers with little chances of accessing inputs (notably fertilisers and mulching materials) can achieve an increase in yield of 18 % and gain 25 % additional gross margin (equivalent to 1005 SDG ha^{-1}), only by applying the recommended spacing and planting density (Table 2). High plant density has been reported to be crucial for increasing yield of rain-fed groundnut grown in the sandy soils of Sudan (Osman & Sid Ahmed, 1993). Farmers can gain 45 % additional revenue, equivalent to 1771 SDG ha^{-1}, by adopting recommended planting density and the improved groundnut variety. They can further improve yields by 46 % by adding seed priming to the previous package, increasing the gross margin by 66 %, equivalent to 2,615 SDG ha^{-1}. Seed priming is a method accessible to the farmers as no external inputs are needed. The effect of mulching was variable in these experiments. In the ladder experiment, mulching increased yield while this effect was not apparent in the experiment with increasing levels of mulching. Though no clear effect of mulching was observed in these experiments, the effect of mulch on improving soil quality is well documented (Buerkert *et al.*, 2002). High cost of sorghum straw due to competitive use of biomass as fodder, fuel and building material has been reported as a constraint limiting the use of sorghum or millet straw for mulching in West Africa (Lal, 2007). Grazing pressure is also high in the off-season, making it difficult to retain mulch on farmers' fields. There is a need to develop new approaches to ensure more mulch retention. One simple approach could be to cut the straw at 10-20 cm above ground level, leaving the bottom part. Introducing mulch-producing trees such as *Guiera senegalen-*

sis, *Piliostigma reticulatum*, *Acacia senegal* and *Faidherbia albida* is another approach (Lahmar *et al.*, 2012). The use of mulch producing trees may be complemented with improved rotations including leguminous crops.

5   Conclusions

This study has shown that farmers can significantly increase groundnut yields and returns by adopting (parts of) a new technological package. The yield and gross margin revenue increased with increasing levels of intensification. Adopting the complete technological package gave 83 % higher gross margin than farmers' package. Resource-limited farmers can choose low-costs, low-risk components of the technological package whereas farmers with better access to resources can adopt the complete technological package.

Acknowledgements

Sincere thanks are due to the Drylands Coordination Group (DCG) for their assistance and support that made this work possible. Our thanks are also extended to the Dry Lands Research Centre (ARC) and ADRA (Sudan) for their cooperation and facilitation. Thanks are due to Mutaz M. Elsadig, Sudan (DCG Coordinator) for his keen follow up and for facilitating the implementation of the project. Sincere thanks are also extended to Khalid Ali Issa and Faiz Ali Ahmed, the drivers who made the journeys to and from trial sites easy and enjoyable. We are very grateful to the members of the communities who participated in the implementation of the activities.

References

Abdalla, E. A. & AlAhmedi, A. B. (1997). A new groundnut cultivar (C36B-25"II"-B) for the sandy rain-fed areas of Western Sudan. 2nd meeting, variety release committee, 15 Sep. 1997, Ministry of Agriculture and forestry, Khartoum, Sudan.

Aune, J. B. & Bationo, A. (2008). Agricultural Intensification in the Sahel The ladder approach. *Agricultural Systems*, 98 (2), 119–125. doi:10.1016/j.agsy.2008.05.002.

Aune, J. B. & Osman, A. (2011). Effect of seed priming and micro dosing of fertilizer on sorghum and pearl millet in western Sudan. *Experimental Agriculture*, 47, 419–430. doi:10.1017/S0014479711000056.

Bagayoko, M., Mason, S. C., Traoré, S. & Eskridge, K. M. (1996). Pearl millet/cowpea cropping system yields and soil nutrient levels. *African Crop Science Journal*, 4 (4), 453–462. Available at: http://hdl.handle.net/1807/20979

Buerkert, A., Piepho, H. P. & Bationo, A. (2002). Multi-site time-trend analysis of soil fertility management effects on crop production in sub-Saharan West Africa. *Experimental Agriculture*, 38 (2), 163–183. doi: 10.1017/S0014479702000236.

Coulibaly, A., Bagayoko, M., Traore, S. & Mason, S. C. (2000). Effect of crop residue management and cropping system on pearl millet and cowpea yield. *African Crop Science Journal*, 8 (4), 411–418. doi: 10.4314/acsj.v8i4.27681.

Lahmar, R., Bationo, B. A., Lamso, N. D., Guéro, Y. & Tittonell, P. (2012). Tailoring conservation agriculture technologies to West Africa semi-arid zones: Building on traditional local practices for soil restoration. *Field Crops Research*, 132, 158–167. doi:10.1016/j.fcr.2011.09.013.

Lal, R. (2007). Constraints to adopting no-till farming in developing countries. *Soil Tillage Research*, 94, 1–3.

Maman, N., Mason, S. C. & Sirifi, S. (2000). Influence of variety and management level on pearl millet production in Niger: I. Grain yield and dry matter accumulation. *African Crop Science Journal*, 8 (1), 24–34. doi: 10.4314/acsj.v8i1.27713.

McCalla, A. F. (1999). Prospects for food security in the 21st Century: with special emphasis on Africa. *Agricultural Economics*, 20 (2), 95–103. doi:10.1016/S0169-5150(98)00080-2.

MSTAT (1993). User's Guide to MSTAT-C. MSTAT Development Team, Michigan State University, USA.

NAPA (2007). National Adaptation Plan of Action (NAPA), Republic of the Sudan. Ministry of Environment and Physical Development, Higher Council for Environment and Natural Resources.

NKSMARD (2014). Annual Survey report 2014. North Kordofan State Ministry of Agriculture and Rural Development (NKSMARD), Department of planning and agricultural statistics, Elobeid, Sudan.

Osman, A. K. & Aune, J. B. (2011). Effect of seed priming and micro dosing of fertilizer on groundnut, sesame and cowpea in western Sudan. *Experimental Agriculture*, 47, 431–443. doi:doi.org/10.1017/S0014479711000068.

Osman, A. K. & Elamin, E. M. (2000). Effect of a production package on yield and profitability in rain-fed groundnut. *Arab Agricultural Research Journal*, 4, 113–122.

Osman, A. K. & Sid Ahmed, S. A. H. (1993). Effect of spacing and plant population on yield and some yield components of groundnut. Crop Husbandry Committee–ARC, Wadmedani, Sudan.

15

Growth performance and digestive tract development of indigenous scavenging chickens under village management

Thomas Raphulu [a,b,*], Christine Jansen van Rensburg [a]

[a]*Department of Animal and Wildlife Sciences, University of Pretoria, Pretoria 0002, South Africa*
[b]*Limpopo Department of Agriculture and Rural Development, Mara Research Station, P/bag x 2467, Makhado, 0920, South Africa*

Abstract

The study was conducted on indigenous scavenging chickens under village management firstly, to evaluate the early development of the digestive tract to 28 days of age and secondly, to determine the growth performance of these chickens up to 20 weeks of age. One hundred and seventeen chicks, 13 chicks per age class (day 1, 4, 7, 10, 14, 17, 21, 24, 28) were randomly purchased from six rural villages in the Vhembe District, Venda, South Africa. The chickens were weighed and sacrificed for measurement of the different parts of its gastrointestinal tract. The liver and pancreas were also weighed. The relative weight of the storage organs and liver peaked at day 4 while that of the small intestine and duodenum peaked at day 10. The relative lengths of the small intestine and jejunum peaked at day 7, duodenum at day 10 and ileum at day 4. Four hundred and forty four (444) chicks from 13 households were recorded at two weekly intervals starting from day old until 20 weeks of age. The mean body weight obtained for males and females were 201.7 and 171.5 g at six weeks of age and 1048.1 and 658.6 g at 20 weeks of age, respectively. The indigenous chickens under village management were characterised by slow digestive tract development, poor growth performance and high mortalities. Further research needs to be conducted to determine the effect of early feed supplementation on the development of the digestive tract and the performance of indigenous chickens under village management.

Keywords: digestive organs, relative weight, relative length, growth performance, rural communities

1 Introduction

The gastrointestinal tract (GIT) of birds undergoes rapid development during the early post-hatch period, which plays a major role in inducing early growth (Lilja, 1983; Sell *et al.*, 1991). Post-hatch development studies have been conducted in broilers (Lija, 1983, Nitsan *et al.*, 1991a,b; Katanbaf *et al.*, 1998; Noy & Sklan, 1998; Ravindran *et al.*, 2006) and in Yangzhou goslings (Liu *et al.*, 2010) and reports indicated that the digestive organs increased more rapidly in weight relative to the whole body mass. The relative weights of these organs were maximal at 6–8 days of age in turkey poults (Sell *et al.*, 1991; Noy & Sklan, 1998) and at 6–10

days of age in broiler chicks (Katanbaf *et al.*, 1998). Kadhim *et al.* (2010) reported that the rate of organ growth relative to the increase of body weight in both Malaysian fowl and broilers fed commercial diets reached a maximum at 10 days post-hatch and after that declined sharply.

Access to nutrients initiates growth about 24 hours after ingestion of exogenous feed for the first time after hatch. Early access to feed resulted in a more rapid post-hatch development of the intestine (Sklan, 2001). Feeding behaviour, rather than differences in individual body weight, accounted for gross anatomical differences in the intestine (Yamauchi & Zhou, 1998). The relative weight of the duodenum, jejunum and pancreas but not ileum was found to be higher in light breeds than heavy breeds (Dror *et al.*, 1977). Breed effect on the development of the digestive tract post-

* Corresponding author – thomas.raphulu@gmail.com

hatch, however, was not noticeable when chickens had full access to feed.

Indigenous chickens are the most common types of poultry raised in the rural communities of Vhembe District, South Africa. Young chicks scavenge with their mothers for food around the household during the day and are provided with shelter at night. Chicks relying on scavenging for their feed might have a low and unbalanced nutrient intake, which could impair growth and the development of the digestive tract. Post-hatch development of the digestive system and growth performance of local chickens under village management have never been documented. Availability of such information might form the basis for improving the productivity of Venda Indigenous Scavenging (VIS) chickens in the rural communities. The study was carried out, firstly, to evaluate changes in the development of the digestive tract up to 28 days of age and, secondly, to determine the growth performance of VIS chickens under village management up to 20 weeks of age.

2 Materials and methods

The study was conducted at 6 adjacent villages, Tshifudi, Tshidzini, Tshamutshedzi, Tshivhilwi, Tshitereke and Makhuvha. All villages are situated between latitude 22°48′ S to 22°53′ S and longitude 30°28′ E to 30°42′ E in the Thulamela Municipality, Vhembe District in the Limpopo Province of South Africa. Vhembe District is the most northern district of the Limpopo Province and shares borders with Botswana, Zimbabwe and Mozambique. The villages are in a summer rainfall area (October to April). The wettest months and the hottest season is between October and March, when the mean maximum temperatures range from 26.7 to 29.1 °C. The coldest season is between May and August, when the minimum temperatures range from 12 to 14 °C. Winter is usually cold but rarely reach freezing point. The main crops cultivated in the area are maize, groundnuts and vegetables e.g spinach, Chinese cabbage (locally known as *mutshaina*), tomato, and beetroot.

2.1 Digestive tract measurements

A hundred and seventeen (117) VIS chicks, 13 chicks per age class (day 1, 4, 7, 10, 14, 17, 21, 24, 28) were randomly purchased from rural villages and were sacrificed in the laboratory through neck cut. The weights of the chickens were recorded before slaughtering. Directly after killing, the abdominal cavity was opened and the digestive tract from the tongue to the cloaca of each bird was removed. The GIT was separated into crop, proventriculus, gizzard, small intestine and caeca. The small intestine was divided into three regions (duodenum, jejunum and ileum) following the demarcation described by Mitchell & Smith (1990). The different segments of the digestive tract were flushed out with water and the empty weights were recorded and length measured. The accessory organs, liver and pancreas, were also removed and weighed.

2.2 Growth performance

Thirteen households participated voluntarily in the study. These households kept chickens under the traditional village management system of allowing the birds to roam free during the day to scavenge for feed, while providing shelter during the night. Shortly after hatch, 444 chicks were tagged for identification. Individual chicken live weights were recorded at two weeks intervals starting from day-old until 20 weeks of age. Prior to data collection, farmers were requested to keep the chickens in the shelter until weighing was completed in the morning.

2.3 Statistics

Data for organ weights were calculated as total or absolute weights (g) and relative weight (g/100 g body weight), absolute (cm) and relative length (cm/100 g body weight) of the digestive tract and growth performance (g). Individual birds were considered as the experimental unit. All the data (growth performance, absolute and relative weights of digestive organs, absolute and relative length of digestive organs) was subjected to General Linear Models procedure of SAS, Version 9.3 (SAS, 2016). The following model was employed on organ weights data: $Y_i = \mu + A_i + E_i$, where Y_i is an observation for a given variable; μ is the general mean common to all observations; A_i is the effect due to ith age and E_i is the random error. On body weight and growth rate data of VIS chickens, the following model was employed : $Y_j = \mu + G_j + E_j$, was employed and a 5 % significant level was used, where Y_j is an observation for a given variable; μ is the general mean common to all observations; G_j is the effect due to jth gender of chickens and E_j is the random error. Differences among means were determined by the least significant difference (LSD) procedure of SAS (2016).

3 Results

As shown in Fig. 1, an increase in mass and length of the different parts of the GIT with age was observed ($P < 0.005$). The relative weights of the storage organs (crop, proventriculus and gizzard) and liver increased rapidly and peaked at day 4 where after it remained more or less constant.

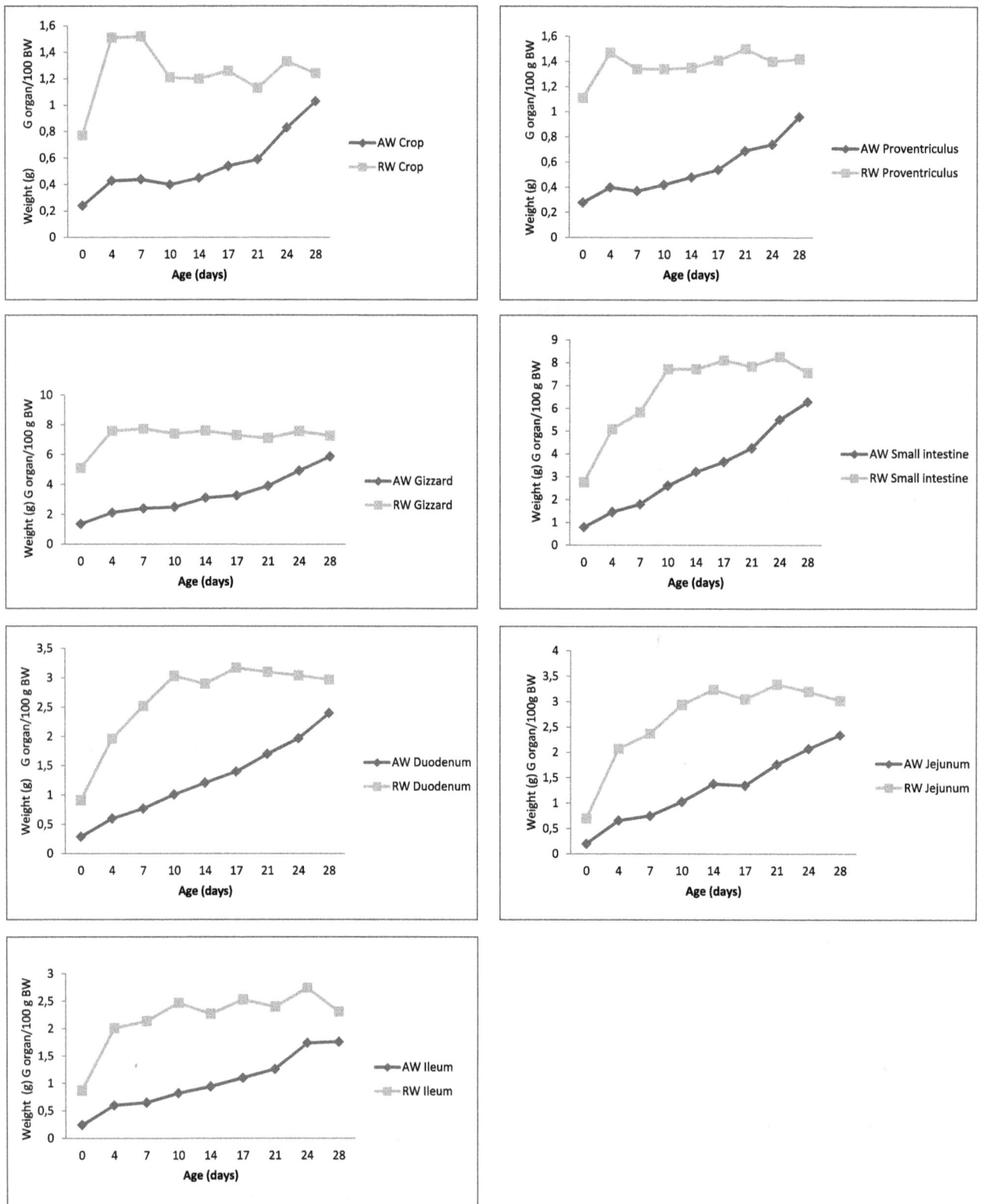

Fig. 1: *Changes in the weight of the gastro-intestinal tract segments of scavenging chicks during the first 28 days after hatch (AW – absolute weight, RW – relative weight).*

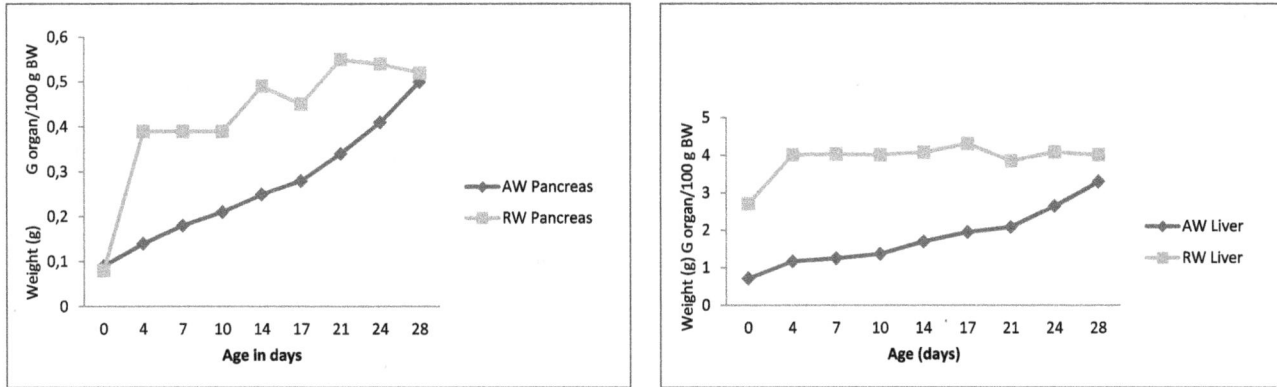

Fig. 2: *Changes in the weight of the pancreas and liver of scavenging chicks during the first 28 days after hatch (AW – absolute weight, RW – relative weight).*

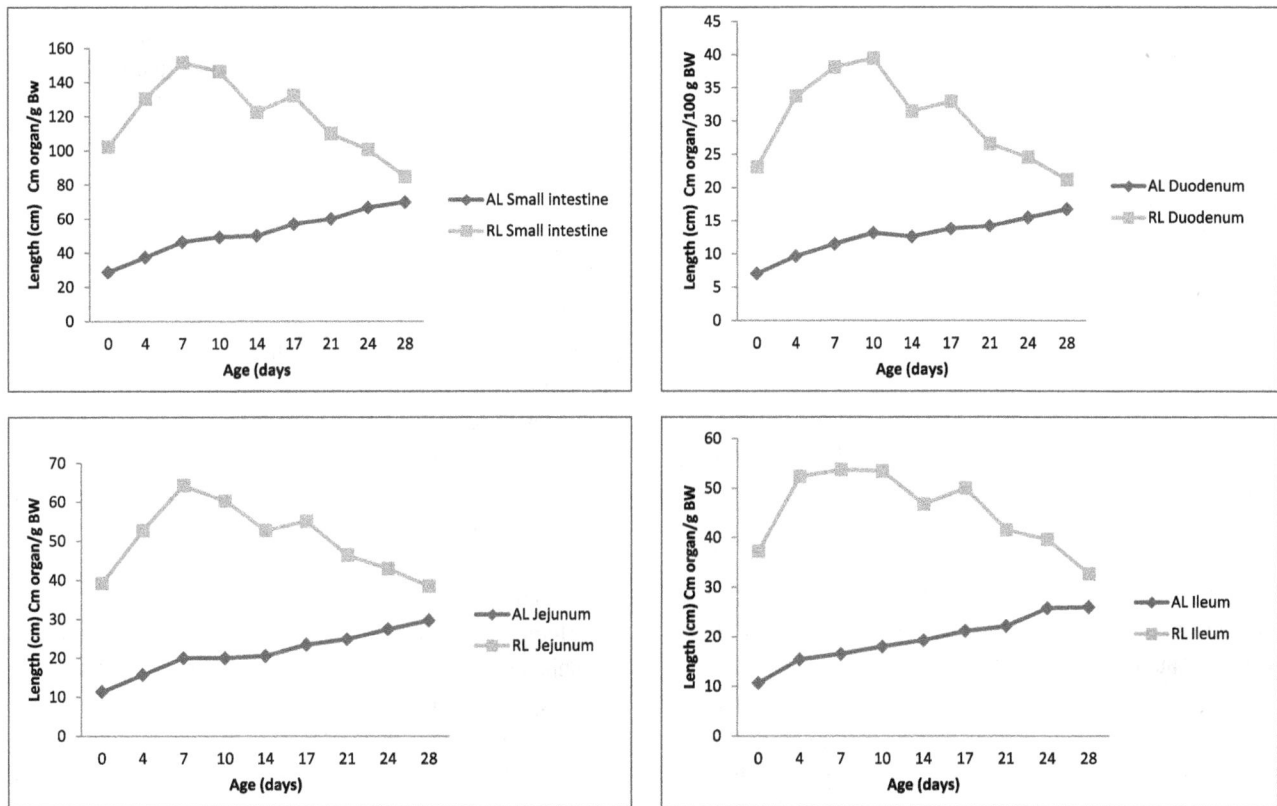

Fig. 3: *Changes in the length of small intestine segments of scavenging chicks during the first 28 days after hatch (AL – absolute length, RW – relative length).*

The relative weight of the small intestine and its separate components rapidly increased with age and peaked at about 10 days of age. The absolute mass of the small intestine increased by 127.8 % during the first 7 days. The relative weight of the pancreas increased rapidly up to day 4, and continued to increase at a slow rate up to about 21 days of age (Fig. 2). The relative length of the small intestine and jejunum peaked at day 7, duodenum at day 10 and ileum at day 4, where after it decreased (Fig. 3).

The mean body weight obtained for males and females were 201.7 and 171.5 g at six weeks of age and 1048.1 and 658.6 g at 20 weeks of age, respectively (Table 1). The cumulative mortality observed in chicks under the age of six weeks and 7–20 weeks was 57.4 and 26.3 %, respectively.

Table 1: *Performance of the VIS chickens under village management.*

Performance parameter	Mean ± standard error
Body weight at 6 weeks (g)	
Males	201.7 ± 5.80^{a}
Females	171.5 ± 6.74^{b}
Growth rate up to 6 weeks (g/day)	
Males	4.1 ± 0.36^{a}
Females	2.9 ± 0.42^{b}
Survival rate (%) up to six weeks	42.6
Body weight at 20 weeks (g)	
Males	1048.1 ± 28.09^{a}
Females	658.6 ± 22.94^{b}
Growth rate 7 to 20 weeks (g/day)	
Males	10.1 ± 0.51^{a}
Females	4.6 ± 0.50^{b}
Survival rate (%) 7 to 20 weeks	73.3

[a,b] Means with different superscripts within a column and a factor differ significantly ($P < 0.05$).

4 Discussion

The development of the GIT during the post-hatch period played a major role in inducing early growth (Sell *et al.*, 1991). The digestive organs of the scavenging, indigenous chicks studied in this trial, followed a similar early growth pattern observed in other chickens (Dror *et al.*,1977; Lilja, 1983), turkeys (Sell *et al.*,1991) and ducks (King *et al.*, 2000). It has been suggested that the accelerated development of the digestive organs immediately after hatching is a prerequisite for sustained post-hatch growth in fast growing poultry (Katanbaf *et al.*,1998).

According to Nitsan *et al.* (1991b), the pancreas of chickens first experiences a rapid growth phase from hatch to day 3 and then a slower growth phase from day 4–8. In this study, however, a rapid increase in the relative weight of the pancreas was noted until 4 days of age and a slower relative growth up to 21 days of age. The different segments of the small intestine developed at slightly different rates in relation to the increase in body weight. The observed results are in accordance with findings of Uni *et al.* (1999) who reported that the temporal increases in intestinal weight and length are not identical for different segments, with the duodenum developing at a faster rate than both the jejunum and ileum.

The absolute growth rate of the small intestine of the chicks in this study was much lower than reported by Noy *et al.* (2001), who found an increase in the mass of the small intestine by nearly 600 % within the first 7 days. Kadhim

et al. (2010) found similar patterns of organ weights relative to body weights in both indigenous breeds (Malaysian local chickens) and broilers fed commercial diets, ruling out the possibility that genotype affects GIT development. The slower development of the small intestine in the current study could rather be attributed to the poor availability of quality feeds to the scavenging chickens in the rural communities. Growth is initiated about 24 hours after first ingestion of exogenous food and it is suggested that early access to nutrients results in the more rapid development of the intestine during the immediate post-hatch period (Sklan, 2001). The withholding of feed and water from birds resulted in reduced growth of all segments of the intestinal tract (Murakami *et al.*, 1992; Uni *et al.*, 1998). It is possible that the development of the digestive tract of the scavenging chickens in this study might have been impaired by a lack of feed and irregular access to water which inhibited the growth of the birds in general. It is known that little care is taken with regard to housing, feeding, breeding or parasite and disease control (Minga *et al.*, 1989). As a result, chicks might survive for a few days post-hatch mainly on nutrients supplied by the yolk. Sell *et al.* (1991) reported that nutrients from the yolk are depleted in broiler chicks and poults within 4–5 days. It has been reported that the yolk is used for maintenance, while exogenous energy is utilised for growth (Anthony *et al.*, 1989). The slower gut development and growth rate and high mortality of the chicks in this study could have been caused by a limited feed intake. There is no planned feeding for scavenging chickens in the rural areas. Chickens are left to scavenge around the homesteads during daytime feeding on household leftovers, waste products and environmental materials such as insects, worms, seeds and green forages (Goromela *et al.*, 2006; Raphulu *et al.*, 2015). Supplementation is rarely done since farmers assume that the chickens scavenging from the natural resource feed base get adequate nutrients to meet their maintenance, reproductive and productive needs (Nzioka *et al.*, 2017). The growth of an animal depends in part on its capacity to digest and assimilate ingested macromolecules (Liu *et al.*, 2010). Results from this study indicated that supplementary feed to chicks in the rural communities might be necessary during the first few weeks to promote the development of the GIT until chicks can scavenge successfully.

The peak of the relative length of the small intestine observed in this study was 2–3 days later than the 5–7 days post-hatch reported by Noy & Sklan (1998). However, our results are in accordance with that of Sell *et al.* (1991), that the process of rapid relative growth was maximal at 6–8 days in the poult and 6–10 days in the chick. Kadhim *et al.* (2010) found that the absolute length of the intestinal segments of the Malaysian local chicken fed a commercial diet were shorter by approximately one fold than those of broilers. Nir *et al*. (1993) suggested that the smaller breeds have relatively lighter and shorter intestines than broilers.

There is a lack of published data on the productivity of local chickens under village management under South African conditions. However, a few studies have been conducted in other parts of Africa. Under village management conditions in Tanzania, Lwelamira *et al*. (2008) reported mean body weights of 1135 and 1240 g for 20 weeks old female and male chickens, respectively. Mwalusanya *et al*. (2002) reported growth rates of chickens up to 10 weeks to be 5.4 and 4.6 g/day for males and females, respectively, whereas chickens aged 10–14 weeks old showed rates of 10.2 and 8.4 g/day for males and females. The obtained body weight gain of the chickens at eight weeks in this study were higher than those described by Mafeni (1995) for Cameroon, Omeje & Nwosu (1984) for Nigeria, and Tadelle & Ogle (2001) for Ethiopia, but less than those obtained by Lwelamira *et al*. (2008). Differences in growth performance of local chickens could be due to genetic differences between birds, climatic condition and local management that determine the availability of feeds between countries. Aini (1990) stated that the productivity of local birds is characteristically very low, but there is large variation in production performance between birds of different localities.

Mortality was high in chicks up to ten weeks of age (73.8 %). The observed results are comparable to the results of Minga *et al*. (1989) who reported 50 % mortality in scavenging chickens during rearing. Mwalusanya *et al*. (2002) reported a mortality of 40.3 % up to 10 weeks of age. Mortality is a serious problem in local chicken production and it needs intervention. It was noted from the famers that the causes of high mortality in local chickens under six weeks of age were lack of quality feeds, theft and predators (dogs and eagles). This was confirmed by Mwalusanya *et al*. (2002) who reported predation to be an important cause of loss in chicken flocks. Chicks can be protected from predators by providing shelters and supplementary feeds can be given to chicks under six weeks of age to improve survivability. It is believed that chicks older than six weeks might be able to escape attacks from the predators and also successfully search for food. The high costs involved in provision of housing and feeds to chicks might be challenging in the poverty restricted rural communities and it might be necessary but feasible to use locally produced feed resources and building materials.

5 Conclusion

The chickens under village management were characterised by slow digestive tract development, growth per-

formance and high mortalities. Dietary supplementation strategies using locally produced feeds, brooding and provision of shelter to newly hatched chicks for the first six weeks might be important tools in improving chicken production in general, through reduced mortality at early age and improved growth rate. Further research needs to be conducted to determine the effect of early feed supplementation on the development of the digestive tract and the performance of chickens under village management.

Acknowledgements

The authors wish to acknowledge the support of National Research Foundation (NRF) for funding the research work. The approval by the local Traditional leaders to conduct research at the rural communities and participation by the community members whose chickens were purchased are gratefully acknowledged. We wish to record our gratitude to Roelof Johannes Coertze for his assistance in data analysis.

References

Aini, I. (1990). Indigenous chicken production in South East Asia. *World's Poultry Science Journal*, 46, 51–57.

Anthony, N., Dunnington, B. E. & Siegel, P. B. (1989). Embryo growth of normal and dwarf chickens from lines selected for high and low 56-day body weight. *Archiv für Geflügelkunde*, 53, 116–122.

Dror, Y., Nir, I. & Nitsan, Z. (1977). The relative growth of internal organs in light and heavy breeds. *British Poultry Science*, 18, 493–496.

Goromela, E. H., Kwakkel, R. P., Verstegen, M. W. A. & Katule, A. M. (2006). Strategies to optimize the use of scavengeable feed resource base by smallholders in traditional poultry production systems in Africa: A review. *African Journal of Agricultural Research*, 1, 91–100.

Kadhim, K. K., Zuki, A. B. Z., Noordin, M. M., Babjee, S. A. & Khamas, W. (2010). Growth evaluation of selected digestive organs from day one to four months post hatch in two breeds of chicken known to differ greatly in growth rate. *Journal of Animal and Veterinary Advances*, 9, 995–1004.

Katanbaf, M. N., Dunnington, E. A. & Siegel, P. B. (1998). Allomorphic relationships from hatching to 56 days in parental lines and F1 crosses of chickens selected 27 generations for high or low body weight. *Growth, Development and Aging*, 52, 11–22.

King, D. E., Asem, E. K. & Adeola, O. (2000). Ontogenetic development of intestinal digestive functions in white Pekin ducks. *Journal of Nutrition*, 130, 57–62.

Lilja, C. (1983). A comparative study of postnatal growth and organ development in some species of birds. *Growth*, 47, 317–339.

Liu, B. Y., Wang, Z. Y., Yang, H. M., Wang, X. B., Hu, P. & Lu, J. (2010). Developmental morphology of small intestine in the Yangzhou gosslings. *African Journal of Biotechnology*, 9 (43), 7392–7400.

Lwelamira, J., Kifaro, G. C. & Gwakisa, P. S. (2008). On station and on-farm evaluation of two Tanzania chicken ecotypes for body weights at different ages and for egg production. *African Journal of Agricultural Research*, 3, 843–851.

Mafeni, M. J. (1995). *Studies of productivity, immuno-competence and genetic diversity of naked neck and normal feathered indigenous Cameroon and German Dahlem Red fowl and their crosses*. Ph.D. thesis, Humboldt University of Berlin, Germany.

Minga, U. M., Katule, A. N., Maeda, T. & Musasa, J. (1989). Potential and problems of traditional chicken industry in Tanzania. *In:* Proceedings of the 7th Tanzania Veterinary Association Scientific Conference, Arusha International Conference Centre, Arusha, Tanzania. pp. 207–215.

Mitchell, M. A. & Smith, M. W. (1990). Jejunal alanine uptake and structural adaptation in response to genetic selection for growth rate in the domestic fowl (*Gallus domesticus*). *Journal of Physiology*, 424, 7–15.

Murakami, H., Akiba, Y. & Horiguchi, M. (1992). Growth and utilization of nutrients in newly hatched chicks with or without removal of residual yolk. *Growth, Development and Aging*, 56, 75–84.

Mwalusanya, N. A., Katule, A. M., Mutayoba, M., Mtambo, M. A., Olsen, J. E. & Minga, U. M. (2002). Productivity of local chickens under village management conditions. *Tropical Animal Health and Production*, 34, 405–416.

Nir, I., Nitsan, Z. & Mahagna, M. (1993). Comparative growth and development of the digestive organs and of some enzymes in broiler and egg type chicks after hatching. *British Poultry Science*, 34, 523–532.

Nitsan, Z., Ben-Avraham, G., Zoref, Z. & Nir, I. (1991a). Growth and development of the digestive organs and some enzymes in broiler chicks after hatching. *British Poultry Science*, 32, 515–523.

Nitsan, Z., Dunnington, E. A. & Siegel, P. B. (1991b). Organ growth and digestive enzyme levels at fifteen days of age in lines of chickens differing in body weight. *Poultry Science*, 70, 2040–2048.

Noy, Y., Geyra, A. & Sklan, D. (2001). The effect of early feeding on growth and small intestine development in the posthatch poult. *Poultry Science*, 80, 912–919.

Noy, Y. & Sklan, D. (1998). Metabolic responses to early nutrition. *Journal of Applied Poultry Research*, 7, 437–451.

Nzioka, S. M., Mungube, E. O., Mwangi, M. D., Muhammed, L. & Wambua, J. M. (2017). The quantity and quality of feed available to indigenous chickens under the scavenging system in semi-arid Eastern Kenya. *East African Agricultural and Forestry Journal*, 82, 57–69.

Omeje, S. S. & Nwosu, C. C. (1984). Hetrosis and superiority in body weight and feed efficiency evaluation of exotic parent stock by local chicken F1 Crossbreds. *Nigerian Journal of Genetics*, 5, 11–26.

Raphulu, T., Jansen van Rensburg, C. & van Ryssen, J. B. J. (2015). Assessing nutrient adequacy from the crop contents of free-ranging indigenous chickens in rural villages of the Venda region of South Africa.

Ravindran, V., Wu, Y. B., Thomas, D. G. & Morel, P. C. H. (2006). Influence of whole wheat feeding on the development of gastrointestinal tract and performance of broiler chickens. *Australian Journal of Agricultural Research*, 57, 21–26.

SAS Institute (2016). *SAS® Statistics Users Guide, Statistical Analysis System, 9.3 version*. SAS Institute Inc., Cary, NC, USA.

Sell, J. L., Angel, C. R., Piquer, F. J., Mallarino, E. G. & Al-Batshan, H. A. (1991). Development patterns of selected characteristics of the gastrointestinal tract of young turkeys. *Poultry Science*, 70 (5), 1200–1205.

Sklan, D. (2001). Development of the digestive tract of poultry. *World's Poultry Science Journal*, 57, 415–428.

Tadelle, D. & Ogle, B. (2001). Village poultry production systems in the Central highlands of Ethiopia. *Tropical Animal Health and Production*, 33, 521–537.

Uni, Z., Ganot, S. & Sklan, D. (1998). Post hatch development of mucosal function in broiler small intestine. *Poultry Science*, 77, 75–82.

Uni, Z., Noy, Y. & Sklan, D. (1999). Post hatch development of small intestine function in the poult. *Poultry Science*, 78, 215–222.

Yamauchi, K. & Zhou, Z. X. (1998). Comparative anatomical observations on small intestine of chickens and water fowls. *In:* Proceedings XVIII World's Poultry Congress, Nagoya, Japan. pp. 1059–1060.

16

Applying phosphorus indices at a small agricultural watershed in Southern Brazil

Josiane C. N. Waltrick [a], Gabriel D. Goularte [b], Nerilde Favaretto [b,*],
Luiz C. P. Souza [b], Jeferson Dieckow [b], Volnei Pauletti [b], Fabiane M. Vezzani [b],
Luciano Almeida [c], Jean P. G. Minella [d]

[a]*Department of Education of Paraná State, Av. Água Verde, 2140, Vila Izabel, CEP 80240-900, Curitiba, Paraná, Brazil*

[b]*Department of Soil Science and Agricultural Engineering, Federal University of Paraná,*
Rua dos Funcionários 1540, CEP 80035-050, Curitiba, Paraná, Brazil

[c]*Department of Agricultural Economics and Extension, Federal University of Paraná,*
Rua dos Funcionários 1540, CEP 80035-050, Curitiba, Paraná, Brazil

[d]*Department of Soil Science, Federal University of Santa Maria, Av. Camobi, 1000, CEP 97105-900, Santa Maria, Rio Grande do Sul, Brazil*

Abstract

Best management practices at watershed scale are essential to mitigate water pollution. The objectives of this study were: (1) to estimate the P-index in a small watershed with intensive agricultural use applying five P-index versions at three scales (watershed, sub-basin and agricultural field); (2) to assess the effect of the connectivity factors (distance between the agricultural field and the stream and width of riparian native vegetation) in estimating the risk of P loss. The five P-index versions resulted in a similar risk of P loss, 75 to 83 % of the whole watershed scale (agricultural plus forest areas) was classified as low or very low risk for P loss. At the agricultural area scale, 79 to 100 % of this area was classed as high and very high risk for P loss. The low risk of P loss at watershed scale is explained by the high occurrence of forest vegetation. The reduced distance between agricultural land and streams and/or the reduced width of riparian native vegetation increased the risk of P loss. Estimated P-index values at a sub-basin scale indicated lower risk of P loss compared to agricultural field scale. In order to better estimate the risk of P loss at an agricultural field scale, we advise using a P-index which considers also connectivity factors.

Keywords: P-index, P risk assessment, water quality, catchment, riparian vegetation, runoff

1 Introduction

Phosphorus (P) is an essential nutrient for plants, and extensively applied on agricultural fields. However, in an aquatic environment P is associated with eutrophication (Correll, 1998; Daniel *et al.*, 1998; Schindler *et al.*, 2008) causing detrimental impacts on aquatic life and derived products for human use (Kay *et al.*, 2009). It is transferred from agricultural fields into streams mainly via surface runoff due to its low mobility in soil (Leinweber *et al.*, 2002; Sharpley & Wang, 2014). Soil P content plays an important role in water pollution, given that high soil P increases the potential for P loss by surface runoff (Pote *et al.*, 1999). To ensure high productivity, ever-growing amounts of organic and inorganic fertilisers have been added to soils in intensive agricultural systems (Hooda *et al.*, 2000; Kleinman *et al.*, 2002). So, especially sloped areas associated with P over-fertilisation become high risk sites of P transport from soil to water (Shigaki *et al.*, 2006; Sharpley *et al.*, 2001; Gburek *et al.*, 2000).

To assess P losses from agricultural land into surface water the P-index was developed as a semiquantitative tool (Sharpley *et al.*, 2001). This would aid farmers to decide about which practices should be applied in soil man-

---
* Corresponding author
Nerilde Favaretto (nfavaretto@ufpr.br)

agement and P fertilisation, considering both agronomic (plant production) and environmental (water quality) aspects (Sharpley et al., 2003; Buczko & Kuchenbuch, 2007).

The P-index is a simple combination of several factors related to P transport and P source. Most versions of P-index consider soil erosion and surface runoff as transport factors, and soil content, amount and method of P application as source factors. Differences among versions are related to factor weight, and variables such as connectivity, presence of animals, and critical areas (Buczko & Kuchenbuch, 2007). Connectivity between fields and receiving water i.e. distance from field to the stream and width of riparian buffer zone, have been included in several P-index versions (Sharpley et al., 2003). Another difference among P indices is the way to calculate the P-index score. Originally published by Lemunyon & Gilbert (1993), the P-index was additive (P-index scores the sum of factors). Later versions were modified utilising a multiplicative approach, where P-index score is calculated by multiplying the transport factor by the source factor. In all versions, the final P-index score is ranked on a scale from very low to excessive, indicating the risk of water contamination by P (Sharpley et al., 2003; Buczko & Kuchenbuch, 2007). In the U.S.A. for example, there are 47 different versions to estimate the P-index (Sharpley et al., 2003). Canada (Reid, 2011) and several European countries (Heathwaite et al., 2003; Bechmann et al., 2005) also developed variations of the P-index. In Brazil, some studies have been done with P indices from other countries (Lopes et al., 2007; Oliveira et al., 2010). Recently, Couto et al. (2015) applied two versions of the P-index (modified from Lemunyon & Gilbert, 1993 and Flynn et al., 2000), to evaluate P loss from agricultural land after long-term application of pig slurry in Southern Brazil. However, so far, there is no Brazilian P-index.

P indices are empirical and developed using the best available professional knowledge from a wide range of scientific literature (Sharpley et al., 2003). Therefore, it is not possible to calibrate the P-index as a mathematical model; however, it is possible and very important to evaluate the sensitivity of this framework using measured data (Buczko & Kuchenbuch, 2007). This sensitive evaluation allows to asses if the framework is correlated with P losses and also to identify which component has the greatest influence on the P-index scores. Several studies comparing estimated and measured values have been conducted at small and large field plots as well as at watershed (Sharpley, 1995; Gburek et al., 2000; Eghball & Gilley, 2001; Veith et al., 2005; Bechmann et al., 2007).

The objectives of this study were: (1) to estimate the P-index in a small watershed with intensive agricultural use (Colombo, Paraná, Brazil) applying five P-index versions obtained by an additive approach at three scales (watershed, sub-basin and agricultural field); (2) to assess the effect of the connectivity factors (distance between the agriculture field and the stream and width of riparian native vegetation) in estimating the risk of P loss.

The Campestre watershed was chosen for this study because it is characterised by intensive vegetable farming using high rates of mineral and organic fertilisers, as well as the presence of shallow soils and sharp slopes. Moreover, almost 50 % of the riparian zone (30 m at each side of the river) is not covered by native vegetation (Ribeiro et al., 2014). In this scenario, a high risk of phosphorus loss from agricultural areas was expected.

2   Materials and methods

2.1   Site description

The study area (1010 ha) was the Campestre watershed, Colombo, northern metropolitan region of Curitiba, Paraná, Brazil (Fig. 1). The climate is classified as Cfb (mesothermal humid subtropical) by Koeppen with cool summers and no dry season. The average of minimum and maximum temperature is 12 and 22 °C, respectively. The average annual rainfall over of the last 22 years amounted 1479 mm (Caviglione et al., 2000).

The slope was determined using topographic data (1 : 10 000 scale, with contour lines every 5 m in digital media). Agriculture occurred predominantly in areas with steep slopes (70 % of agriculture occurred with slope > 13 %) (Table 1). The predominant slope was 20–45 % followed by 13–20 % (representing 45 % and 24 % of the watershed, respectively).

Land cover and soil use was obtained from an aerial photography at a scale of 1 : 30 000 (Suderhsa, 2000) and revised by field survey (Ribeiro et al., 2014), describe as follow: (a) 44 % was covered by native vegetation, predominantly characterised by secondary forest at different stages of regeneration; (b) 23 % was covered with reforested wood species such as Mimosa scabrella Benth. and Eucalyptus spec.; (c) 19 % was used for agriculture predominantly for vegetable production including lettuce (Lactuca sativa L.), broccoli (Brassica oleracea var. italica L.), cauliflower (Brassica oleracea var. botrytis L.), squash (Cucurbita pepo L.), beetroot (Beta vulgaris L.), swiss chard (Beta vulgaris L.), cucumber (Cucumis sativus L.), tomato (Lycopersicon esculentum Mill.), pepper (Capsicum annuum L.), green beans (Phaseolus vulgaris L.); and (d) 14 % was used as animal pasture or unused.

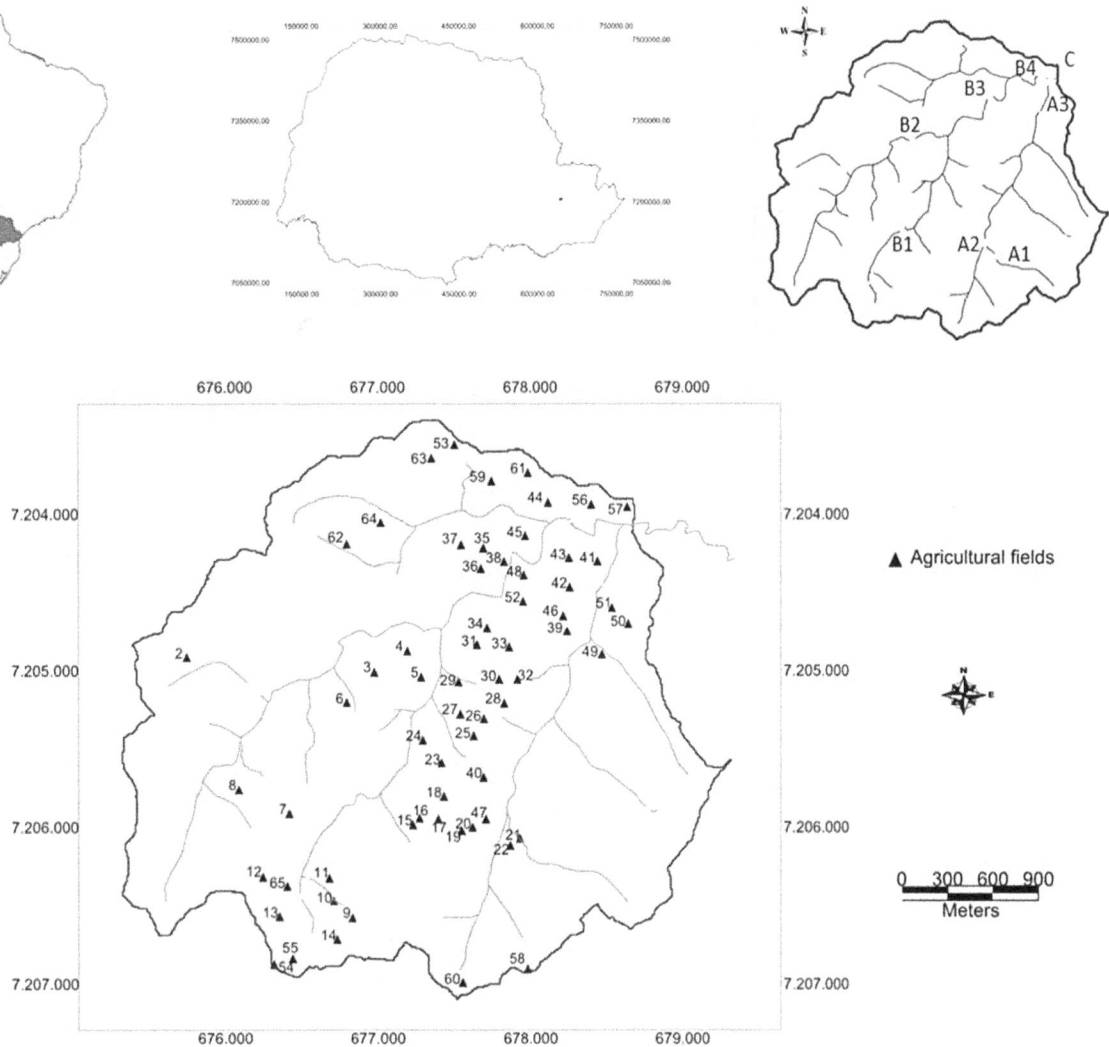

Fig. 1: *Localisation of the Campestre watershed, the sub-basins and the agricultural fields, Colombo, Paraná, Brazil.*

The soil management was based on conventional tillage (ploughing and harrowing predominantly with animal traction) with heavy use of mineral and organic fertilisers (poultry litter). However, some farmers practiced a conservation tillage based on the organic production.

Further information on Campestre watershed characterisation can be found in Ribeiro *et al.* (2014) and Ramos *et al.* (2014).

The soil survey was performed on 14 soil profiles coming from different landscape positions (summit, shoulder, backslope and footslope) based on Embrapa (2006). There were 11 soil units (loam and clay loam texture) distributed among Cambisol (50 %), Association of Cambisol and Leptosol (42 %), Leptosol (4 %), Ferralsol (3 %) and Gleysol (1 %) (FAO-World Reference Base for Soil Resources). Further information on soil classification of the watershed can be found in Ribeiro *et al.* (2014).

2.2 Versions of P-index

As no Brazilian P-index exists, five different versions from the U.S.A. were applied (Original – Lemunyon & Gilbert, 1993; New Mexico – Flynn *et al.*, 2000; Alabama – NRCS, 2001; Nebraska – Eghball & Gilley, 2001; Montana – Fasching, 2006). These five versions were chosen because they all used an additive framework, which means the final P-index is calculated from the sum of the site characteristics (factors). P-index with multiplicative frameworks were not chosen to facilitate the comparison using only additive framework. Table 2 describes the framework of the original version (Lemunyon & Gilbert, 1993) and Table 3 describes the contribution of each factor in the five different versions of P-index used in our study. The original and Nebraska versions consider the same factors (erosion and surface runoff as transport factors; soil P, rate and application method of mineral and organic fertilisers as source factors) but with different weights. The Alabama, Montana and New Mexico versions have all these factors plus the connectivity factors

Table 1: *Land use (ha) in each slope class in the Campestre watershed, Colombo, Paraná, Brazil.*

Sub-basin	Land use	Slope (%)							Σ
		0–3	>3–8	>8–13	>13–20	>20–45	>45–75	>75	
A1	Agriculture	0	0	0	0	0	0	0	0
	Whole	0	0.2	0.8	4.3	20.2	7.3	2.7	35
A2	Agriculture	0.1	0.6	2.9	2.0	2.9	0.2	0.0	8
	Whole	0.2	2.4	12.3	25.1	69.2	19.7	5.8	134
A3 (A1+A2)	Agriculture	0.1	1.1	7.7	7.6	10.6	1.3	0.2	28
	Whole	0.5	6.6	36.3	68.2	164.3	43.1	12.1	331
B1	Agriculture	0.1	4.2	7.4	10.5	8.8	0.9	0.1	32
	Whole	0.7	17.0	20.4	27.8	28.8	4.2	0.9	99
B2	Agriculture	0.1	1.3	3.4	4.2	3.5	0.3	0.0	13
	Whole	0.8	10.0	36.3	59.9	123.0	30.2	6.8	267
B3	Agriculture	0.5	9.9	25.2	39.6	39.7	3.1	0.5	118
	Whole	2.2	35	81.5	133.0	211.6	41.9	9.8	515
B4 (B1+B2+B3)	Agriculture	0.5	14.0	32.2	51.0	58.0	5.8	1.3	163
	Whole	2.5	44.0	104	170	284.4	56.6	14.1	675
C (A3 + B4)	Agriculture	0.6	15.0	40.4	58.9	68.8	7.2	1.6	192
	Whole	3	50	141	240	450	100	26	1010

Table 2: *Framework of the original P-index (after Lemunyon & Gilbert, 1993).*

Site characteristics (factors)	Weight	Phosphorus loss rating (value)				
		None (0)	Low (1)	Medium (2)	High (4)	Very High (8)
Transport factors:						
1. Soil erosion (t ha^{-1} year^{-1})[†]	1.5	Not applicable	<12	12–25	25–37	>37
2. Runoff class	0.5	Negligible	Very Low or Low	Medium	High	Very High
Source factors:						
1. Soil P test	1.0	Not applicable	Low	Medium	High	Very High
2. P fertiliser (mineral) application rate (kg ha^{-1} year^{-1} P$_2$O$_5$)[‡]	0.75	None applied	1–34	35–100	101–168	>168
3. P fertiliser (mineral) application method	0.5	None applied	Placed with planter deeper than 5 cm[§]	Incorporated immediately before crop	Incorporated >3 months before crop or surface applied <3 months before crop	Surface applied > 3 months before crop
4. P fertiliser (organic) application rate (kg ha^{-1} year^{-1} P$_2$O$_5$)[‡]	1.0	None applied	1–34	35–67	68–100	>100
5. P fertiliser (organic) application method	1.0	None applied	Injected deeper than 5 cm[§]	Incorporated imediatley before crop	Incorporated >3 months before crop or surface applied <3 months before crop	Surface applied to pasture or >3 months before crop

[†] Unit transformed from t ac^{-1} to t ha^{-1}; [‡] Unit transformed from lbs ac^{-1} to kg ha^{-1}; [§] Unit transformed from inches (in.) to centimetres (cm).
Final P-index = $\sum$ (factor $*$ weight).

Table 3: *Contribution of each factor in the five versions of P-index used in our study[†].*

Site characteristics (factors)	Original Weight	(%)	Nebraska Weight	(%)	Alabama in Weight	(%)	out Weight	(%)	Montana in Weight	(%)	out Weight	(%)	New Mexico in Weight	(%)	out Weight	(%)
Transport factors:																
1. Soil erosion	1.5	24	4.0	50	3.0	16	3.0	21	1.5	19	1.5	21.4	1.5	15	1.5	21.4
2. Runoff class	0.5	8	0.5	6	4.0	21	4.0	30	0.5	6	0.5	7.1	1.5	15	1.5	21.4
Source factors:																
1. Soil P test	1.0	16	0.5	6	1.0	5	1.0	7	1.0	12.5	1.0	14.3	1.0	10	1.0	14.3
2. P fertiliser (mineral or mineral + organic) application rate	0.75	12	0.5	6	3.0	16	3.0	21	1.0	12.5	1.0	14.3	1.0	10	1.0	14.3
3. P fertiliser (mineral or mineral + organic) application method	0.5	8	1.0	13	3.0	16	3.0	21	1.0	12.5	1.0	14.3	1.0	10	1.0	14.3
4. P fertiliser (organic) application rate	1.0	16	0.5	6					1.0	12.5	1.0	14.3				
5. P fertiliser (organic) application method	1.0	16	1.0	13					1.0	12.5	1.0	14.3	1.0	10	1.0	14.3
Connectivity factors:																
1. Distance between agricultural field and streams					3.0	16			1.0	12.5			1.5	15		
2. Riparian filter strip width					2.0	10							1.5	15		

[†] *in* = with connectivity factors; *out* = without connectivity factors.

Table 4: *Interpretation of risk of P loss for the different P-index versions.*

Risk of P loss	Original (Lemunyon & Gilbert, 1993)	Nebraska (Eghball & Gilley, 2001)	Alabama (NRCS, 2001)	Montana (Fasching, 2006)	New Mexico (Flynn et al., 2000)
Very low					0–10
Low	<8	<3	<65	<11	11–17
Medium	8–14	3–6.5	66–75	11–21	18–27
High	15–32	6.6–10	76–85	22–43	28–37
Very high	>32	>10	86–95	>43	38–47
Extremely high			>95		>47

(distance between agricultural field and the stream and/or width of the filter strip) in its framework. The final P-index score indicates the relative vulnerability to P loss from a field following each version (Table 4).

The P-index was applied at three scales: watershed (C); sub-basin (seven sub-basins: A1, A2, A3, B1, B2, B3 and B4) (Fig. 1 and Table 1) and field (65 agricultural fields) (Fig. 1). The agricultural fields represented all hillslopes with vegetables production existent in the watershed.

The similarity of P-index at sub-basin and agricultural fields was analysed by the Euclidian distance and the cluster analysis was applied using the MATLAB software (Math-Works, 2017).

2.3 Estimation of P-index

The P-index was estimated using IDRISI 15.0 software (Eastman, 1999). In the following, the estimation of the source, transport and connectivity factors used for the calculation of the P-index is briefly described. More detailed information on these factors can be found in Waltrick (2011).

2.3.1 Estimation of source factors

(a) Soil P content: Soil samples (73 samples composed of 20 sub-samples) were taken at a depth of 0–20 cm on 65 agricultural fields, three representative grassland fields and five representative forest fields. The samples were air dried,

homogenized and passed through a 2 mm mesh. The P Mehlich I (the standard soil test for farmers in Paraná State) was determined according to Sparks (1996) and Pavan *et al.* (1992). The soil P status (low, medium, high, very high) was classified following the recommendation for Paraná state (SBCS, 2004).

(b) Amount and method of organic and mineral P application: Information was obtained through interviews with the farmers of all agricultural fields. Cauliflower and Swiss chard were regarded as the main crops as these were most prevalent and received the larger quantity of mineral and organic fertilisers.

2.3.2 Estimation of transport factors

(a) Erosion: The risk of soil loss was not available from any national source. Therefore, soil loss was estimated using the Revised Universal Soil Loss Equation ($A = R K LS C P$) (Renard *et al.*, 1997) with IDRISI 15.0 software (Eastman, 1999). Rainfall erosivity (R) was calculated according to Rufino *et al.* (1993), using a historic series from 1988 to 2009. Soil erodibility (K) was calculated according to Roloff & Denardin (1994). Slope (LS) was calculated according to Moore & Burch (1986) and Engel & Mohtar (2006). Input data on cover management (C) and conservation practices (P) was obtained from Bertoni & Lombardi Neto (1999).

(b) Surface runoff: Surface runoff (water loss) was estimated according to soil permeability and slope based on Fasching (2006). Permeability was determined in the field for each soil class as reported by Santos *et al.* (2005) while slope was defined using the SPRING 15.0 software with topographic data in digital media (1 : 10 000 scale, with contour lines every 5 m).

2.3.3 Estimation of connectivity factors

Riparian filter strip width and distance between agricultural land and the watercourse (in meters) was determined using the IDRISI 15.0 software (Eastman, 1999).

3 Results

3.1 Evaluation of source, transport and connectivity factors

The average soil P Mehlich I in 0-20 cm was 2.5 mg dm^{-3} in the forest vegetation areas ($n = 3$), 2.7 mg dm^{-3} in the grassland ($n = 5$) and 120.7 mg dm^{-3} in the agriculture fields ($n = 65$; varying from 9.1 to 325.2 mg dm^{-3} P). In loam and clay loam texture, soil P Mehlich I above 24.0 mg dm^{-3} is classified as very high in Paraná state (SBCS, 2004). According to all five P-index estimation

versions most results of the soil P analyses for agricultural land in the Campestre watershed were classified as very high risk of P loss (data not shown). The high soil P content is due to the large quantities of P being applied as fertilisers to crops in the summer and winter seasons. An estimated 48 Mg ha^{-1} year^{-1} of poultry litter was applied, a total of 1.152 kg P$_2$O$_5$ ha^{-1} year^{-1} (503 kg P ha^{-1} year^{-1}) of organic fertilization. In the conventional system, farmers applied organic fertiliser (48 Mg ha^{-1} year^{-1}) plus 4 Mg ha^{-1} year^{-1} of mineral fertiliser (10-10-10), 400 kg P$_2$O$_5$ ha^{-1} year^{-1} (175 kg P ha^{-1} year^{-1}) of mineral fertilization (Ribeiro *et al.*, 2014). 120 to 450 kg of P$_2$O$_5$ ha^{-1} depending on soil P content are the amounts recommended for most vegetable crops in Paraná State (SBCS, 2004). The application of organic and mineral fertilisers occurs immediately before manual planting, which gives a medium risk of P loss in all P-index versions.

The highest estimated soil loss occurred on cultivated land. Out of the 192 ha used for agriculture (Table 1), 175 ha showed soil losses of above 37 t ha^{-1} year^{-1}. This rate is classified as very high risk of P loss in all five P-index versions. Surface runoff was classified as high on more than 52 % of the total watershed. High runoff associated with high soil loss indicates an increased risk of P loss, especially on the cultivated fields.

According to Brazilian law (Brasil, 2012), 30 m on each side of a stream should be preserved with native vegetation as riparian buffer. For the Campestre watershed, using the buffer routine from IDRISI 15.0, this would mean 135 ha of riparian zone, however, only 57 % of this riparian zone was covered with native forest (Fig. 2). Sub-basin A was the most protected by riparian vegetation: 60.1 % was covered by native forest, 22.3 % was covered with reforested wood species, 11.2 % was used as animal pasture or fallow land and only 6.4 % was agricultural area. On the other hand 19 % of the riparian zone in sub-basin B was agricultural area. Also, most agricultural production occurred very near or inside the riparian zone (Fig. 2). Of the 192 ha under cultivation, 20 ha were located inside the riparian zone and only 50 ha were more than 300 m away from a watercourse.

3.2 P-index at watershed, sub-basin and agricultural field scale

The five P-index versions resulted in a similar risk of P loss, 75 % to 83 % of the whole watershed was classified as low or very low risk of P loss. Additionally, while analysing the P-index for the agricultural area, it was noted that 79 to 100 % of this area was classed as high and very high risk of P loss (Table 5). Exemplarily, Fig. 3 illustrates the results of the original P-index (Lemunyon & Gilbert, 1993).

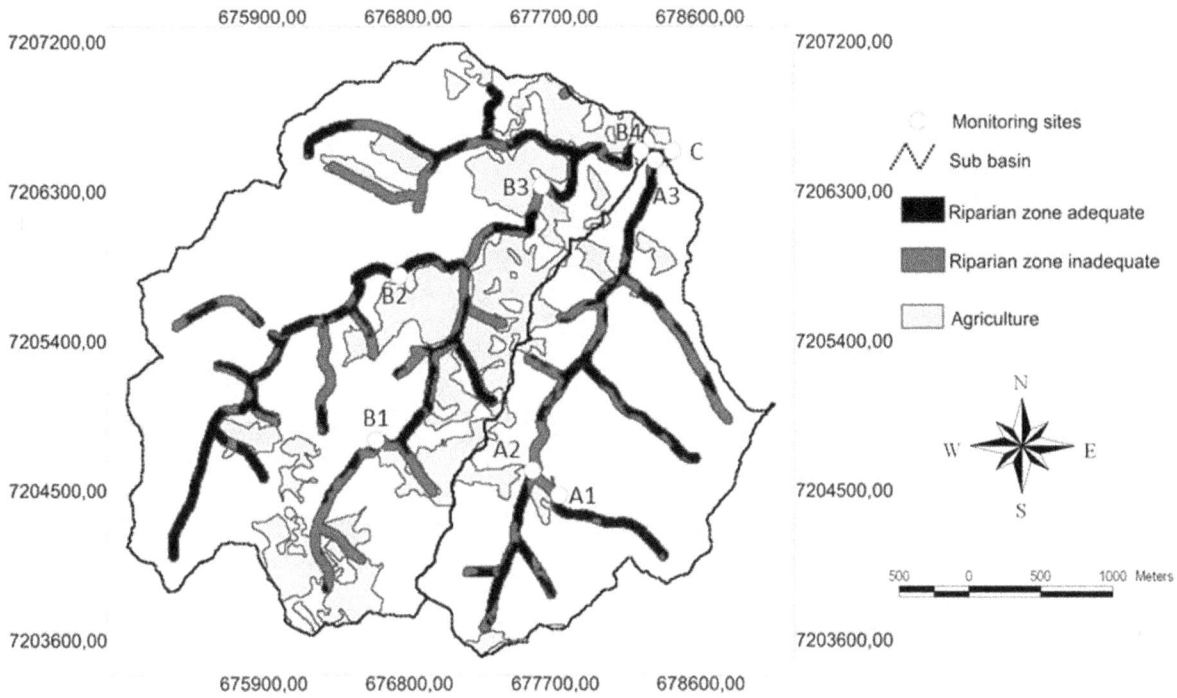

Fig. 2: *Agricultural area and riparian zone in the Campestre watershed, Colombo, Paraná, Brazil.*

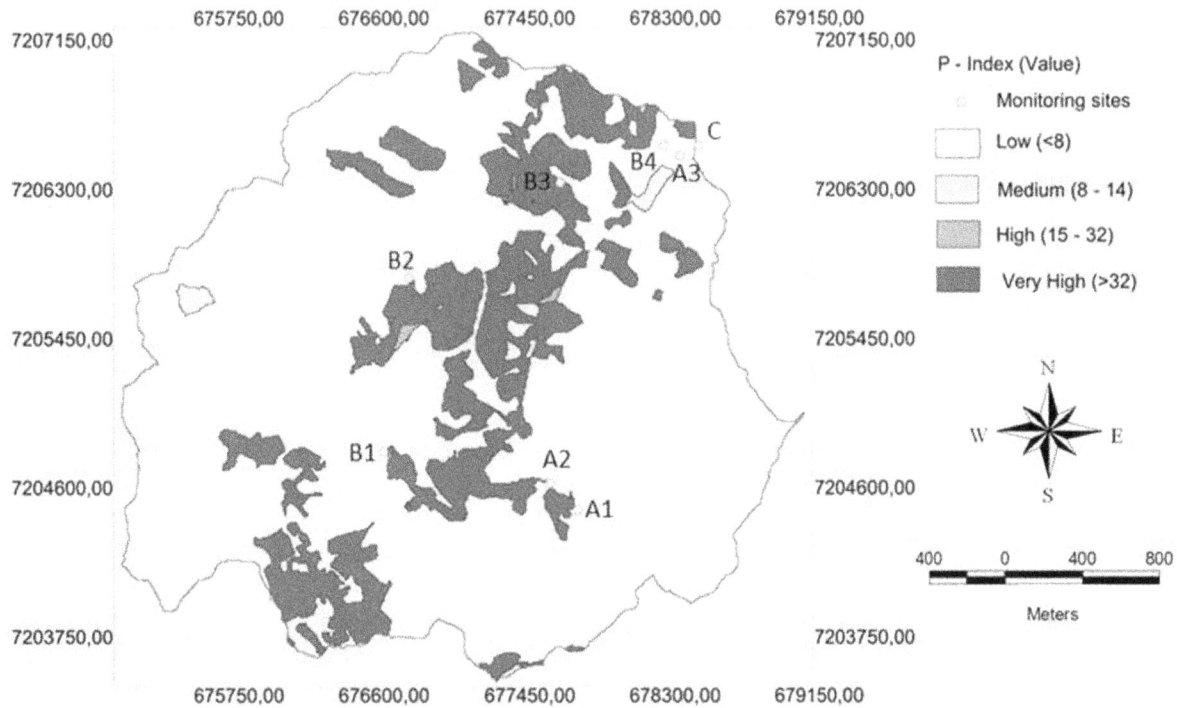

Fig. 3: *P-index estimation based on Lemunyon & Gilbert (1993) for the different sub-basins of the Campestre watershed, Colombo, Paraná, Brazil.*

Table 5: *Estimation of the risk of P loss using different P-index versions for the whole watershed (Whole) and for the agricultural area (Agric.) in the Campestre watershed, Colombo, Paraná, Brazil[†].*

Risk of P loss (P-index class)	% of the P-index class									
	Original (Lemunyon & Gilbert, 1993)		Nebraska (Eghball & Gilley, 2001)		Alabama (NRCS, 2001)		Montana (Fasching, 2006)		New Mexico (Flynn et al., 2000)	
	Whole	Agric.	Whole	Agric.	Whole	Agric.	Whole	Agric.	Whole	Agric.
Very Low									80.9	
Low	80.9	–	74.5	0.3	82.5	5.5			0.1	–
Medium	0.1	0.1	7.0	3.1	2.9	15.6	80.9	–	0.5	1.6
High	2.6	11.7	0.6	0.3	3.1	16.8	0.5	–	2.4	13.1
Very High	16.4	88.2	17.9	96.3	1.0	5.5	0.2	1.2	4.8	24.7
Extremely High					10.5	56.5	18.4	98.8	11.3	60.6

[†] Alabama, Montana and New Mexico P-index with connectivity factors.

Table 6: *Estimated P-index (mean value) at sub-basin scale (whole sub-basin (Whole) and for the agricultural area (Agric.) in the Campestre watershed, Colombo, Paraná, Brazil[†].*

Sub-basin	Original (Lemunyon & Gilbert, 1993)		Nebraska (Eghball & Gilley, 2001)		Alabama (NRCS, 2001)		Montana (Fasching, 2006)		New Mexico (Flynn et al., 2000)	
	Whole	Agric.	Whole	Agric.	Whole	Agric.	Whole	Agric.	Whole	Agric.
A1	4	–	2	–	11	–	5	–	5	–
A2	6	29	2	13	16	85	9	56	8	46
A3	6	39	2	13	16	86	8	57	8	46
B1	15	29	5	13	38	99	21	58	20	53
B2	5	39	2	13	15	96	7	56	8	52
B3	12	30	4	13	30	95	16	57	16	52
B4	12	31	4	13	30	94	17	57	16	51
C (A+B)	11	39	3	13	25	93	14	57	14	50

[†] Alabama, Montana and New Mexico P-index with connectivity factors.

At a sub-basin scale, there was a greater risk of P loss in sub-basin B than in sub-basin A (Table 6).

We compared the mean P-index across seven sub-basins and between 65 agricultural plots. We observed a greater value as well as a greater variation in P loss risk across fields than at the sub-basin scale. For example, the result (mean value) of the original P-index at sub-basin scale (whole area of the sub-basin) varied from 4 to 15 and at agricultural fields varied from 29 to 39 (Table 6). For the 65 agricultural fields, the minimum P-index was 14 and the maximum was 40 (data not shown). Bechmann et al. (2007), testing the Norwegian P-index across 50 fields and 9 sub-basins also observed a greater difference in risk of P loss at the field scale.

The Euclidian distance and cluster analysis showed the difference of scales between the cluster "agricultural fields" and the cluster "sub-basins" (Fig. 4a).

3.3 Effect of the connectivity factors

Connectivity factors such as distance between cropped fields and water streams, which effectively means the distance between P application and water course and/or riparian filter strip width are included in most P-index versions (Sharpley et al., 2003; Buczko and Kuchenbuch, 2007). From the five versions applied in this study, three of these included the connectivity factor in their frameworks (Table 3). According to the New Mexico P-index version (Flynn et al., 2000), the risk of P loss is very low with riparian buffer > 30 m. The Alabama P-index (NRCS, 2001) is less restrictive; the risk of P loss is classified as very

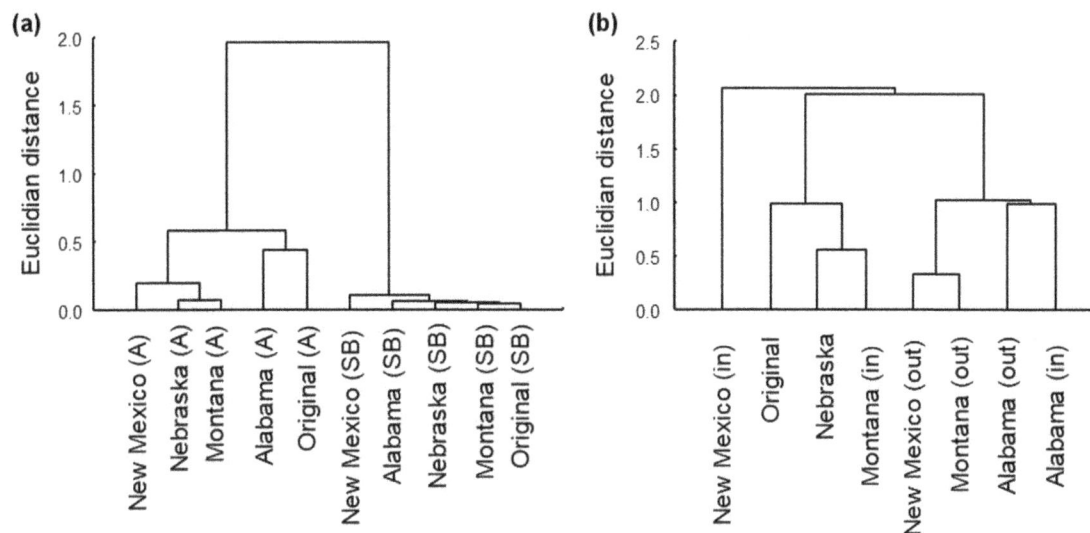

Fig. 4: *Dendrogram of the P-index versions in the Campestre watershed, Colombo, Paraná, Brazil: (a) agricultural fields (A) and sub-basins (SB); (b) agricultural fields with (in) and without (out) the connectivity factors.*

low with riparian buffer >15 m. According to the Montana P-index (Fasching, 2006), the risk of P loss and distance between watercourse and agriculture was: < 30 m, very high and > 600 m, very low. The New Mexico (Flynn *et al.*, 2000) is less restrictive (very low risk of P loss with distance from the agriculture field to the stream > 300 m and very high risk with distance < 9 m). The Alabama (NRCS, 2001) also is less restrictive (very low risk of P loss with distance from the agriculture field to the stream > 122 m and very high risk with distance < 5 m).

The mean P-index calculated for a whole watershed in the Alabama version varied from 24 to 25 (without and with the connectivity factors, respectively), remaining in the same risk of P loss (low). However, the mean P-index of agricultural area varied from 84 to 93 (without and with the connectivity factors, respectively), increasing the risk of P loss from high to very high (Table 7).

Taking into account the distance between cropped fields and water streams in the Montana version, the P-index values for the agricultural area were also modified (Table 7). In this version, without considering the connectivity factor, the average P-index was 27 (high risk of P loss). However, when this factor was considered, the average P-index increased to 57 (very high risk of P loss).

In the New Mexico version, there was an increase in the estimated P-index value with the connectivity factors. The average P-index across the whole watershed increased from 10 to 14 and for agricultural area from 35 to 50 (Table 7), modifying the risk from high to extremely high.

3.4   P indices at agricultural field scale

The cluster analysis identified the similarity among the P-index versions tested at agricultural field scale (Fig. 4b). In general, the versions Alabama (*in* and *out*), Montana (*out*) and New Mexico (*out*) formed a well-defined cluster (Fig. 4b) with 23 to 62 % of agricultural fields classified as very high and extremely high risk of P loss (Fig. 5). The Original, Nebraska, and Montana (*in*) formed the second cluster with 88 to 99 %, and New Mexico (*in*) individually formed a third cluster (Fig. 4b) with 85 % within the very high and extremely high risk of P loss category (Fig. 5). The Montana and New Mexico versions were sensitive to the connectivity factors (Fig. 4b). When the distance between agriculture field and the stream and/or the riparian filter strip were considered, the classification of very high and extremely high risk of P loss decreased from 99 to 49 % for the Montana version and from 85 to 23 % for the New Mexico version (Fig. 5). The Alabama version also decreased the proportion of areas classified as very high and extremely high risk of P loss when the connectivity factor was included (Fig. 5), however, the cluster analysis identified still similarities between both cases (Fig. 4b). The Original and Nebraska versions do not consider connectivity factors in their frameworks, but these versions also showed high percentage of areas classified as very high and extremely high risk of P loss. These results are explained by the soil erosion factor weighting 24 and 50 %, respectively (Table 3). Even without considering the distance between agriculture field and the stream and the riparian filter strip, 92 % (on average) of the agricultural fields was classified as very high and extremely high risk of P loss (Fig. 5).

Table 7: *P-index (mean value and risk) at whole watershed and at agricultural area with (in) and without (out) the connectivity factor in the Campestre watershed, Colombo, Paraná, Brazil.*

Alabama (NRCS, 2001)				*Montana* (Fasching, 2006)				*New Mexico* (Flynn *et al.*, 2000)			
Whole		Agric.		Whole		Agric.		Whole		Agric.	
in	*out*	*in*	*out*	*in*	*out*	*in*	*out*	*in*	*out*	*in*	*out*
25	24	93	84	14	11	57	27	14	10	50	35
low	low	very high	high	medium	medium	very high	high	low	very low	extremely high	high

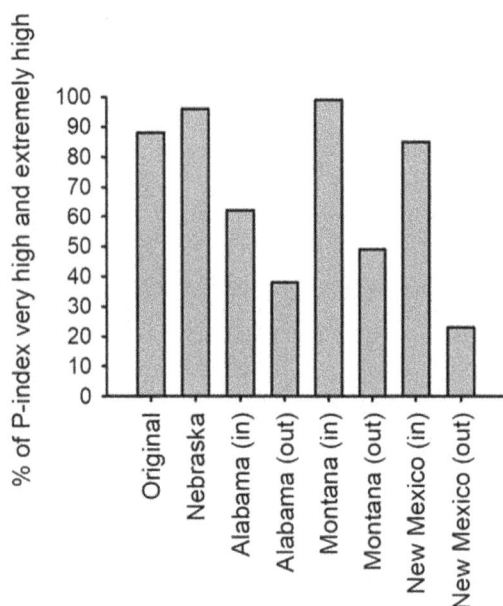

Fig. 5: *Percentage of areas for the agricultural fields with P-index very high and extremely high of versions with (in) and without (out) the connectivity factor.*

4 Discussion

The low risk of P loss on the whole watershed is explained by the presence of forest vegetation (67 % of the area is occupied by forest vegetation and only 19 % by agriculture). On the other hand, the high risk of P loss for the agricultural area is explained by the steep slopes (more than 70 % of the cropped fields are located on slopes steeper than 13 %). Soil loss and surface runoff are greatly affected by slope, which strongly influences P transport (Eghball & Gilley, 2001). Delaune *et al.* (2004) found a greater effect of surface runoff on P loss in pastures systems. However, in conventional systems, the contribution of soil loss (erosion) is usually higher than water loss (surface runoff) (Gburek *et al.*, 2000). In addition to slope, the high soil P content (Pote *et al.*, 1999) and rates and source of P application (Shigaki *et al.*, 2007) contribute to high risk of P loss.

The greater P-index at sub-basin B than sub-basin A could be explained by land use and connectivity; 24 % of sub-basin B is intensively used for vegetable cultivation, while this only accounts for 8 % in sub-basin A; 41 % of agriculture in sub-basin B occurs on slopes > 20 % and 19 % of the riparian zone is covered by agriculture. In sub-basin A only 6 % of the riparian zone is used for agriculture, on the other hand, 60 % of agriculture occurred on slopes > 20 %. Ribeiro *et al.* (2014), studying the same sub-basins, observed a better water quality at sub-basin A. The transport of pollutants from soil to water is highly related to the land use and soil management (Sharpley *et al.*, 2014).

The connectivity factor (such as distance between cropped fields and water streams and riparian filter strip width) was more important when determining P-index for agricultural areas than at a whole watershed scale, especially because the Campestre watershed has high forest coverage. The increased P-index for versions with connectivity factor can be explained by the lack of riparian native vegetation and by the short distance between cultivated fields and watercourse. Agricultural land closer to the stream and without required riparian vegetation have greater soil and water loss (Nair & Graetz, 2004) and consequently a greater risk of P loss (Flynn *et al.*, 2000; NRCS, 2001; Fasching, 2006).

High vulnerability for P loss (high P-index) on agricultural land in the Campestre watershed highlights the need for improved management practices to control soil erosion and surface runoff (transport factors). It is also necessary to reduce P fertilisation (factor source) avoiding future problems with water contamination (Sharpley & Wang, 2014). In freshwaters, phosphorus is the main nutrient associated with eutrophication, so conservation measures across the watersheds need to be implemented (Sharpley *et al.*, 2014). The P-index was developed to estimate the vulnerability of P loss from soil to waters (Lemunyon & Gilbert, 1993) and it is an important tool for farmers, field staff and watershed planners (Sharpley *et al.*, 2003; Sharpley *et al.*, 2001)

So, the five P-index versions can be recommended as a tool to rank the vulnerability to P loss in surface runoff and subsequently to recommend best management practices. However, as the P-index increased when connectivity factors (distance between cropped fields and the stream and/or riparian filter strip) was included, we suggest to use at agricultural field scale a P-index incorporating connectivity in its framework.

Acknowledgements

The authors thank the National Council for Scientific and Technological Development (CNPq), the farmers from the watershed, and colleagues who helped with the field and laboratory work.

References

Bechmann, M., Krogstad, T. & Sharpley, A. N. (2005). A phosphorus index for Norway. *Acta Agriculturae Scandinavica, Section B – Soil & Plant Science*, 55 (3), 205–213.

Bechmann, M. E., Stalnacke, P. & Kvaerno, S. H. (2007). Testing the Norwegian phosphorus index at the field and subcatchment scale. *Agricultural Ecosystems & Environment*, 120, 117–128.

Bertoni, J. & Lombardi Neto, F. (1999). *Conservação do solo*. (4th ed.). Editora Ícone, São Paulo. 355 pp. (in Portuguese)

Brasil (2012). Lei Federal 12.651 Novo Código Florestal Brasileiro. Available at: http://www.planalto.gov.br/ccivil_03/_ato2011-2014/2012/lei/l12651.htm (accessed on: 26 March 2018).

Buczko, U. & Kuchenbuch, R. O. (2007). Phosphorus indices as risk-assessment tools in the U.S.A. and Europe- a review. *Journal of Plant Nutrition and Soil Science*, 170, 445–460.

Caviglione, J. H., Kiihl, L. R. B., Caramori, P. H. & Oliveira, D. (2000). Cartas climáticas do Paraná. Iapar, Londrina, CD.

Correll, D. L. (1998). The role of phosphorus in the eutrophication of receiving waters: A review. *Journal of Environmental Quality*, 27, 261–266.

Couto, R. R., Santos, M., Comin, J., Martini, L. C. P., Gatiboni, L. C., Martins, S. R., Belli Filho, P. & Brunetto, G. (2015). Environmental vulnerability and phosphorus fractions of areas with pig slurry applied to the soil. *Journal of Environmental Quality*, 44, 162–173.

Daniel, T. C., Sharpley, A. N. & Lemunyon, J. L. (1998). Agricultural phosphorus and eutrophication: A symposium overview. *Journal of Environmental Quality*, 27, 251–257.

Delaune, P. B., Moore, P. A., Carman, D. K., Sharpley, A. N., Haggard, B. E. & Daniel, T. (2004). Development of a phosphorus index for pastures fertilized with poultry litter – Factors affecting phosphorus runoff. *Journal of Environmental Quality*, 33, 2183–2191.

Eastman, J. R. (1999). *Guide to GIS and image processing*. Clark University, Worcester. 193 pp.

Eghball, B. & Gilley, J. E. (2001). Phosphorus risk assessment index evaluation using runoff measurements. *Journal of Soil and Water Conservation*, 56, 202–207.

Embrapa (2006). Sistema Brasileiro de Classificação de Solos. 2 ed., Embrapa, Rio de Janeiro, Brasília, DF. Available at: http://livraria.sct.embrapa.br/liv_resumos/pdf/00053080.pdf (accessed on: 26 March 2018).

Engel, B. & Mohtar, R. (2006). Estimating soil erosion using RUSLE and the ArcView GIS. Available at: http://pasture.ecn.purdue.edu/~abe526 (accessed on: 1 June 2010).

Fasching, R. A. (2006). Phosphorus index assessment for Montana. Agronomy Technical Note 77. United State Department of Agriculture – Natural Resources Conservation Service.

Flynn, R., Sporcic, M. & Scheffe, L. (2000). Phosphorus assessment tool for New Mexico. Agronomy Technical Note 57. United State Department of Agriculture – Natural Resources Conservation Service.

Gburek, W. J., Sharpley, A. N., Heathwaite, A. L. & Folmar, G. J. (2000). Phosphorus management at the watershed scale: a modification of the phosphorus Index. *Journal of Environmental Quality*, 29, 130–144.

Heathwaite, L., Sharpley, A. & Bechmann, M. (2003). The conceptual basis for a decision support framework to assess the risk of phosphorus loss at the field scale across Europe. *Journal of Plant Nutrition and Soil Science*, 166, 447–458.

Hooda, P. S., Edwards, A. C., Anderson, H. A. & Miller, A. (2000). A review of water quality concerns in livestock farming areas. *Science of the Total Environment*, 25, 143–167.

Kay, P., Edwards, A. C. & Foulger, M. (2009). A review of the efficacy of contemporary agricultural stewardship measures for ameliorating water pollution problems of key concern to the UK water industry. *Agricultural Systems*, 99, 67–75.

Kleinman, P. J. A., Sharpley, A. N., Noyer, B. G. & Elwinger, G. (2002). Effect of mineral and manure phosphorus sources on runoff phosphorus. *Journal of Environmental Quality*, 31, 2026–2033.

Leinweber, P., Turner, B. L. & Meissner, R. (2002). Phosphorus. *In:* Haygarth, P. M. & Jarvis, S. C. (eds.), *Agriculture, hydrology and water quality.* pp. 29–55, CAB International, Cambridge.

Lemunyon, J. L. & Gilbert, R. G. (1993). The concept and need for a phosphorus assessment tool. *Journal of Production Agriculture*, 6, 483–486.

Lopes, F., Merten, G. H., Franzen, M., Giasson, E., Helfer, F. & Cybis, L. F. A. (2007). Utilização de P-index em uma bacia hidrográfica através de técnicas de geoprocessamento. *Revista Brasileira de Engenharia Agrícola e Ambiental*, 11, 312–317. (in Portuguese)

MathWorks (2017). MATLAB. The language of technical computing. Available at: https://www.mathworks.com (accessed on: 26 March 2018).

Moore, I. & Burch, G. (1986). Physical basis of the length-slope factor in the universal soil loss equation. *Soil Science Society of America Journal*, 29, 1624–1630.

Nair, V. D. & Graetz, D. A. (2004). Agroforestry as an approach to minimizing nutrient loss from heavily fertilized soils: the Florida experience. *Agroforestry Systems*, 61, 269–279.

NRCS (2001). Phosphorus index for Alabama: A planning tool to asses and manage P movement. Agronomy Technical Note 72. United State Department of Agriculture – Natural Resources Conservation Service. United State Dep.

Oliveira, M. F. M., Favaretto, N., Roloff, G. & Fernandes, C. V. S. (2010). Estimativa do potencial de perda de P através da metodologia "P-index". *Revista Brasileira de Engenharia Agrícola e Ambiental*, 14, 267–273. (in Portuguese)

Pavan, M. A., Bloch, M. F., Zempulski, H. C., Miyazawa, M. & Zocoler, D. C. (1992). Manual de análise química de solo e controle de qualidade. Londrina, IAPAR, Circular 76. (in Portuguese)

Pote, D. H., Daniel, T. C., Nichols, D. J., Sharpley, A. N., Moore, P. A., Miller, D. M. & Edwards, D. R. (1999). Relationship between phosphorus levels in three ultisols and phosphorus concentrations in runoff. *Journal of Environmental Quality*, 28, 170–175.

Ramos, M. R., Favaretto, N., Dieckow, J., Dedeck, R. A., Vezzani, F. M., Almeida, L. & Sperrin, M. (2014). Soil, water and nutrient loss under conventional and organic vegetable production managed in small farms versus forest system. *Journal of Agriculture and Rural Development in the Tropics and Subtropics*, 115 (1), 31–40.

Reid, D. K. (2011). A modified Ontario P-index as a tool for on-farm phosphorus management. *Canadian Journal of Soil Science*, 91, 455–466.

Renard, K. G., Foster, G. R., Weesies, G. A., Mccoll, D. L. & Yoder, D. C. (1997). Predicting soil erosion by water. A guide to conservation planning with the Revised Universal Soil Loss Equation. USDA. Agricultural Handbook 703.

Ribeiro, K. H., Favaretto, N., Dieckow, J., de Paula Souza, L. C., Minella, J. P. G., de Almeida, L. & Ramos, M. R. (2014). Quality of surface water related to land use: a case study in a catchment with small farms and intensive vegetables production in southern Brazil. *Revista Brasileira de Ciência do Solo*, 38 (2), 656–668.

Roloff, G. & Denardin, J. E. (1994). Estimativa simplificada da erodibilidade do solo. *In:* Resumos da X Reunião Brasileira de Manejo e Conservação do Solo e Água. Florianópolis, SBCS. (in Portuguese)

Rufino, R. L., Biscaia, R. C. M. & Merten, G. H. (1993). Determinação do potencial erosivo da chuva do estado do Paraná através da pluviometria: terceira aproximação. *Revista Brasileira de Ciência do Solo*, 17, 439–444. (in Portuguese)

Santos, R. D., Lemos, R. C., Santos, H. G., Ker, J. C. & Anjos, L. H. C. (2005). *Manual de descrição e coleta de solo no campo.* (5th ed.). Viçosa, Sociedade Brasileira de Ciência do Solo. (in Portuguese)

SBCS (2004). *Manual de adubação e de calagem para os Estados do Rio Grande do Sul e de Santa Catarina.* (10th ed.). Sociedade Brasileira de Ciência do Solo (SBCS), Comissão de Química e Fertilidade do Solo, Porto Alegre. Available at: http://www.sbcs-nrs.org.br/docs/manual_de_adubacao_2004_versao_internet.pdf

Schindler, D. W., Hecky, R. E., Findlay, D. L., Stainton, M. P., Parker, B. R. & Paterson, M. J. (2008). Eutrophication of lakes cannot be controlled by reducing nitrogen input: results of a 37-year whole-ecosystem experiment. *Proceedings of the National Academy of Sciences USA*, 105 (32), 11254–11258.

Sharpley, A. (1995). Identifying sites vulnerable to phosphorus loss in agricultural runoff. *Journal of Environmental Quality*, 24, 947–951.

Sharpley, A., Jarvie, H. P., Buda, A., May, L., Spears, B. & Kleinman, P. (2014). Phosphorus legacy: overcoming the effects of past management practices to mitigate future water quality impairment. *Journal of Environmental Quality*, 42, 1308–1326.

Sharpley, A. & Wang, X. (2014). Managing agricultural phosphorus for water quality: Lessons from the USA and China. *Journal of Environmental Sciences*, 26 (9), 1770–1782.

Sharpley, A. N., Mcdowell, R. W. & Kleinman, J. A. (2001). Phosphorus loss from land to water: Integrating agricultural and environmental management. *Plant and Soil*, 237, 287–307.

Sharpley, A. N., Weld, J. L., Beegle, D. B., Kleinman, P. J. A., Gburek, W. J., Moore Jr, P. A. & Mullins, G. (2003). Development of phosphorus indices for nutrient management planning strategies in the United States. *Journal of Soil and Water Conservation*, 58, 137–153.

Shigaki, F., Sharpley, A. N. & Prochnow, L. I. (2006). Animal-based agriculture, phosphorus management and water quality in Brazil: Options for the future. *Scientia Agricola*, 63 (2), 194–209.

Shigaki, F., Sharpley, A. N. & Prochnow, L. I. (2007). Rainfall intensity and phosphorus source effects on phosphorus transport in surface runoff from soil trays. *Science of the Total Environment*, 373, 334–343.

Sparks, D. L. (1996). *Methods of soil analysis. Part 3 Chemical Methods*. Soil Science Society of America (SSSA), Madison.

Suderhsa (2000). *Fotografias aéreas coloridas. Escala 1:30.000*. Instituto das Águas do Paraná. Available at: http://www.aguasparana.pr.gov.br/modules/conteudo/conteudo.php?conteudo=78

Veith, T. L., Sharpley, A. N., Weld, J. L. & Gburek, W. J. (2005). Comparison of measured and simulated phosphorus losses with indexed site vulnerability. *Transactions of the ASAE*, 48, 557–565.

Waltrick, J. C. N. (2011). *Aplicação da metodologia P-index na bacia hidrográfica do campestre-Colombo, PR*. Master's thesis, Universidade Federal do Paraná, Curitiba, Brazil. (in Portuguese).

17

Physical and chemical optimisation of the seedball technology addressing pearl millet under Sahelian conditions

Charles I. Nwankwo [a,*], Jan Mühlena [a], Konni Biegert [a], Diana Butzer [a], Günter Neumann [b], Ousmane Sy [c], Ludger Herrmann [a]

[a] *Institute of Soil Science and Land Evaluation, University of Hohenheim, Emil-Wolff Str. 27, 70599 Stuttgart, Germany*
[b] *Nutritional Crop Physiology, Institute of Crop Science, University of Hohenheim, Fruwirth Str. 20, 70599 Stuttgart Germany*
[c] *Institut Sénégalais de Recherches Agricoles (ISRA), Bambey, Senegal*

Abstract

This study deals with the development of the seedball technology in particular for dry sowing under Sahelian conditions and pearl millet as crop. At first, our participatory evaluation in Senegal showed that (i) local materials needed for seedball production are locally available, (ii) the technology conforms to the existing management systems in the Sahel, and (iii) socio-economic conditions do not hinder seedball adoption. Afterwards, seedball was mechanically and chemically optimised. Pearl millet seedlings derived from the seedball variants were grown and compared to the control under greenhouse conditions. Our results showed that the combination of 80 g sand + 50 g loam + 25 ml water is the standard seedball dough, which produces about ten 2 cm diameter-sized seedballs. Either 1 g NPK fertiliser or 3 g wood ash can be added as nutrient additive to enhance early biomass of pearl millet seedlings. Ammonium fertiliser, urea and gum arabic as seedball components hampered seedlings emergence. Seedball + 3 g wood ash and seedball + 1 g NPK-treatments enhanced shoot biomass by 60 % and 75 %, root biomass by 36 % and 94 %, and root length density by 14 % and 28 %, respectively, relative to the control. Shoot nutrient content was not greatly influenced by treatment. However, multiplying biomass yield with nutrient content indicates that nutrient extraction was higher in nutrient-amended seedballs. On-station field tests in Senegal showed over 95 % emergence under real Sahelian conditions. Since early seedlings enhancement is decisive for pearl millet panicle yield under the Sahelian conditions, on-farm trials in the Sahel are recommended.

Keywords: Pearl millet early growth, seedball technology, local resources, dry sowing, seedling emergence, subsistence farming, smallholder farmer, cheap seed pelleting technique

1 Introduction

Under Sahelian conditions, seedling establishment is a major yield factor in pearl millet (*Pennisetum glaucum* [L.] R. Br.) production systems (Rebafka, 1993). This fact, apart from low seed quality and limited water supply, is mainly explained by low chemical fertility (Valluru *et al.*, 2010) of the widespread sandy soils (Arenosols according to the World Reference Base for soil resources – WRB; IUSS, 2014). Arenosols are characterised by low phosphorus (Rebafka, 1993; Muehlig-Versen *et al.*, 2003), nitrogen and organic matter content (Bationo & Buerkert, 2001). Unfortunately, the improvement of seedling establishment through seed coating (Rebafka *et al.*, 1993; Karanam & Vadez, 2010) and the application of commercial mineral fertiliser (NPK) (Bationo *et al.*, 1993; Bationo & Ntare, 2000; Bationo & Buerkert, 2001; Aune & Ousman, 2011; El-Lattief, 2011) is hardly feasible for smallholder farmers. This is due to lack of skills and financial resources (van der Pol & Traore, 1993; Cooper *et al.*, 2008), as well as lacking infrastructure.

* Corresponding author – c.nwankwo@uni-hohenheim.de;
charsile2000@yahoo.com

In the Sahel, farmers partly practice dry sowing with un-coated seeds to optimally use the vegetative period in order to ensure higher yield (Bationo & Buerkert, 2001). How-ever, uncoated seeds bear high risks of loss through preda-tion and early season droughts. In contrast, seed coating has the potential to improve seedling establishment. E.g., it mitigates high seed size variation when lack of uniformity poses challenges (Peske & Novembre, 2011), controls seed predation (Overdyck et al., 2013), and ensures early nutri-ent supply (Rebafka et al., 1993; Karanam & Vadez, 2010). Small seeded species have more advantages from seed coat-ing relative to large seeds due to nutrient addition through the coating materials. Pearl millet is such a crop, having 7–10 mg weight per seed (Rebafka et al., 1993), only.

The seedball technology, invented by Fukuoka (1978) in the frame of the permaculture concept, was introduced to improve rice seedling establishment under dry sowing con-ditions. It has been used in Australia for rangeland im-provement (Atkinson & Atkinson, 2003). Apart, hardly any research ever addressed this technology. It combines local materials such as sand, loam and seeds. Other addi-tives such as nutrients or pesticides can potentially be ad-ded, depending on local needs. NPK, organic compost (Ba-diane et al., 2001) and wood ash (Saarsalmi et al., 2012) can play significant roles in increasing the nutrient supply of plants. E.g., wood ash can serve the dual function of P nutrient release and low soil pH amelioration (Nkana et al., 2002). These materials can be incorporated into the seedball coating materials as additives, addressing the often observed soil-related plant growth limitations in the Sahel (Herrmann et al., 1994). However, their content needs to be optimised in order to avoid any effects that hinder seed germination, e.g. through high osmotic pressure.

The low reserve of 20 μg P per seed (Rebafka et al., 1993) qualifies the pearl millet crop for nutrient supplementation at emergence. Pearl millet seeds have been successfully coated (Rebafka et al., 1993; Karanam & Vadez, 2010; Peske & Novembre, 2011). However, a technology is lack-ing that is based on local resources and affordable to sub-sistence farmers in the Sahel. Therefore, the present study describes the development of the seedball technology for Sahelian subsistence pearl millet production systems and its potential for seedling improvement under poor soil condi-tions. The main objectives of this study were to physically (materials, size) and chemically (nutrients, osmotic pres-sure) optimise seedballs in order to improve early seedling performance (biomass, nutrient content) and prepare on-site testing.

2 Materials and methods

A participatory discussion with farmers in Louga, Senegal on coating pearl millet seeds with local materials and five greenhouse experiments are reported. The green-house experiments were conducted at the University of Ho-henheim, Germany and at ISRA experimental station in Bambey, Senegal. We report here only the key methodolo-gies and findings. The presentation is chronological, with later experimental layouts depending on the previous re-sults.

2.1 Participatory approach on seedball testing and adop-tion in the Sahel

In Louga, Senegal, a participatory study was conducted with the Louga federation of farmers associations (FAPAL: Fédération des Associations Paysannes de Louga) for a period of four weeks. A workshop on seedball produc-tion was carried out to practically demonstrate to the farm-ers, how seedballs are produced. Sand, loam, water, and seeds were used as basic constituents, wood ash and ani-mal dung as nutrient additives, charcoal and termite soil as conditioner, and chili pepper (Capsicum annuum) as repel-lent. Doughs were formed from gravimetric mixtures of these materials. Afterwards, seedballs of about 2 cm dia-meter size were handmade and dried in < 24 hours (h) to avoid unwanted germination. Every step taken in seedball production was carefully explained to farmers. Expert in-terviews, based on social status and gender, were conducted in Wolof language with the help of a translator. Data on the cultivation methods and management norms in the inter-vention zone Louga, as well as on the potential benefits and limitations of seedballs, were collected. An open-discussion class that allowed the farmers to freely interact about seed-ball was conducted. The opinions and perceptions of the farmers on seedball usage and applicability were evaluated. 20 female and 25 male farmers participated in this study.

The qualitative outcomes of our participatory study clearly indicated that the materials necessary for seedball production are freely available in the farm households, i.e. wood ash, charcoal, animal dung, sand as well as seeds and water. Loam and termite soil can be sought less than 4 km away from the settlements. Seedball sowing appeared to be simple using the "drop-and-match" technique in partic-ular in the predominant sandy fields. Farmers stated that seed wastage could be minimised since a known number of seeds is inserted into the seedballs. As a compromise, seed-ball development (mechanical as well as chemical optimisa-tion) using these farmers' affordable local recourses became a main task.

2.2   Seed pre-germination test and material preparation

Local seed varieties collected from the Bambey area, Senegal were used for this study. Seed quality plays a vital role in crop establishment. Thus, checking the viability of any seed lot through a germination test is essential (Meyer & Schmid, 1999). Germination tests for the available seeds were conducted as reported by Throneberry & Smith (1955), but slightly modified. Fifty seeds were randomly selected from the seed lot and placed into 9.0 cm diameter by 1.8 cm height petri-dishes, each, in 12 repetitions. Whatman™ filter paper, 47 mm diameter was soaked in distilled water up to saturation. The water-saturated filter papers were placed into the petri-dishes and, after seed addition, inside a germination chamber. The germination conditions were set at 29.4 °C average temperature, 62 % relative humidity and 12 h / 12 h day/night cycle. On the 7th day, the germinated seeds were counted for each petri-dish.

Cheap and potentially locally available materials such as sand, organic compost, charcoal, animal manure, cattle urine and wood ash were identified as potential seedball components. NPK in minute quantity as a non-local resource was identified, too. The "local materials" used were classified into three groups: matrix, fillers and nutrient additives. Sand was used as matrix since it mimics the major soil property and is available everywhere, where millet is cropped in the Sahel. Loam, gum arabic and termite soil served as potential fillers. Loam is frequently available for free at least in Sahelian subsoils and characterised by higher cation exchange capacity relative to sand (Lorenz, 1999). Compost, charcoal, sheep and goat dung, cattle urine, wood ash and NPK as well as calcium nitrate tetrahydrate (CNT) served as potential nutrient additives. All materials, except urine and CNT, were air-dried at ambient temperature as well as hand-crushed or grinded with a mortar where necessary. Afterwards, these materials were sieved through a 2 mm mesh to remove over-sized particles.

2.3   Laboratory analyses

The pH, electrical conductivity, soluble cations, total P and N as well as organic carbon (C) were measured in all the tested seedball materials, except gum arabic. The pH of the materials was measured using a glass electrode pH-meter (1 : 20 H_2O). Electrical conductivity (1 : 20 H_2O) was measured using a portable electrical conductivity meter, Model 3320 obtained from Xylem Analytics Germany Sales GmbH & Co. KG, Germany (www.wtw.com). Water soluble cations (1 : 20 ratio wt./wt.) – Ca, K and Na were photometrically measured using an Elex 6361 flame photometer (Eppendorf, Hamburg, Germany). Water-soluble magnesium (Mg) was measured with a Perkin-Elmer Model 3100 AAS PerkinElmer, Norwalk, CT, USA. Total C and N

were measured from finely ground sample materials using VarioMacro EL instrument (Elementar, Hanau, Germany). Plant available P was extracted with calcium acetate lactate and determined colometrically based on the molybdenum blue method (Rodriguez et al., 1994). It was measured with a Cary 50 UV-Visible Spectrophotometer (Varian, Mulgrave, Australia) at 710 nm wavelength.

In addition, the shoot P, Mg and K contents were measured in pearl millet seedlings after harvest. Finely ground shoot samples were digested with a HNO_3/H_2O_2 solution (10 minutes, 105 °C temperature, ventilated) in a microwave (MLS 1200 mega, Leutkirch, Germany). Afterwards, the extract was filtered using blueband filter paper. For K and Mg content determination, 25 ml from the filtrate was transferred to 50 ml volumetric flask and after, filled to 50 ml mark with distilled water. For P determination, 10 ml of the extract was mixed with 8 ml of John solution in a 50 ml volumetric flask that was then filled to the 50 ml mark with distilled water. Nutrients in the solution were measured as described above.

Table 1: *Chemical properties of the seedball components and nutrient additives used in this study.*

Component	$pH_{1:20 H_2O}$	$EC_{1:20}$ ($\mu S\,cm^{-1}$)	C_{org} (%)	C : N
Loam	5.9	11	0.8	12
Charcoal	8.1	143	78.9	114
Manure	8.3	2560	32.3	19
Termite soil	8.3	55	0.1	7
Wood ash	11.6	8430	1.1	35
Mineral fertiliser	4.8	5160	0.3	–

Table 2: *Total nitrogen and phosphorus as well as the cation content of wood ash and mineral fertiliser used in this study.*

Content ($mg\,kg^{-1}$)	Wood ash	Mineral fertiliser
N_{total}	326	151,100
P_{total}	1880	67,200
K^+	65,100	152 500
Ca^{2+}	683	27,500
Mg^{2+}	860	7430
Na^+	2190	2470

2.4   Experiment 1: Mechanical optimisation of seedballs

Seed germination is often related to sowing depth (Chen & Maun, 1999; Benvenuti et al., 2001). Shallow sowing depth stimulates more germination than surface placement (Benvenuti et al., 2001). A sowing depth of 2–4 cm is considered optimum for the emergence of *Calligonum* L. species (Ren et al., 2002). This depth is exactly applicable for pearl millet seeds with similar seed size. Bearing this in mind, two major factors: the (i) diameter of the seedball and (ii) location of seeds inside the seedballs, were considered during the seedball development. Where sowing depth is influential, seeds emerging from higher diameter seedballs or the core centre of seedballs might differ from those emerging from near the seedball's surface. On the other hand, randomised seed placement distributes germination failure risk and eases production.

The reason behind the mechanical optimisation is to determine the optimum seedball diameter and the best seed placement position that will not hamper seedlings emergence. In addition, an ideal seedball, after drying, will not break when dropped from about 2 m above the soil surface, i.e. the height when sown by an adult person.

The first part of the mechanical optimisation study was conducted at ISRA/CNRA research station, Bambey, Senegal, to observe seedling (i) emergence and (ii) development. Sandy topsoil material and loam were collected from an uncultivated area inside the station. These materials were prepared as described in section 2.2 (see above). Loam, termite mound material, and gum arabic were separately and permutatively combined with sand. Each combination was mixed with water to point of dough formation. Seedballs were manually moulded from the dough. The seedballs were of four different diameters: 1.0, 1.5, 2.0, 2.5, and 3.0 cm. Afterwards, they were dried under ambient temperature (25–30 °C). For the seedling emergence experiment, seven treatments were tested: (i) conventional sowing without seedballs served as absolute control. Otherwise NPK- and wood ash-amended seedballs of random and central seed placement formed from 80 g sand + 50 g loam + 25 ml water and 1–3 cm diameter range served as seedball treatments, labelled as (ii) Sball+3gAsh+1cmdiam, (iii) Sball+3gAsh+2cmdiam, (iv) Sball+3gAsh+3cmdiam, (v) Sball+1gNPK+1cmdiam, (vi) Sball+1gNPK+2cmdiam, and (vii) Sball+1gNPK+3cmdiam. Each seedball contained 15 seeds, placed in two different positions: (i) random and (ii) central placement. Number of repetitions was 6. Seedlings emergence, only, was counted on the 7th day after planting (DAP).

The second part of the mechanical optimisation study assessed pearl millet seedling height development, only. Three diameters (1.0, 2.0 and 2.5 cm) and six treatments were tested: (i) conventional sowing as absolute control, (ii) seedballs without amendments and amended ones with (iii) charcoal (Sball+30gCha), (iv) compost (Sball+30mlComp), (v) animal manure (Sball+4gMan) and (vi) termite soil (Sball+30gTerm). Each seedball contained 6 seeds. Number of repetitions was 6. Seedlings height and leaf development were measured on the 9th DAP.

Day and night temperatures of 36 and 23 °C respectively, were observed in the greenhouse throughout the emergence period. Seedballs were sown with the physical centre at 3.0 cm depth, i.e. approximately the depth at which pearl millet is sown by farmers. Each experimental unit consisted of a black 2-liter polyethylene bag, filled with sand at a bulk density of 1.6 g cm^{-3}. Each treatment was repeated six times in a completely randomised design. Soil moisture of 60 % field capacity was adjusted in each experimental unit every 24 h throughout the experiment.

2.5 Experiment 2: Chemical optimisation of seedballs

The objective of seedball chemical optimisation was to identify the optimum contents of seedball additives that conserve germination rates and at the same time enhance biomass development. Charcoal, wood ash, termite soil and gum arabic collected nearby Bambey, Senegal were tested as seedball additives. The experimental conditions (temperature, soil water content, sand-substrate, bulk density and germination bags) were same as in the mechanical optimisation. This study concentrated on wood ash, CNT and exclusively NPK fertiliser as nutrient additives. The intention was to optimise the nutrient content in a way that negative osmotic effects are avoided but maximum nutrient amounts incorporated into the seedball.

In the first part of the chemical optimisation study, only the seedlings emergence was assessed. Wood ash, cattle urine, charcoal and NPK served as nutrient additives. NPK 17:17:17, manufactured by Green Partners International GmbH & Co. KG, Germany was used. It contained 4.1 %, 4.8 % and 8.1 % ammonium, nitrate and carbamide N, respectively. So-called quartz sand, i.e. sieved alluvial sand from SW-Germany, was used as growth medium. It contained 2 wt.% coarse sand (630–2000 µm), 60 wt.% medium sand (200–630 µm) and 38 wt.% fine sand (63–200 µm). The intention to use this sand was to mimic the sandy soil textures as reported to be typical for Sahelian pearl millet sites by Hebel (1995). The loam for seedball production was collected from the subsoil of a field called "Goldener Acker" located at the University campus of Hohenheim, Germany. According to WRB classification system, the reference soil group there is a Luvisol. Wood ash, cattle urine, charcoal and NPK were added in variable quantities. Where urine was used as nutrient additive, no further water was added to produce the seedball dough. 2 cm diameter-sized seedballs were formed from 80 g sand + 50 g loam + 25 ml water. Seed number was adjusted to 6 and 10 per seedball. Seeds were randomly placed. Conventional sowing served as control. All treatments were sown at 3 cm depth. The experimental design was a randomised complete block with six replications per treatment. Seedling number was counted on the 7th DAP.

In the second part of the chemical optimisation study, sandy subsoil, collected from Rastatt (48° 49′ N, 8° 11′ E) in Germany, was used as substrate. The intention was to mimic the typical Sahelian pearl millet soils. The collected soil material was air-dried and passed through a 2 mm sieve to remove coarser particles. The soil is characterised by > 90 % sand, a pH_{CaCl_2} of 4.5, < 1 wt. % organic matter, a C : N ratio of 23 and a potential cation exchange capacity of 39 mmol kg^{-1} at around 0.7 m depth. Further properties of this soil can be accessed from Stahr *et al.* (2009). The seven tested treatments were: (i) conventional sowing as absolute *Control*; (ii) seedballs without amendments as *Sball* control; NPK-containing seedballs at two levels (iii) Sball+0.5gNPK (iv) Sball+1gNPK; (v) one wood ash-containing seedball variant (Sball+3gAsh); and CNT-containing seedballs at two levels (vi) Sball+0.1gCNT and (vii) Sball+0.5gCNT. Each seedball as well as the control contained ten seeds. The used fertiliser was NPK 15 : 15 : 15, 2–5 mm granular sized, white-coloured, containing < 2.0 wt. % water. CNT, ≥ 99 % pure, obtained from Carl Roth GmbH, Germany (www.carlroth.com), was used in addition as ammonia-free N-source.

Seedballs of 2 cm diameter size were formed and air-dried in < 24 h. The sowing depth for all treatments was 3.0 cm. Plastic containers of 12.0 cm in diameter and 14.0 cm in height were used. Each container was filled with sieved-sand at a bulk density of 1.6 g cm^{-3}. At the bottom of each plastic container, Whatman™ filter paper was installed to avoid sand materials from sipping through. The soil was air-dried before the treatments were sown. This was intended to mimic dry sowing as often practiced by Sahelian farmers. The experimental design was fully randomised, comprising six treatment replications. About 2.5 mm sized gravels covered the topmost 2.0 cm of each plastic container. This was to reduce soil water loss via evaporation. Water sprinkler was used for watering the experimental containers throughout this study. Watering started 48 h after sowing the treatments. 16 wt. % soil moisture content was adjusted daily using a weigh balance until harvest. A day/night cycle of 10/14 h was ensured. Day and night temperatures of 32 and 26 °C, respectively, with a relative air humidity of 48.5 % were maintained throughout the growth period. Seedling height and leaf number were repeatedly measured per week.

On the 28th DAP, the seedlings were harvested. Root (weight, length) and shoot (weight, leaf number) variables were measured. The dry matter was obtained after drying to a constant weight in an oven at 58–60 °C temperature range for 48 h. Dried shoots were analysed for P, Mg and K using the method described in section 2.3. Total nutrient uptake was calculated as: root biomass × nutrient content + shoot biomass × nutrient content. The roots were obtained by carefully washing the materials through a 2 mm sieve and were cut into lengths of about 1.0 cm with a clean pair of scissors to minimise root inter-twisting particularly during scanning. Measurement errors in root diameter as well as root length measurement are often associated with long (> 3 cm) root sections during scanning (Nwankwo *et al.*, 2013 – unpublished). Half of each sample was used for dry matter and the other half for root length determination. The stored roots were scanned with an EPSON Perfection V700 PHOTO dual lens scanner and the values of the root length and diameter were measured using WinRhizo® V2009c software (Regent Instruments, Nepean, Canada).

2.6 Experiment 3: Seedball storage effects and on-station testing

Seedball production is time demanding, without mechanisation requiring approximately 40 h for 10,000 units per person. This means quite an investment for the farmer. For the applicability of the technology, it is important to know whether the seedballs can be manufactured before the season (when labour load is low) and stored without causing decreasing germination rates.

The objective of the first experiment was, therefore, to identify whether storage time has an effect on number of germinated seeds. A second experiment was dedicated to the question whether seedballs function also under real Sahelian conditions using local materials. For this purpose, Sball+3gAsh and Sball+1gNPK treatments, containing 25 seeds per seedball were produced according to the same recipe as used in the two preceding experiments and tested against conventional sowing. The reason to constrain to wood ash and NPK as nutrient carrier was that these materials showed a positive growth effect in the chemical optimisation study and are available to Sahelian farmers, at least in small amounts. In contrast, CNT that showed the best biomass results is a pure chemical not available and affordable to these farmers.

For the first experiment, about 200 seedballs, per treatment, were produced at once and stored in the greenhouse of University of Hohenheim. Every week, germination tests were carried out with 20 seedballs per treatment for nine weeks period. Same greenhouse experimental conditions as stated in mechanical and chemical optimisation section (see sections 2.4 and 2.5), except average temperature that was 28.6 °C, were maintained. Average number of germinated seeds per seedball after eight days are reported.

For the field test under Sahelian conditions, Control, Sball, Sball+3gAsh and Sball+1gNPK treatments were tested for germination inside the ISRA/CNRA station, Bambey, Senegal. Seedballs were produced with local materials collected from the locality of Bambey. Each seedball contained 15 seeds; same seed number was inserted per planting pocket in the control i.e., the conventional sowing. The experimental site (14° 42′ N, −16° 28′ W) was characterised by a brown coloured sandy-loam soil characterised by a pH of 6.0 and a C:N ratio of 48. The exchangeable cations as extracted by NH_4-acetate from the first 0.3 m topsoil revealed in g kg^{-1}: 24.8 Ca, 5.1 Mg, 1.0 K, and 0.7 Na, as well as 26.5 mg kg^{-1} of plant available P as extracted by the Bray1 method.

Dried stands of *Vetiveria nigritana* and scarcely located seedlings of *Balanites aegyptiaca* were manually cleared off the site, which was fallowed for two years before this study. The planting area was 14 m × 15 m. Sowing was 3 cm deep, at a spacing of 1 m × 1 m, in a completely randomised block design of four treatment replications. No form of fertilisation was applied since we did not intend for any harvest. It was an off-season experiment; therefore, water was supplied via irrigation. Water equivalent to 20 mm rain every four days was supplied using sprinklers starting from the 2nd DAP. The intention of sowing before watering was to mimic dry sowing as often practiced by Sahelian farmers. Throughout the study, the observed average day and night temperatures were 27 and 19 °C, respectively. For reasons of genetic variation and environmental conditions, pearl millet seedlings may not survive after emergence under field conditions (Peacock *et al.*, 1993). Therefore, to check for seedlings survival after germination, repeated germination counts per planting pocket were conducted every week in a four weeks period. The number of emerged seedlings per planting pocket was noted.

2.7 Statistical analysis

Where statistical analysis was applicable, normal distribution and variance homogeneity were tested based on the Shapiro-Wilk test. As data were not evenly distributed, Welch's one-way analysis of variance using Proc. GLM was performed for all data sets of one-time measurements (e.g. biomass and dry matter). Proc. MIXED was performed for repeated measurements (e.g. plant height and leaf count). The treatment means were compared for significant differences at $p < 0.05$. Results are presented as means (± standard deviations) of the measured variables while the mean values represent the treatment means. All analyses were performed with SAS version 9.4 while Sigma Plot version 13.0 was used to plot all graphs.

3 Results

The used seed lot showed a germination rate of over 90 % in the greenhouse. Thus, the seeds were accepted as viable for the experiments.

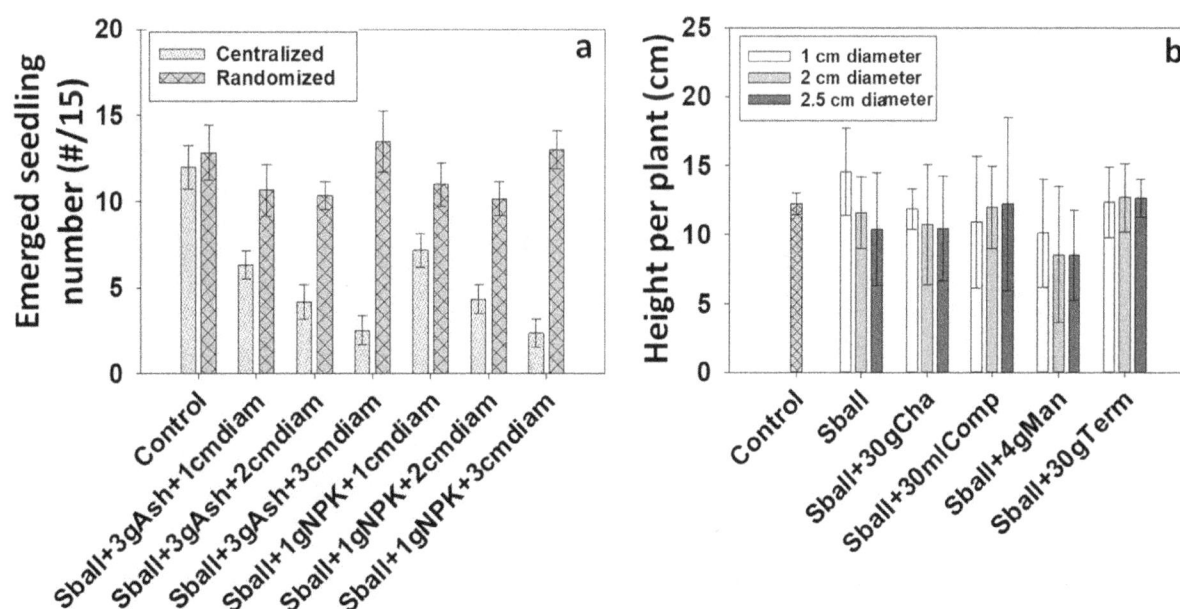

Fig. 1: *Treatment effects on pearl millet: (a) number of plants at day 7 after sowing, (b) plant height at day 9 after sowing at the greenhouse of ISRA/CNRA research station, Bambey, Senegal. Symbols show arithmetic means (n = 6) and error bars indicate standard deviations (±). Control = non-pelleted seeds, Sball = 80 g sand + 50 g loam + 25 ml water, Ash = wood ash, NPK = 15 : 15 : 15 mineral fertiliser, Cha = charcoal, Comp = compost, Term = termite soil, Man = manure, diam = diameter size in cm, centralized = seed placement at the core centre of the seedball i.e. seeds were inserted into the seedballs after the seedball was moulded, and randomized = scattered seed placement in seedball i.e. seeds were mixed with the seedball components before the seedball was moulded.*

3.1 Experiments 1 and 2: Chemical and mechanical optimisation of seedballs

Pre-trials (data not presented) have shown that the best base recipe for seedball dough is derived from a mixture of 80 g sand + 50 g loam + 25 ml water and that germination is best at a shallow sowing depth of about 3 cm as practiced by farmers. The greater the seedball diameter for centrally placed seeds in particular, the less the number of emerged plants one week after sowing (Fig. 1a). With respect to seed placement within seedballs, randomised placement showed an overall good performance, while central placement reduced number of emerged seeds in dependence of seedball diameter. There was no significant effect of seedball diameter on biomass development as presented by the plant height at day 9 after sowing. However, the manure-amended seedballs showed in general lower means (Fig. 1b). 2 cm diameter appears as good compromise between material needed, nutrient amount added and emergence rate.

Two striking effects of additives can be observed: gum arabic as well as urine heavily depress pearl millet emergence from seedballs (Fig. 2).

Seedling emergence was affected by treatment. Wood ash and CNT at high application rates of 3 and 0.5 g per standard seedball recipe significantly reduced seedlings emer-

gence by 41 % and 64 %, respectively (Fig. 3a). Treatments showed effects on shoot and root variables at harvest. Sball+3gAsh, Sball+0.5gCNT and Sball+1gNPK treatments were 60 %, 202 % and 75 % higher in shoot and 36 %, 154 % and 94 % in root biomass compared to the control (3b and 3c). The root length density repeats these trends. Sball+3gAsh, Sball+0.5gCNT and Sball+1gNPK treatments showed 14 %, 12 % and 28 % increment in root length density relative to the control (Fig. 3d). Higher root diameter was observed in Sball+1gNPK treatment relative to the control. Root to shoot ratio did not respond to treatment (data not shown).

Treatment did not clearly influence the nutrient content of the seedlings (Fig. 3e). Sball+3gAsh, Sball+0.5gCNT and Sball+1gNPK treatments showed 82 %, 440 % and 193 % more P uptake, respectively, than the control (Fig. 3f). The total nutrient uptake is more indicative than the nutrient content. The former shows the already known pattern for all three investigated nutrients (Fig. 3f). In particular, K uptake was affected. It was 127 %, 380 % and 82 % higher in Sball+3gAsh, Sball+0.5gCNT and Sball+1gNPK treatments, respectively (Fig. 3f). Mg uptake was influenced by treatment as well. Sball+3gAsh, Sball+0.5gCNT and Sball+1gNPK treatments showed 66 %, 367 % and 68 % more Mg uptake than the control.

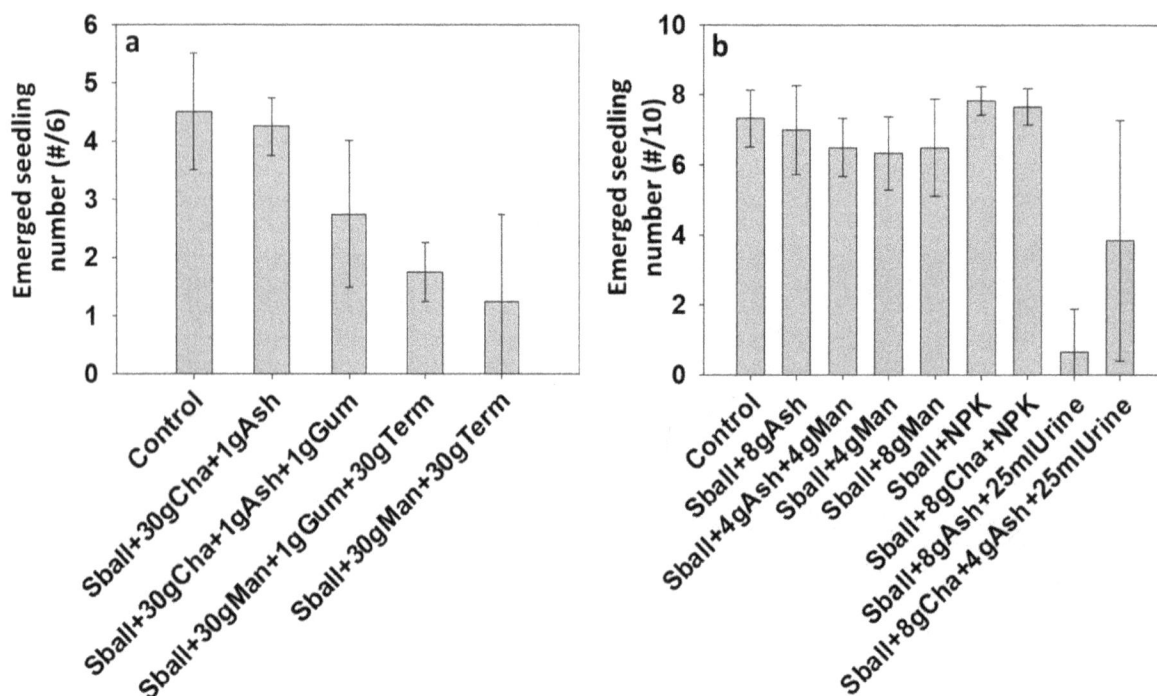

Fig. 2: *Treatment effects on pearl millet seedling number at the 7th day after sowing for (a) six and (b) ten seeds per seedball, observed at the greenhouse of University of Hohenheim, Germany. Bars represent arithmetic means (n = 6) and error bars indicate standard deviations (±). Control = non-pelleted seeds, Sball = 2 cm diameter sized-seedball made from a mixture of 80 g sand + 50 g loam + 25 ml water. Cha = charcoal, Ash = wood ash, Gum = gum arabic, Man = manure, Term = termite soil, NPK = 25 ml 17 : 17 : 17 mineral fertiliser in 200 ml g^{-1} solution and Urine = cattle urine.*

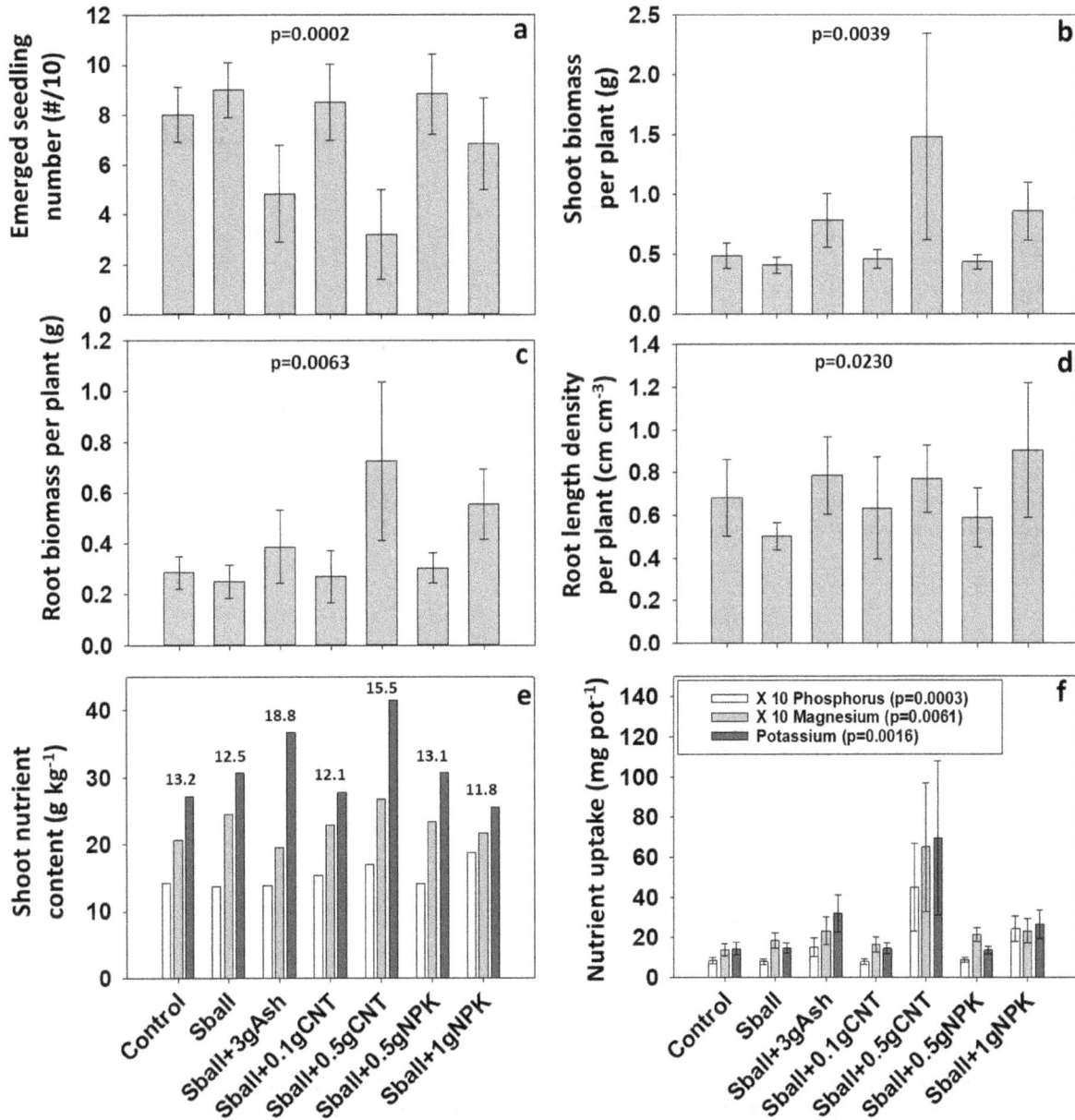

Fig. 3: *Treatment effects on pearl millet (**a**) emergence at 7th DAP, and (**b**) shoot biomass (**c**) root biomass (**d**) root length density (**e**) shoot nutrient content as well as (**f**) total nutrient uptake, at 28th DAP, observed at the greenhouse of University of Hohenheim, Germany. Numbers in (**e**) indicate the ratio of K to Mg content of the shoot. Symbols show arithmetic means (n = 6) and error bars indicate standard deviations (±), except for (**e**) where biomass was pooled due to small sample sizes. p = probability value, Control = non-pelleted seeds, Sball = 2 cm diameter sized-seedball made from a mixture of 80 g sand + 50 g loam + 25 ml water, Ash = wood ash, CNT = calcium nitrate tetrahydrate and NPK = 15 : 15 : 15 mineral fertiliser.*

3.2 Seedball storage effects and on-station testing

Plant height and leaf number as biomass proxies responded to treatments. Relative to the control, 29 % and 18 % increments in height were observed in Sball+1gNPK on the 12th and 16th DAP, respectively (Fig. 4a). Within 24 DAP, Sball+3gAsh, Sball+0.5gCNT and Sball+1gNPK treatments showed 11 % and 18 % and 17 % height increment, compared to the control. Sball+1gNPK treatment in

particular enhanced leaf development, particularly between 15th and 25th DAP (Fig. 4b).

Seedlings emergence slightly declined in Sball+3gAsh and Sball+1gNPK treatments after about six weeks of storage (Fig. 5). As for the on-station seedball germination in Senegal, the germination rates for the Control (98.2 %), Sball (97.4 %), Sball+3gAsh (97.1 %) and Sball+1gNPK (96.1 %) treatments were comparable under field conditions. Over 20 seedlings per germination pocket was ob-

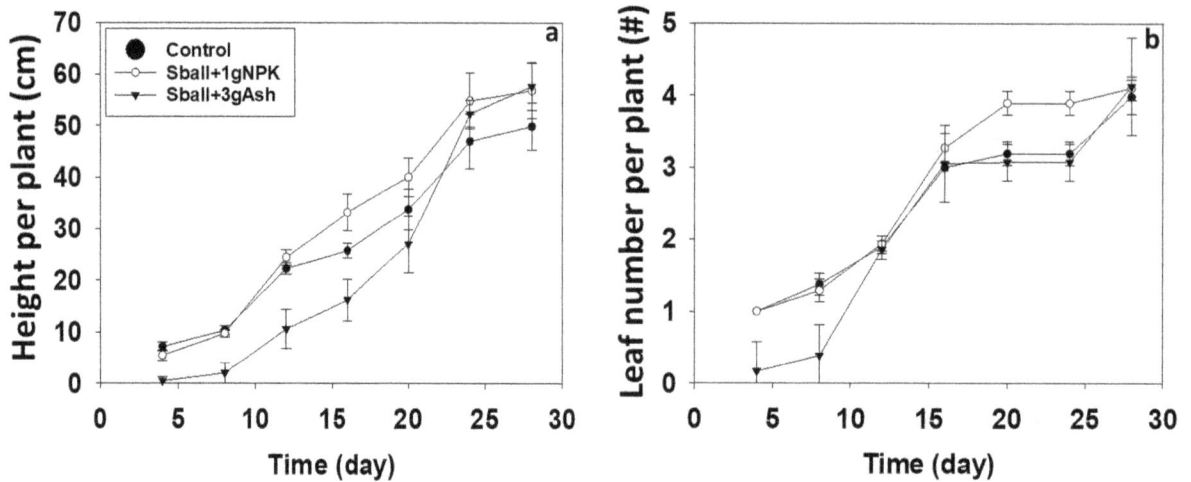

Fig. 4: *Pearl millet (**a**) shoot height and (**b**) leaf number development for different treatments at the greenhouse of University of Hohenheim, Germany. Symbols show arithmetic means (n = 6) and error bars indicate standard deviations (±). Control = non-pelleted seeds, Sball = 2 cm diameter sized-seedball made from a mixture of = 80 g sand + 50 g loam + 25 ml water, Ash = wood ash, and NPK = 15 : 15 : 15 mineral fertiliser.*

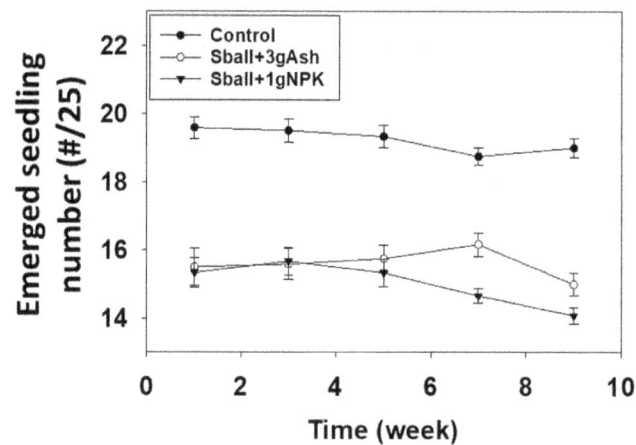

Fig. 5: *Absolute number of emerged pearl millet seedlings 8 DAP for three treatments as affected by storage time at the greenhouse of University of Hohenheim, Germany. Symbols show arithmetic means (n = 20) and error bars indicate standard deviations (±). Control = non-pelleted seeds, Sball = 2 cm diameter sized-seedball made from a mixture of = 80 g sand + 50 g loam + 25 ml water, Ash = wood ash and NPK = 15 : 15 : 15 mineral fertiliser.*

served for all treatments. The seedlings of Sball+3gAsh and Sball+1gNPK treatments in particular were more vigorous than those of the conventional sowing.

4   Discussion

4.1   Participatory approach on seedball testing and adoption in the Sahel

Technology adaptation and sustainable adoption are faster reached in a development context if target farmers are involved in all steps of technology development (Herrmann et al., 2013). Therefore, Sahelian local farmers participated

in this study as early as possible, i.e. identifying potential constraints to adoption of the seedball technology at pearl millet cropping sites with their predominant sandy soils.

Seedball technology seems to reduce seed usage per hectare at sowing. Seed wastage poses in particular a problem when children do the sowing and get physically exhausted. Then more than 300 seeds can be found in single sowing pockets (Klaij & Hoogmoed, 1993). Farmers preferred dry sowing practice because it prolongs the vegetation period and thus, potentially increases yield. Hand-sowing was preferred to the use of local sowing machines that demand cattle, horses or donkeys for traction and present a limitation.

The seedball technology absolutely conforms to the already established pearl millet management systems in the Sahel. Its application does not pose any form of disadvantage if the production is done during the dry season when opportunity costs are low, i.e. labour is not a limiting factor. Neither material availability nor social factors (gender and religion) per se, seem to hamper the adoption of the seedball technology at the Sahelian site investigated.

Female farmers appeared to be more interested in the technology, though. This might be explained by the fact that female farmers have less access to sowing machines and do need to support their husbands during the major sowing time at the beginning of the rainy season and are, thus, more keen to apply dry sowing.

4.2 Experiment 1: Mechanical optimisation of seedballs

Most likely, pearl millet seeds in central placement were unable to mechanically make their way through the substrate (Fig. 1a). With respect to the diameter, the biomass experiment did not provide a final argument (Fig. 1b). However, in a seedball of 1 cm diameter only a limited amount of seeds and nutrients can be incorporated. On the other hand, seedballs of 3 cm diameter need a lot of material (about threefold the amount of 2 cm diameter seedballs) that needs to be transported to the production site and afterwards to the fields, meaning elevated costs. Therefore, as a compromise, a diameter of 2 cm was considered optimum for seedball production.

4.3 Experiment 2: Chemical optimisation of seedballs

The negative effect of gum arabic on seedlings emergence (Fig. 2a) can potentially be explained by its strong tendency to absorb water itself. In consequence, if only low amounts of water are added, the seeds cannot compete. For urine and Macrotermes termite mound material (Fig. 2b), another explanation is necessary. The liberation of ammonia (Bremner & Krogmeier, 1989; Haden et al., 2011) or similar ammonium compounds (Pan et al., 2016) can intoxicate cereal seeds at direct contact. The urea compound in urine decomposes into ammonia. Own observations in other trials not reported here showed the negative effects of any ammonia containing fertiliser on pearl millet emergence. Macrotermes mound material can also contain relative high amounts ($> 50\,mg\,kg^{-1}$) of ammonia (Hebel, 1995).

The probable reasons for failed emergence in the wood ash and CNT amended seedballs (Fig. 3a) are osmotic effects since both components have a very high solubility in water. All other treatments with lower share of osmotic compounds did not significantly reduce emergence compared to the control. It is well established that germination can be impaired in the case pearl millet seeds are coated with P at higher concentration (Rebafka et al., 1993) or other materials (Peske & Novembre, 2011).

We did not assess the temporal root development in this study. However, positive effects were observed in the early root development of Sball+1gNPK treatment in other experiments, using computer tomography (Nwankwo et al., 2018). N and P addition through NPK and wood ash, (Table 2) can be suspected to cause this effect. It is well documented that local nutrient supply as early as emergence influences early root development in pearl millet (Rebafka et al., 1993; Karanam & Vadez, 2010; Valluru et al., 2010). This is particularly true for phosphorus.

On the other hand, seedlings potassium content could be increased by ash and CNT application. While the ash effect can be explained by the high content of water soluble potassium in the ash itself, the CNT effect must be indirect, i.e. by better extraction of potassium from the soil mediated by a longer root network (Fig. 3c). The same argument can be applied for magnesium, since the 0.5 g CNT application yields highest content for magnesium as well. The potassium in the plant can contribute to biomass production (Fig. 3b and c) by increasing drought tolerance through better water use efficiency by effective regulation of the stomata. This is in agreement with the findings of Ashraf et al. (1994) on dry matter and biomass production of pearl millet shoot and root systems under drought conditions. Since wood ash ($860\,mg\,kg^{-1}$) and the NPK fertiliser ($7430\,mg\,kg^{-1}$) contained this nutrient (Table 2), only for CNT the effect needs to be explained by extraction from the growth medium (Cummins & Perkins, 1974). The seedlings of all our treatments contained Mg in higher amount than the range (0.102–0.126 %) reported as deficient by Embleton (1966).

In the acidic soils of the African Sahel where potentially plant available P can be fixed by soil aluminium (Scott-Wendt et al., 1988), wood ash- and NPK-amended seedballs can potentially enhance P uptake in pearl millet. This is in particular true for the wood ash that locally increases the soil pH and thus counteracts Al-toxicity. This can be of great advantage, since early P uptake in pearl millet is decisive for higher dry matter and panicle yield under Sahelian conditions (Rebafka et al., 1993; Buerkert, 1995; Karanam & Vadez, 2010).

Poor pearl millet seedling performance (Fig. 3b and c), as observed in our absolute control (conventional sowing) and seedball control (no nutrient amendment) treatment is often caused by low P and K nutrient uptake (Scott-Wendt et al., 1988). The non-nutrient amended seedball treatment, Sball, showed similar K, Mg and P uptake as the control, indicating the importance of nutrient additives for the success of this technology. Nutrients positively influence biomass development and allocation in plants (Poorter & Nagel, 2000; Hermans *et al.*, 2006) in particularly if water is not limiting – as in this study. Conversely, low nutrient availability decreases plant nutrient uptake and consequently reduces leaf dry mass (Evans, 1996).

Similar observations have been reported on pearl millet when nutrients were supplied as early as the establishment stage (Rebafka *et al.*, 1993; Karanam & Vadez, 2010; Valluru *et al.*, 2010). Excessive shoot development in Sball+0.5gCNT treatment led to a lodging effect. Speculatively, the high N content of the CNT most likely triggered this effect. In rice, excessive N content was responsible for seedlings lodging (Mannan *et al.*, 2010). In this experiment, the wood ash treatment shows slow early development, but overtakes the control in the last phase. Possible reasons for the first effect is the high osmotic pressure exerted by the soluble components of the wood ash, and for the second effect the equilibrated nutrient supply, since wood ash – that derives from plant materials – is the most complex fertiliser that can be imagined.

The marginal plant height difference observed from 24th DAP onwards indicates nutrient depletion in the limited rooting volume by the well-established seedlings. Therefore, nutrient supplementation to maintain the seedlings is necessary, precisely three weeks after planting. This is exactly the time when the local farmers carry out weeding and thinning. Fertilisation can be supplemented at this stage to ensure a continuation of the already established seedling. This could be in form of animal manure, considering its availability as well as affordability in the Sahel.

4.4   Experiment 3: Seedball storage effects and on-station testing

Long-term cumulative osmotic effect arising from the wood ash and NPK contents of the seedball can be suspected for the declined seedlings emergence six weeks after storage (Fig. 5). Emergence rate was lower in the nutrient amended treatments, but still high enough with respect to farmer needs. In addition, number of emerged seeds per seedball can be adjusted by the number of seeds inserted. As a common practice, Sahelian farmers often thin down to 2–3 plants per pocket from > 30 emerged seedlings. Seedlings emergence rates as observed in this experiment of about 14 seedlings per pocket is, therefore, acceptable.

In the on-station test, the > 96 % seedlings emergence rate observed in all the treatments is a clear indication that seedballs are a viable option in sandy Sahelian fields. Since here no quantitative biomass variables were assessed, testing should be continued for whole cropping seasons.

5   Conclusions on the applicability and optimised formula of the seedball technology under Sahelian conditions

Opportunities exist, through the seedball technology, to improve the performance of pearl millet production under Sahelian conditions (poor soil + erratic rainfall). Farmers' perception about the technology was in general positive. The standard base dough consists of 80 g sand + 50 g loam + 25 ml water (+ 1 g NPK or 3 g wood ash). All components that potentially contain ammonia (urea, urine, manure) should be avoided since they consistently reduce germination rates. Once confectioned and dried, seedballs can be stored for prolonged periods (at least two months), showing only a slight trend of decreasing germination over time. Number of germinated seeds can be adjusted via the seed number per seedball. In farmers' environment, the production of seedballs can be based on simple volumetric ratios using traditional bins, plastic cups or bottle caps. A topic is workload, but this can be circumvented by seedball manufacturing before the planting season.

In this study, nutrient amended seedballs have proven to enhance seedling performance in the early growth stages (first 3–4 weeks). From a farmer perspective, as soon as crop establishment is guaranteed, further fertilisation (e.g. with organic manure) needs to be applied. Seedballs reduce the risk of loss on investment in particular with respect to fertiliser, requiring < 2 kg NPK per hectare, and lower seed number per pocket. On-station in Senegal, first tests have shown that seedballs can be produced with the indigenous local materials, and that germination is sufficient under the given Sahelian environmental conditions. The next steps are now to quantitatively test performance effects on-station and on-farm.

Acknowledgements

This study was conducted within the framework of the "Anton & Petra Ehrmann Research Training Group" in Water – People – Agriculture – College at the University of Hohenheim, and was made possible through the generous support by the American People provided to the Feed the Future Innovation Lab for Collabourative Research on Sorghum and Millet through the United States Agency for International Development (USAID). The contents are the responsibility of the authors and do not necessarily reflect the views of USAID or the United States Government. Program activities are funded by the United States Agency for International Development (USAID) under Cooperative Agreement No. AID–OAA–A–13–00047.

References

Ashraf, M., Zafar, Z. & Cheema, Z. (1994). Effect of low potassium regimes on some salt-and drought-tolerant lines of pearl millet. *Phyton*, 34, 219–227.

Atkinson, V. L. & Atkinson, V. (2003). *Mine and industrial site revegetation in the semi-arid zone, North-Eastern Eyre Peninsula, South Australia*. Master's thesis, University of South Australia, Australia.

Aune, J. B. & Ousman, A. (2011). Effect of seed priming and micro-dosing of fertiliser on sorghum and pearl millet in Western Sudan. *Experimental Agriculture*, 47, 419–430.

Badiane, A., Faye, A., Yamoah, C. F. & Dick, R. (2001). Use of compost and mineral fertilisers for millet production by farmers in the semiarid region of Senegal. *Biological Agriculture & Horticulture*, 19, 219–230.

Bationo, A. & Buerkert, A. (2001). Soil organic carbon management for sustainable land use in Sudano-Sahelian West Africa. *Nutrient Cycling in Agroecosystems*, 61, 131–142.

Bationo, A., Christianson, C. & Klaij, M. (1993). The effect of crop residue and fertiliser use on pearl millet yields in Niger. *Fertiliser Research*, 34, 251–258.

Bationo, A. & Ntare, B. (2000). Rotation and nitrogen fertiliser effects on pearl millet, cowpea and groundnut yield and soil chemical properties in a sandy soil in the semi-arid tropics, West Africa. *The Journal of Agricultural Science*, 134, 277–284.

Benvenuti, S., Macchia, M. & Miele, S. (2001). Light, temperature and burial depth effects on Rumex obtusifolius seed germination and emergence. *Weed Research*, 41, 177–186.

Bremner, J. M. & Krogmeier, M. J. (1989). Evidence that the adverse effect of urea fertiliser on seed germination in soil is due to ammonia formed through hydrolysis of urea by soil urease. *Proceedings of the National Academy of Sciences*, 86, 8185–8188.

Buerkert, A. (1995). *Effects of crop residues, phosphorus, and spatial soil variability on yield and nutrient uptake of pearl millet (Pennisetum glaucum L.) in southwest Niger*. Ph.D. thesis, University of Hohenheim, Germany.

Chen, H. & Maun, M. (1999). Effects of sand burial depth on seed germination and seedling emergence of *Cirsium pitcheri*. *Plant Ecology*, 140, 53–60.

Cooper, P., Dimes, J., Rao, K., Shapiro, B., Shiferaw, B. & Twomlow, S. (2008). Coping better with current climatic variability in the rain-fed farming systems of sub-Saharan Africa: an essential first step in adapting to future climate change? *Agriculture, Ecosystems & Environment*, 126, 24–35.

Cummins, D. & Perkins, H. (1974). Response of Millet and Sorghum × Sudangrass Crosses to Magnesium Fertilization 1. *Agronomy Journal*, 66, 809–812.

El-Lattief, E. A. (2011). Growth and fodder yield of forage pearl millet in newly cultivated land as affected by date of planting and integrated use of mineral and organic fertilisers. *Asian Journal of Crop Science*, 3, 35–42.

Embleton, T. W. (1966). Magnesium. *In:* Chapman, H. D. (ed.), *Diagnostic Criteria for Plants and Soils*. erkeley, University of California, USA. Pp. 793

Evans, J. R. (1996). Developmental constraints on photosynthesis: effects of light and nutrition. *In:* Baker, N. R. (ed.), *Photosynthesis and the Environment. Advances in Photosynthesis and Respiration, vol 5.*. Springer, Dordrecht, Netherlands.

Fukuoka, M. (1978). *The one-straw revolution: an Introduction to natural farming*. New York Review of Books, United States.

Haden, V. R., Xiang, J., Peng, S., Bouman, B. A., Visperas, R., Ketterings, Q. M. & Duxbury, J. M. (2011). Relative effects of ammonia and nitrite on the germination and early growth of aerobic rice. *Journal of Plant Nutrition and Soil Science*, 174, 292–300.

Hebel, A. (1995). *Einfluß der organischen Substanz auf die räumliche und zeitliche Variabilität des Perlhirse-Wachstums auf Luvic Arenosolen des Sahel (Sadoré, Niger)*. Ph.D. thesis, University of Hohenheim, Germany.

Hermans, C., Hammond, J. P., White, P. J. & Verbruggen, N. (2006). How do plants respond to nutrient shortage by biomass allocation? *Trends in Plant Science*, 11, 610–617.

Herrmann, L., Haussmann, B. I. G., van Mourik, T., Traoré, P. S., Oumarou, H. M., Traoré, K. & Naab, J. (2013). Coping with climate variability and change in research for development targeting West Africa: need for paradigm changes. *Science et changements planétaires / Sécheresse*, 24 (4), 294–303.

Herrmann, L., Hebel, A. & Stahr, K. (1994). Influence of microvariability in sandy Sahelian soils on millet growth. *Journal of Plant Nutrition and Soil Science*, 157, 111–115.

IUSS Working Group WRB (2014). *World Reference Base for Soil Resources 2014, International soil classification system for naming soils and creating legends for soil maps.*. FAO, Rome.

Karanam, P. & Vadez, V. (2010). Phosphorus coating on pearl millet seed in low P Alfisol improves plant establishment and increases stover more than seed yield. *Experimental Agriculture*, 46, 457–469.

Klaij, M. & Hoogmoed, W. (1993). Soil management for crop production in the West African Sahel. II. Emergence, establishment, and yield of pearl millet. *Soil and Tillage Research*, 25, 301–315.

Lorenz, P. M. (1999). Determination of the cation exchange capacity (CEC) of clay minerals using the complexes of copper(II) ion with triethylenetetramine and tetraethylenepentamine. *Clays and Clay Minerals*, 47, 386–388.

Mannan, M., Bhuiya, M., Hossain, H. & Akhand, M. (2010). Optimisation of nitrogen rate for aromatic Basmati rice (*Oriza sativa* L.). *Bangladesh Journal of Agricultural Research*, 35, 157–165.

Meyer, A. & Schmid, B. (1999). Seed dynamics and seedling establishment in the invading perennial *Solidago altissima* under different experimental treatments. *Journal of Ecology*, 87, 28–41.

Muehlig-Versen, B., Buerkert, A., Bationo, A. & Roemheld, V. (2003). Phosphorus placement on acid Arenosols of the West African Sahel. *Experimental Agriculture*, 39, 307–325.

Nkana, J. V., Demeyer, A. & Verloo, M. (2002). Effect of wood ash application on soil solution chemistry of tropical acid soils: incubation study. *Bioresource Technology*, 85, 323–325.

Nwankwo, C. I., Blaser, S. R., Vetterlein, D., Neumann, G. & Herrmann, L. (2018). Seedball-induced changes of root growth and physico-chemical properties—a case study with pearl millet. *Journal of Plant Nutrition and Soil Science*, 181, 768–776.

Overdyck, E., Clarkson, B. D., Laughlin, D. C. & Gemmill, C. E. (2013). Testing broadcast seeding methods to restore urban forests in the presence of seed predators. *Restoration Ecology*, 21, 763–769.

Pan, W. L., Madsen, I. J., Bolton, R. P., Graves, L. & Sistrunk, T. (2016). Ammonia/ammonium toxicity root symptoms induced by inorganic and organic fertilisers and placement. *Agronomy Journal*, 108, 2485–2492.

Peacock, J., Soman, P., Jayachandran, R., Rani, A., Howarth, C. & Thomas, A. (1993). Effects of high soil surface temperature on seedling survival in pearl millet. *Experimental Agriculture*, 29, 215–225.

Peske, F. B. & Novembre, A. D. L. (2011). Pearl millet seed pelleting. *Revista Brasileira de Sementes*, 33, 352–362.

van der Pol, F. & Traore, B. (1993). Soil nutrient depletion by agricultural production in southern Mali. *Fertilizer Research*, 36, 79–90.

Poorter, H. & Nagel, O. (2000). The role of biomass allocation in the growth response of plants to different levels of light, CO_2, nutrients and water: a quantitative review. *Functional Plant Biology*, 27, 1191–1191.

Rebafka, F. P. (1993). *Deficiency of phosphorus and molybdenum as major growth limiting factors of pearl millet and groundnut on an acid sandy soil in Niger, West Africa.* Ph.D. thesis, University of Hohenheim, Germany.

Rebafka, F. P., Bationo, A. & Marschner, H. (1993). Phosphorus seed coating increases phosphorus uptake, early growth and yield of pearl millet (*Pennisetum glaucum* (L.) R. Br.) grown on an acid sandy soil in Niger, West Africa. *Nutrient Cycling in Agroecosystems*, 35, 151–160.

Ren, J., Tao, L. & Liu, X.-M. (2002). Effect of sand burial depth on seed germination and seedling emergence of *Calligonum* L. species. *Journal of Arid Environments*, 51, 603–611.

Rodriguez, J., Self, J. & Soltanpour, P. (1994). Optimal conditions for phosphorus analysis by the ascorbic acid-molybdenum blue method. *Soil Science Society of America Journal*, 58, 866–870.

Saarsalmi, A., Smolander, A., Kukkola, M., Moilanen, M. & Saramäki, J. (2012). 30-Year effects of wood ash and nitrogen fertilization on soil chemical properties, soil microbial processes and stand growth in a Scots pine stand. *Forest Ecology and Management*, 278, 63–70.

Scott-Wendt, J., Hossner, L. & Chase, R. (1988). Variability in pearl millet (*Pennisetum americanum*) fields in semi-arid West Africa. *Arid Land Research and Management*, 2, 49–58.

Stahr, K., Böcker, H. & Fischer, H. (2009). *Exkursionsführer – Landschaften und Standorte Baden-Württembergs.* Institut für Bodenkunde und Standortslehre (310), University of Hohenheim, Germany.

Throneberry, G. O. & Smith, F. G. (1955). Relation of respiratory and enzymatic activity to corn seed viability. *Plant Physiology*, 30, 337–343.

Valluru, R., Vadez, V., Hash, C. & Karanam, P. (2010). A minute P application contributes to a better establishment of pearl millet (*Pennisetum glaucum* (L.) R. Br.) seedling in P deficient soils. *Soil Use and Management*, 26, 36–43.

18

Understanding the emergence and evolution of pastoral community groups from the perspective of community members and external development actors in northern Kenya

Raphael Lotira Arasio [a,*], Brigitte Kaufmann [b],
David Jakinda Otieno [c], Oliver Vivian Wasonga [a]

[a]Department of Land Resource Management and Agricultural Technology, College of Agriculture and Veterinary Sciences, University of Nairobi,
P.O. Box 29053, Loresho Ridge, Nairobi, Kenya

[b]German Institute for Tropical and Subtropical Agriculture (DITSL), Steinstrasse 19, 37213 Witzenhausen, and
Institute of Agricultural Sciences in the Tropics (Hans-Ruthenberg-Institute 490g), University of Hohenheim, 70599 Stuttgart, Germany

[c]Department of Agricultural Economics, College of Agriculture and Veterinary Sciences, University of Nairobi,
P.O. Box 29053, Loresho Ridge, Nairobi, Kenya

Abstract

Whereas there is abundance of information on community groups that engage in income generation in rural agricultural and peri-urban areas, information on community groups in pastoral areas still remains scarce. However, in the recent past, a growing trend of such groups has been observed in the pastoral areas in northern Kenya. This study therefore explores how these groups have emerged since Kenya's independence in 1963 to date, and which factors have contributed to their evolution. A full survey on all income-generating community groups was conducted and different types of interviews were used to elicit the perspectives of members of the community and external development actors.

The findings on the history of group formation show the roles played by different entities over time and reveal how and why various factors influenced group formation. The characterisation of all 153 income-generating groups found in Marsabit South showed the diversity of the different group activities and yielded information on the reasons why usually a combination of different income-generating activities is practised. The collective group activities offer a possibility for income diversification for pastoralists despite labour constraints posed by key domestic and livestock-management tasks. The findings explain why community groups are increasingly gaining importance in pastoral areas, as a means to solve problems and fulfil diverse needs at household and community level.

Keywords: community organisations, history of formation, income-generating activities, pastoralists, Kenya

1 Introduction

In sub-Saharan Africa, since the early 20th century, colonial governments (Hussi *et al.*, 1993; Lele, 1981; Lele & Christiansen, 1989; Porter & Lyon, 2006), African post-colonial governments (Lele & Christiansen, 1989) and donors such as the World Bank and International Monetary Fund (Dorward *et al.*, 2005; Jayne *et al.*, 2002; Porter & Lyon, 2006; Shiferaw *et al.*, 2006) have promoted co-operatives, rural organisations and community groups as an option for socio-economic development particularly among the rural poor. Community groups have been variedly referred in development studies literature as: membership-based organisations (Chen *et al.*, 2006), local development groups (Porter & Lyon, 2006), grass root organisations or community-based organisations (De Weerdt, 2001). The contexts within which these terminologies are used differ (social advocacy, provision of services, income generation, among others) but the meaning of the different terminologies is the same – the use of group-based approach at community level to undertake activities for the benefit of members and by extension the entire community.

* Corresponding author – rlotira@yahoo.co.uk

Specific country examples indicate that in Ghana the history of community groups can be traced back to colonial times where traditional authorities and groups based on chieftaincy system were used to pursue commercial interests (Porter & Lyon, 2006). In Kenya, the history of community groups also dates back to colonial and post-colonial times (Bender, 1986; Chitere, 1988; Holmquist, 1984; Ngau, 1987; Thomas, 1988; Winans & Haugerud, 1977). The church, government and non-governmental organisations (NGOs) have been instrumental in supporting the emergence and evolution of groups (Coppock & Desta, 2013). Groups are formed for different purposes, for instance natural resource management (Kariuki & Place, 2005; Meinzen-Dick et al., 2002; White & Runge, 1995), smallholder agricultural marketing (Adong et al., 2013; Asante et al., 2011; Bernard & Spielman, 2009; Coulter, 2007; Markelova & Mwangi, 2010; Meier zu Selhausen, 2015) and income generation (Gugerty, 2007), among others.

The emergence and evolution of community groups has been associated with various factors, which can be categorised as endogenous and exogenous factors, and have also been differentiated into micro (household) and macro (community) level factors (White & Runge, 1995). These factors are context-specific (Markelova & Mwangi, 2010) and have been studied in domains such as natural resource management (White & Runge, 1995) and smallholder agricultural marketing (Meier zu Selhausen, 2015). For community groups whose purpose is income generation, much information is available for the urban and/or peri-urban (Abala, 2013; Fafchamps & La Ferrara, 2012), rural and farming communities (Gugerty, 2007; Jivetti & Edwards, 2009; Kabugua, 2014; Koech, 2014; Molesworth et al., 2017; Place et al., 2004; Shiferaw et al., 2006).

In pastoral areas, substantial information can be found for community groups engaged in natural resource management in Mongolia (Schmidt, 2006; Undargaa, 2017) and in Africa (Nganga & Robinson, 2016). However, fewer examples are found for communal breeding (Ouma & Abdulai, 2009), animal health (Badejo et al., 2017; Leyland et al., 2014) or income generation. For the latter, there are a few such studies for southern Ethiopia (Aklilu, 2004; Coppock et al., 2011) and eastern Kenya (Mutinda, 2017). In northern Kenya, the first study on community groups in pastoral areas was conducted in the frame of the Pastoral Risk Management (PARIMA) project in 2006 (Coppock & Desta, 2013). Other than this, there is only the work of Ngutu et al. (2011), who studied livestock marketing activities of two community groups in northern Kenya and the work of Aklilu & Catley (2009), who reviewed one pastoral livestock marketing group in north-eastern Kenya. These groups differ in objective from informal socio-cultural collective actions or groupings within the community that form part of the pastoral culture whose purpose is to pool efforts to address challenges such as insecurity or pool labour in cultural ceremonies, among other cultural uses.

Up to now, no comprehensive historical and situational analysis has been undertaken to understand the diversity and evolution of community groups in the pastoral areas of northern Kenya.

This article follows the conceptual framework used by White & Runge (1995) for understanding emergence and evolution of groups. Using a combination of different kinds of interviews, the article explores in detail why and how pastoral community groups emerge and evolve from the perspectives of the involved actors, i.e. group members and officials, community leaders, and staff of governmental and non-governmental organisations. It specifically analyses the history of group formation, establishes the prevalence and activities of existing groups and assesses factors influencing emergence and evolution of groups. An analysis of the history of groups and of the drivers of group formation serves as a starting point to understand the context within which groups emerged over time and to draw lessons from history for promotion of sustainable community groups in future.

2   Methods

2.1   Study area

This study was conducted in Marsabit South District of Laisamis Sub-County, located to the south of the greater Marsabit County. Marsabit South District (Fig. 1) is arid with an annual average temperature of 20.1 °C and average annual rainfall is around 200 mm (County Government of Marsabit, 2014). Laisamis Sub-County has a land area of 20,290.5 km^2 and human population of 65,669 persons, predominantly Rendille/Ariaal pastoralists (County Government of Marsabit, 2014). Marsabit South District is prone to droughts and also characterised by high illiteracy and poverty levels, and poor infrastructure (roads, telecommunication, markets, water) (County Government of Marsabit, 2014). On account of the spatial and temporal variation in resources and unpredictable climatic conditions, pastoralism is the dominant livelihood activity and livestock production system in these areas (Krätli et al., 2013). Pastoralists raise livestock for income through sale of livestock and livestock products, and for food (milk, meat), asset accumulation, and cultural obligations (Ministry of Livestock Development, 1991).

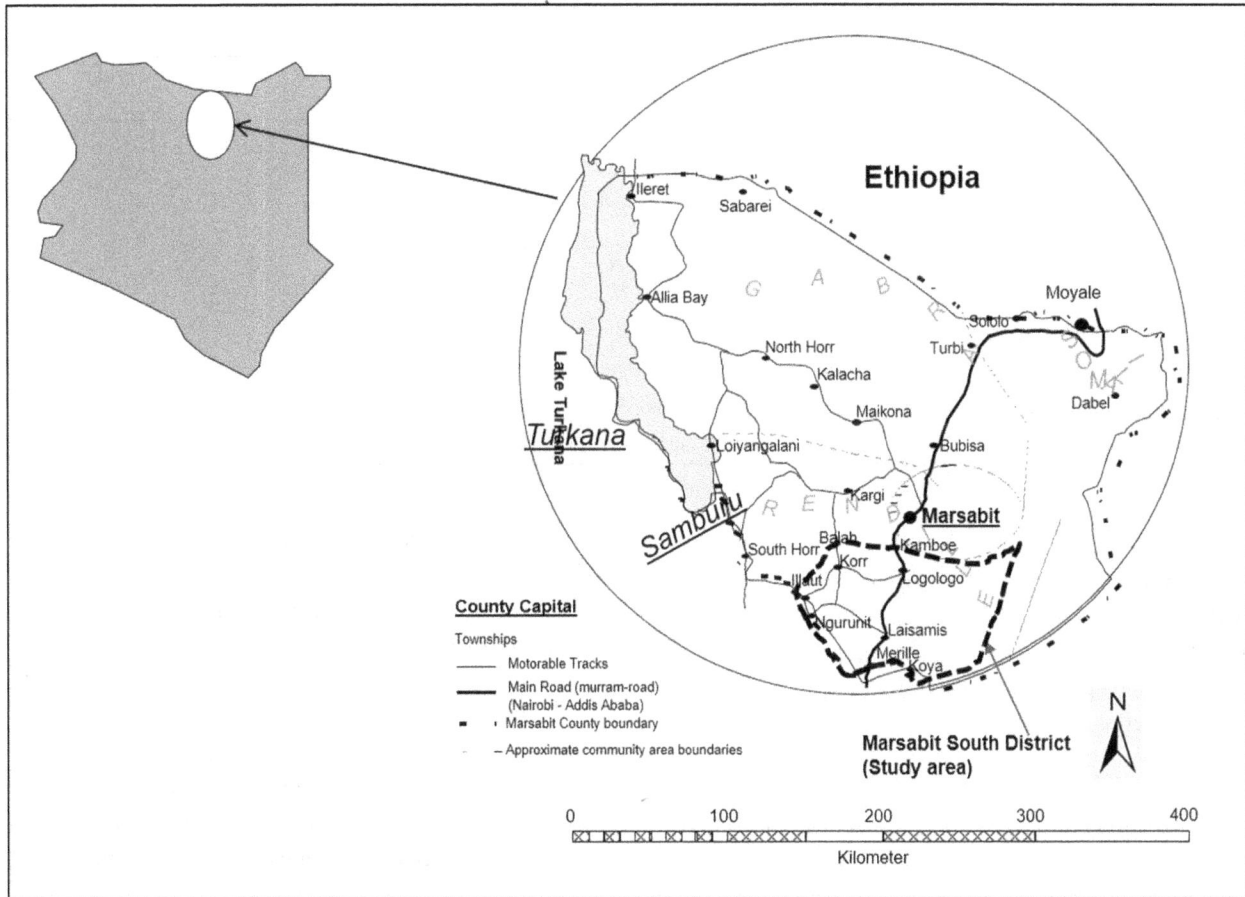

Fig. 1: *A map of the study area in Marsabit County, Kenya.*

2.2 Data collection and analysis

2.2.1 Group identification and selection for narrative interviews

A preliminary list of registered groups was obtained from the Department of Social Services at the Sub-County and County headquarters. All the 10 chiefs and 24 sub-chiefs in Marsabit South District were consulted by the research team to get additional names (including contacts) of both formal and informal groups in their areas of jurisdiction. Staff of six government departments/ministries, four private sector groups, two church organisations and five NGOs that deal with groups in the study area were also consulted to get a list of groups they currently support or have supported in the past or that they know exist in the area. In this study, the term "NGO" includes also bilateral projects supported by governments of other countries e.g. German Agency for International Cooperation (GIZ).

Using a snowball sampling approach, once a group was identified, its members were asked to identify other groups within and outside their area and for their contacts. Three livestock markets (Merille, Ilaut and Korr) were visited to identify existing groups operating in the markets. It was not possible to capture information on groups that were discontinued or collapsed. The records obtained from government and NGOs, did not show whether the groups were existent or not. Community members in particular were reluctant to disclose information on collapsed groups for fear of being labelled as traitors. Among the existing groups, the oldest were formed in the early 1980s.

The compiled list contained groups dealing with either community social-support activities only (these include activities such as environmental awareness/conservation, girl child education, campaigns against female genital mutilation, HIV/AIDS awareness, promotion of culture and heritage, hygiene and sanitation, and management of community water services), income-generating activities (IGAs) only or both. A total of 153 groups that dealt with only income generation were identified.

In each of the 153 identified income-generating groups (IGGs), a group official was asked about gender composition of the group members, administrative division, geographical location, year of formation, entity involved in formation, registration status, size of membership disaggregated by gender, and types of activities undertaken.

Table 1: *Profile of groups selected for narrative interviews.*

Name of group	Division	Gender category	Geographical location	Year of formation	Current group size
Merille livestock traders group	Laisamis	Mixed	Urban	1998	17
Nairibi Morans	Laisamis	Youth	Rural	2012	20
Natala women's group	Laisamis	Women	Peri-urban	2009	14
Nkupulu women's group	Laisamis	Women	Rural	2009	19
Pambazuko women's group	Laisamis	Women	Rural	2005	15
Salamis women's group	Laisamis	Women	Peri-urban	2007	21
Balah self-help group	Korr	Mixed	Rural	2012	28
Jitto Jille women's group	Korr	Women	Rural	2011	30
Meriki women's group	Korr	Women	Rural	2014	10
Lorugo women's group	Korr	Women	Rural	2008	20

2.2.2 Narrative interview with group officials

For narrative interviews, 10 groups were purposively selected from the 153 groups. The main selection criterion was their involvement in the trade of sheep and goats – this was due to the specific interest of the wider study in collective marketing of sheep and goats. Further selection criteria used yielded a sample covering different administrative divisions (Laisamis and Korr), geographical locations (urban, peri-urban and rural) and gender categories (women's groups, mixed-gender groups, men-only groups, and youth groups). The profiles of selected groups are shown in Table 1.

In each group, the interview was done with a group official, either the chairperson or the secretary, to gather information on the history of group formation. Before starting the interview, the interviewee was asked consent for audio recording of the interview. The initial question asked was: *"Can you please give us the story of your group from the time you conceived the idea until now?"* Further steps of the interview were conducted according to Bauer (1996). The interviews took between 45 and 90 minutes.

2.2.3 In-depth interviews with group officials

Using a guiding checklist, ten in-depth interviews were conducted with officials of the same groups involved in the narrative interviews to gather additional information on factors and key personalities/institutions that influenced the emergence and evolution of groups in the study area in the last 20 years. This was the time (between 1996–2015) that the large majority of the existing groups were formed in the study area, including much involvement of NGOs, community elites/leaders and formation of self-initiated groups. For ease of recall of information, reference was made to a major event (the El Nino floods) that took place in the study area in 1997/98. In order to prevent potential response duplication, in each group, the interview was con-

ducted with an official who had not participated in the narrative interview. Depending on the knowledge of a respondent, attempts were also made to gather information on group formation before the 1990s.

2.2.4 Key-informant interviews with other stakeholders

Nine chiefs, staff of six government departments/ministries and staff of eight NGOs, church organisations and members of the private sector dealing with groups in the study area were asked the same questions as in the in-depth interviews with group officials.

2.2.5 Data analysis

Qualitative content analysis was done for the narrative interviews, in-depth interviews and key-informant interviews (Mayring, 2014). The interviews were coded in a step-wise manner according to categories representing key themes (for instance, history of group emergence, drivers of group emergence) and sub-categories (for instance, for history: involvement of churches, for drivers: internal factors such as distance to markets). Data from the survey (on characteristics and activities of existing groups) and some data from in-depth and key-informant interviews (on drivers of emergence and types and weaknesses of external support) were entered into MS Excel and frequencies and proportions were generated.

3 Results

3.1 History of group emergence and evolution in Marsabit South District in northern Kenya

The analysis of the history of group formation provides a timeline entailing entities that have supported formation and development of groups in northern Kenya from the 1960s to the present day and the type of support provided. Characteristics of groups differ between older and new generation of groups in Marsabit South.

Table 2: *Historical trends in emergence of and institutional support to pastoral community groups in the study area (1963–2015).*

Institutions and personalities	Type of support to groups	Starting period*
Catholic Church	Evangelical services to prayer groups at chapels	1963–1965
	Income generation	1976–1980
African Inland Church	Income generation	1991–1995
Government	Registration of groups in Marsabit town	1981–1985
	Formation of groups	1981–1985
	Women Enterprise Fund; Youth Fund	2002
	Financial support: Uwezo Fund	2013
	Transfer of registration to Marsabit South	2009
Non-governmental organisations	Humanitarian aid; income generation	1991–1995
Community elites	Formation; funding	1991–1995
Ordinary community members	Formation of self-initiated groups; income generation	1996–2000

* All types of institutional support persist until present.
Source: Narrative and in-depth interviews (*n* = 10); Key informant interviews.

3.1.1   Involvement of churches

In northern Kenya, group formation started just after Kenyan independence in 1963 with first groups initiated by the Catholic Church (Table 2). The Catholic Church organised the Christian community into prayer groups, Catholic women's associations, adult literacy and choir groups with a membership of about 100 members each. These groups were composed mainly of poor households who settled around the Catholic chapels in pursuit of external aid and alternative livelihoods. In addition to evangelism, the Catholic Church provided relief assistance and social amenities (such as hospitals and schools) to these poorer households. Between the late 1970s and the early 1980s, besides promotion of evangelical work through groups, the missionaries (mainly the Catholic nuns) also introduced and supported IGAs in these groups as a way to sustain evangelical connections. The first set of IGAs supported were the sale of traditional artefacts (beadwork) and micro-credit schemes (merry-go-round). Other churches such as the African Inland Church (AIC) started supporting IGGs in Marsabit South in the early 1990s.

3.1.2   Involvement of the Kenyan government

Kenyan government involvement in group development in Marsabit South District began in 1978. Riding on the precedent set by the Catholic Church as well as its own efforts to form groups, the government through the Department of Social Services began to formally register groups as a way to attract and maintain recognition by external agencies (churches, government, NGOs and private sector).

Between the early 1980s and 2008, registration of groups by the government was centralized at the headquarters of the parent district (Marsabit town in the then greater

Marsabit District). Registration of groups was moved to Marsabit South District (now part of Laisamis Sub-County) headquarters only in 2009, when the new district was created, but still as a national government function. At the beginning, the government's roles through the Department of Social Services were confined to registration and training in group dynamics and development. Other services such as financial support to groups in the form of loans and grants began in 2002, with priority given to women's groups (Women Enterprise Fund) and youth groups (Youth Fund) and scaled up from 2013 to have one fund that supports women's groups, youth groups and groups of disabled persons (*Uwezo* fund).

3.1.3   Support by non-governmental organisations (NGOs) and bilateral projects

In Marsabit South District, various NGOs and bilateral projects supported different livelihood activities through groups since the early 1990s but with mixed outcomes. Some of the earlier projects and the group activities they supported include activities on environmental management and sheep and goats trade supported by the German Development Cooperation (formerly GTZ, now GIZ), Arid Lands Resource Management Project (ALRMP) (goat restocking) and Heifer Project International (HPI) (camel restocking). These organisations and projects supported large groups of about 100 members each so as to spread benefits to a wider spectrum of the community. They also targeted mainly the poor and women, with a precondition that all activities must be done collectively by members. Emergence of groups was further influenced by the precondition that any outside support to livelihood activities can only be undertaken through an organised group.

In common with church-supported and government-supported groups, most NGO/project-supported groups malfunctioned soon after external support ended. The malfunctioning was mainly attributed to large size of groups, mismanagement and inconsistent benefits. Some groups adjusted their names, number of members and objectives just to take advantage of the prevailing opportunities of working with an external agency, as illustrated by a government administrator:

> *In this Ngurunit area, St. Mary's women's group*
> *and Umoja women's group initially supported*
> *by the Catholic Church and the African Inland*
> *Church (AIC) respectively changed their names to*
> *Salama, Saidia and Salato women's groups to tar-*
> *get camel restocking supported by the HPI*
> (Chief, Ngurunit Location, 11 July 2015).

Learning from weaknesses of externally supported groups, some of this older generation of groups were later (in the late 2000s) reorganised by members themselves into small manageable groups of 10–20 people, with a focus on immediate benefits for members. Additionally, these groups adapted their organisational and governance features to the community's socio-cultural norms and practices such as belonging to the same clan, village and age-set, trust and behaviour/conduct of a community member. Instead of group members doing group activities collectively, some groups opted for group loans to support individual members' businesses. Another interesting case is where, contrary to the guideline that women's groups should include only women, some women's groups innovatively co-opted one to three men for difficult tasks such as trekking animals to the market.

3.1.4   Community elites' and leaders' involvement

From the 1990s, having seen the importance of groups, community elites – for instance, some specific primary school leavers, secondary school leavers, primary school teachers and adult literacy teachers – started guiding the community in the formation of groups and fundraising for them. Local political leaders were particularly involved in fundraising for these groups. Groups formed through guidance of community elites/leaders were small (composed of 15–30 members) and performed fairly well as compared to the older generation of groups.

3.1.5   Self-initiated groups

Learning from organisational and governance weaknesses of externally supported groups, in the past two decades, fully self-initiated groups were formed. Usually this process has been started by a potential group initiator acquiring information from neighbouring existing groups, relatives or friends. The information gathered includes group formation process, activities and implementation procedures, group size, performance, benefits and challenges of group work, among others. The idea to form a group is first shared with close friends to check its utility and then spread to the rest of the community members. Potential members are gradually mobilized under an agreed common felt need until the desired membership size is attained (10–20 members). After formation, the new groups continue to consult with older groups until they are able to stand on their own. Although formed on their own, some of the groups later develop relationships with external agencies mainly for training support. However, in the absence of opportunities for external training, groups seek advice from well-functioning groups (peers) and community elites in case of need. Some community members have a long experience and knowledge of undertaking some of the activities e.g. livestock trade hence are a source of knowledge for groups. In most cases, the person who initiates the group ends up being the chairperson. These self-initiated groups take advantage of the socio-cultural fabric and dynamics of the community to organise themselves and set up their governance structures. Moreover, they take into consideration the strengths and weaknesses of existing groups to guide their path towards a conceivably better-functioning model.

3.1.6   Income-generating groups established between 1981 and 2015

The records from the Department of Social Services as well as the survey of groups conducted in the study area showed that, by 2015, a total of 300 groups in a population of approximately 12,000 households had been established, out of which 153 participated in diverse IGAs. These groups were formed between 1981 and 2015.

Generally, there is an increasing trend of IGGs, with the majority being women's groups (Fig. 2). A remarkable increase over time is seen in groups supported by community elites and leaders and in self-initiated groups.

3.2   Existing income-generating groups and their activities

Most of the surveyed IGGs were registered (130 out of 153), and more than half were found in *Manyattas* (clan-based "villages") (86 out of 153) with the majority being women's groups (69 out of 86). The most common group size was 11–20 members (85 out of 153) with a few cases of groups (7 out of 153) with more than 40 members, indicating a shift towards smaller groups.

These groups undertake two categories of IGAs, i.e. livestock-related and non-livestock activities (Table 3).

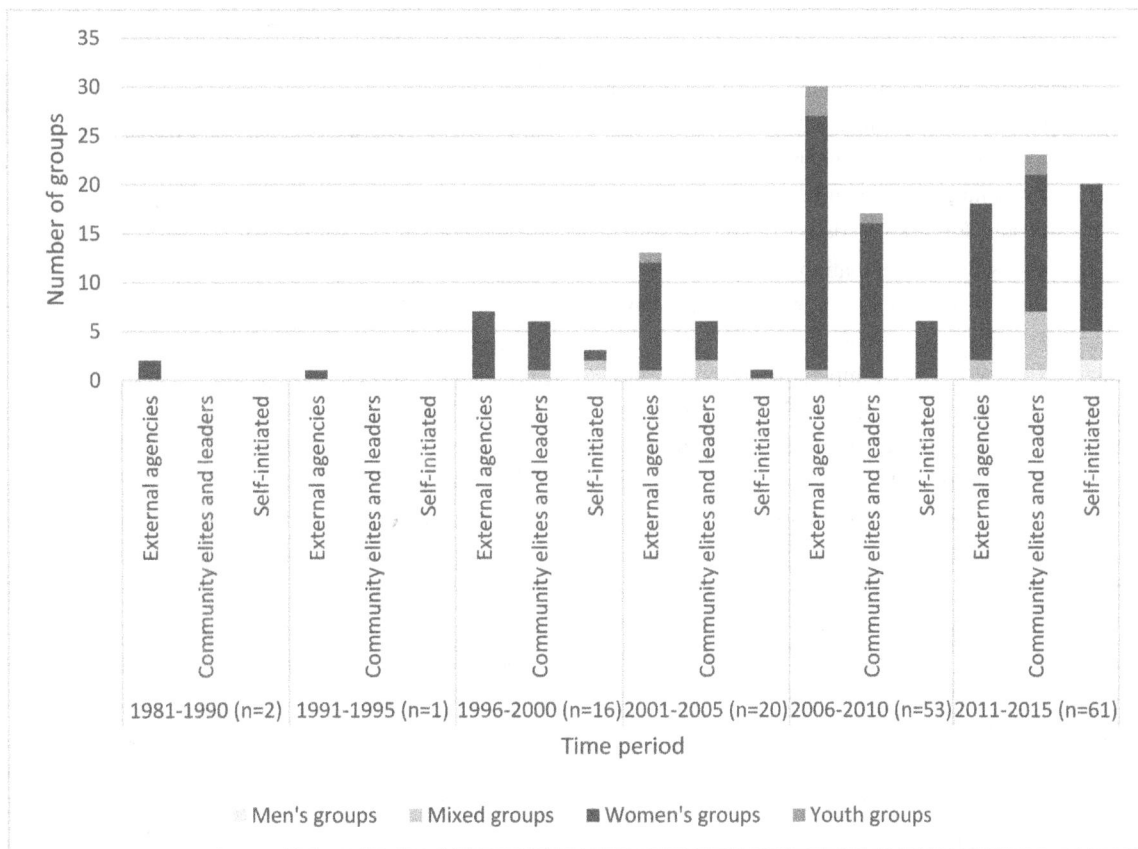

Source: Survey (*n* = 153)

Fig. 2: *Period of formation of community groups in the study area between 1981 and 2015.*

Table 3: *Activities undertaken by income-generating groups (n = 153).*

Non-livestock business activities		Livestock-based business activities	
Type of business (*n* = 18)	No. of practising groups (out of 153 groups)	Type of business (*n* = 13)	No. of practising groups (out of 153 groups)
Sale of household items	89	Sheep and goat trade	100
Micro-credit	32	Skins trade	27
Beadwork	25	Cattle trade	15
Vegetables	15	Veterinary medicines	9
Rental premises	8	Meat trade (butchery)	8
Honey trade	4	Camel trade	4
Water kiosk	3	Goat breeding	4
Tailoring	2	Milk trade	3
Natural forest products: gums and resin	2	Poultry production	3
Motor vehicle repair (garage)	2	Meat processing	2
Tree nurseries (seedlings)	2	Camel breeding	2
Bread making	2	Fodder production	2
Others: mill (2), fuel (1), gravel/ballast (1), khat/miraa (1), blacksmith (1), motor cycle hire (1)		Leather processing	1

Source: Survey (*n* = 153)

Overall, most groups practised trade in sheep and goats and in combination with other activities, majorly sale of household items, micro-credit schemes, skins trade and cattle trade. With regard to micro-credit schemes, through further probing of group officials, the groups that did not mention it as an activity ($n = 121$) said that, though they did not practise it in a structured manner, group members and other community members could still ask for loans from the groups.

Reasons for practising a combination of different IGAs were to enhance income generation, to insure against business risks and to achieve sustainability of the group, as group officials explained. For example, groups commonly use part of income from the sale of household items to buy sheep and goats for trade. They do this as a strategy to stock their cash in the absence of formal banking systems and as they wait to replenish stocks of household items. Furthermore, profits from these businesses are often moved to the micro-credit scheme to allow group members and the entire community better access to loans on interest. Hence, an important objective of engaging in IGAs including trade in sheep and goats is to generate money to facilitate (emergency) loans for members and the entire community.

Several factors determine the choice of business activities undertaken by groups or the shift from one business activity to the other. These include profitability, the need to fill the commodity supply gap in the community, seasonality of demand, challenges such as debts, and advice from an external agency.

3.3 Drivers of emergence of income-generating groups

It was established from community members (group members and officials, and chiefs) and external agencies (staff of government departments, NGOs, church organisations and private sector) that both factors external and internal to the community have influenced the emergence of the existing groups over time (Fig. 3). Community members provided more factors than the external agencies and indicated the different factors more often: five factors were mentioned by more than 50 % of the community members. The community members explained why and how these factors influenced group emergence and continuation, as outlined in the paragraphs below.

3.3.1 Distance to market

In the 1990s, formal livestock and commodity markets were fewer or non-existent in Marsabit South and people moved long distances (> 100 km) to access livestock markets and purchase household items. On account of high transport costs and high capital outlays, it was challenging to acquire household commodities that were highly demanded, particularly in rural villages. In places where formal markets (able to attract many buyers) did not exist and the dominant buyers of livestock were either restaurants in town centres, mobile itinerant traders or rural shopkeepers, groups were formed to increase access to distant markets. As access to distant livestock and commodity markets was costly in terms of time and transport, people pooled human

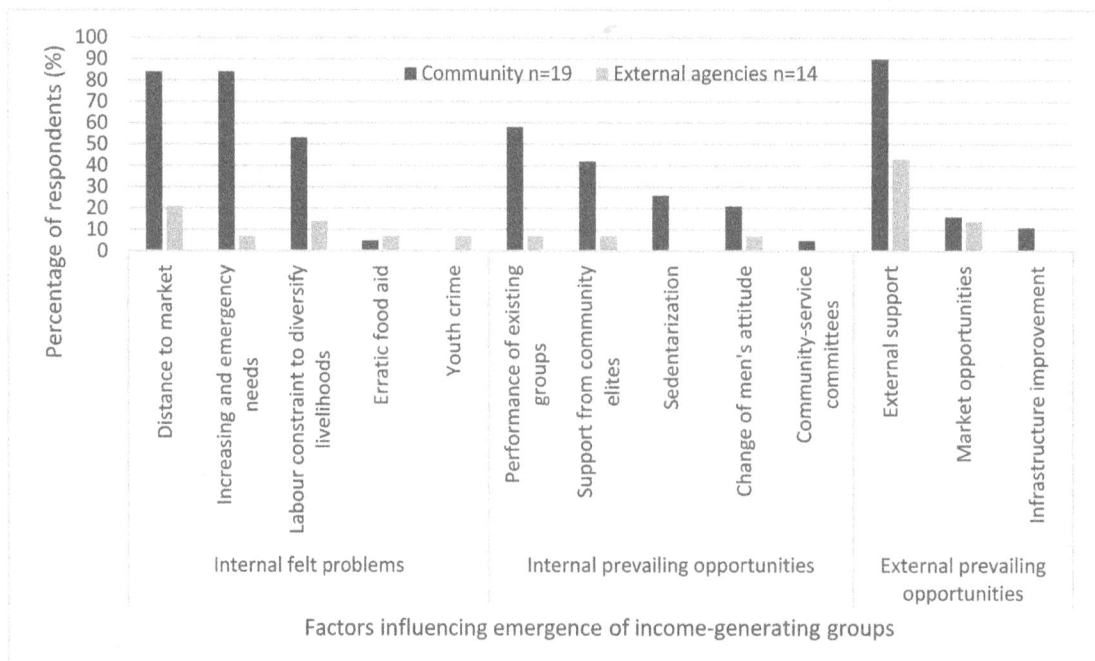

Source: In-depth and key informant interviews (for community members $n = 19$; for external agencies $n = 14$)

Fig. 3: *Factors that influence emergence of income-generating groups.*

and financial resources together to bring services, goods and market outlets closer to the village. However, distance to livestock and commodity markets has reduced in the recent past as a result of the establishment of formal markets in settlements within the community area such as those in Korr and Ilaut, which operate on a weekly and bi-weekly basis, respectively.

3.3.2 Increasing and emergency needs amid decreasing social support

Droughts, livestock diseases and increasing household needs are putting pressure on the available livestock resources. This makes it increasingly difficult to get emergency support from neighbours, relatives or friends. The situation is especially difficult for single-parent families (widows, divorcees).

Among others, one of the priority needs was (and is) payment of school fees and other school-related expenses because of the increasing number of school-going children. Consequently, parents moved long distances, mainly between villages and town centres, to seek loans from distant traders or relatives to cater for school needs. However, there was no guarantee that the required support would be availed. This situation compelled households to form groups at the village level to access emergency loans for school fees, among other benefits.

Women were (and are) more involved in the search for money to educate children, particularly through groups. Compared to pastoralist men who still prefer children to take up livestock-management roles, women have complied more with the government's advice to take children to school. Other motivating/driving factors include seeing families with educated children leading a better life and the wish to be able to cover the increasing household needs. Single-parent families (for instance widows), particularly those with limited or no support from relatives, have to struggle for the education of their children. Although nowadays there are clan-based education funds, they cater only for some of the school needs for secondary and tertiary education; thus parents still struggle to cover the remaining costs.

3.3.3 Labour constraint to diversify livelihoods individually

Pastoral livestock-management and domestic tasks require keen attention from household members. Household participation in additional livelihood activities such as an additional business put a strain on the available labour and time, particularly in labour-deficient households. Furthermore, some of these additional livelihood activities are difficult to undertake individually. For instance, the nature of

livestock trade makes it difficult to effectively organise the activities (buying, trekking and selling), particularly when source and destination markets are far apart. Being in a group enables the households to more easily combine business and domestic tasks. The pooling of labour by households to diversify livelihoods becomes particularly useful in the dry season and during droughts, when livestock-management tasks increase, yet household needs such as provision of food are critical and need concerted attention at the same time. During the dry season, livestock are away from home for a longer period and are not easily accessible for sale (i.e. income generation); thus the group becomes a strategic alternative to cover food and cash needs through loans.

3.3.4 Performance of existing groups

Positive experiences of existing groups including accruing benefits to members and the entire community, and good cooperation between members and officials of a group can influence other people to form or join a group. Other positive experiences include recognition by government through representation and praise at public gatherings as well as winning tenders to supply goods and do construction works. Such positive experiences can also influence a poorly performing group (poorly performing due to misunderstanding, unmanageable large size, little benefits, poor management) to be reorganised into an outfit in which members have more influence on leadership, activities and outcomes.

3.3.5 External support

External agencies prefer to use groups to deliver livelihood-support interventions so to spread the benefits to the wider community. External support ranges from mere awareness about the group approach to direct support. The types of external support mentioned by chiefs, government staff, staff of NGOs and church organisations, and members of the private sector are shown in Table 4.

When asked what they see as shortfalls of external support provided by their peers, government staff and staff of NGOs, church organisations and the private sector gave their views as shown in Table 4. In addition, community members criticized external support, particularly expressing that they found the training and mentoring inadequate either because the content did not respond to their training needs or because the duration was too short for them to develop the respective capacities. They also pointed to the lack of attention to adult literacy. Groups were divided over the need for external funding – a majority of them approved of it while two of the ten interviewed said it is not necessary. Groups that were against external funding referred to cases of groups that malfunctioned once external funding

Table 4: *Types and weaknesses of external support.*

Types of external support			Weaknesses of external support		
Type	Frequency (n = 23)	Type of respondents*	Weakness	Frequency (n = 33)	Type of respondents*
Training	23	9 chiefs, 6 Govt, 8 NGO	Inadequate training and mentorship	19	**10 IGG**, 4 chiefs, 2 Govt, 3 NGO
Funding	23	9 chiefs, 6 Govt, 8 NGO	Inadequate funding	10	**8 IGG**, 1 chief, 1 Govt
Material inputs	5	4 chiefs, 1 Govt	Encourage vested interests and dependency of beneficiaries	4	2 chiefs, 2 NGO
Monitoring/follow-up	4	2 chiefs, 1 Govt, 1 NGO	Inadequate monitoring	11	5 chiefs, 3 Govt, 3 NGO
Link to other service providers	4	4 Govt	Wrong selection of beneficiary groups	8	2 chiefs, 4 Govt, 2 NGO
Registration of groups	2	1 chief, 1 Govt	External agencies prioritizing own interests	4	1 chief, 1 Govt, 2 NGO
Infrastructure development	2	2 Govt	Less attention to adult literacy education	2	**2 IGG**
Commodity supply contracts	1	1 chief	Inadequate stakeholder participation and coordination	4	2 chiefs, 1 Govt, 1 NGO
Humanitarian assistance during disasters	1	1 chief	Short project duration and poor exit strategies	3	2 chiefs, 1 Govt
Funding proposal development	1	1 Govt	Inappropriateness of some interventions	3	2 Chiefs, 1 Govt
Link to markets	1	1 NGO	Less attention to thorough feasibility studies	3	1 chief, 1 Govt, 1 NGO

* IGG: officials and members of income-generating groups; Govt: government staff; NGO: staff of non-governmental organisations, bilateral projects and church organisations, and private sector
Source: In-depth and key informant interviews (n = 23 for types- 9 chiefs, 6 Govt and 8 NGO; n = 33 for weaknesses- 10 IGG, 9 chiefs, 6 Govt and 8 NGO)

ended. IGGs were also not spared in the criticism – staff of NGOs (including bilateral projects) in particular complained of vested interests and lack of seriousness of some groups to put external funding to the right use.

External agencies use certain criteria to select groups for support. Over 20 % of interviewed external agencies mentioned the following as important criteria: registration status, interest of external agency, vulnerability of potential beneficiaries, gender composition of a group, history of a group handling grants as well as its current performance, and the geographical location of interest to an external agency. Other criteria mentioned include records of business and financial transactions, results of site verification, size of membership, business acumen and interest of the group, duration of group and specific activities, religious affiliation, and having a written proposal. Some of these criteria such as group records and written proposals were viewed by government officials (n = 3) as strict and therefore need to be reconsidered given the high rate of illiteracy among the pastoralists.

4 Discussion

In this study, we explored from the perspectives of community group officials and other relevant stakeholders why and how pastoral community groups for income generation emerged and evolved in northern Kenya. We focussed particularly on the history of group formation, prevalence of existing group activities and factors influencing emergence and evolution of such groups.

The detailed examination of IGGs revealed that the emergence of groups was initially externally supported. It was only from the late 1990s onwards that self-initiated groups were formed. The Catholic Church was instrumental in the formation of the first groups in Marsabit South District and this illustrates their involvement in social and economic development in pastoral areas. The Kenyan government's involvement in group affairs in Marsabit South District started only two decades after independence, which is late considering that cooperatives and rural organisations in other parts of Kenya were promoted by the government since colonial times (Chitere, 1988; Thomas, 1988). NGOs used group formation as a strategy to implement their livelihood support programmes from early 1990s onwards. However, many of those groups malfunctioned after external support ended, as was also observed in Ghana by Porter & Lyon (2006). The malfunctioning was mainly attributed to organisational and governance weaknesses. Interestingly, these weaknesses of externally supported groups informed the formation of self-organised groups with modified characteristics, such as favouring small group sizes of 10–20 people for effective management and adapting their organisational and governance features to the socio-cultural norms and practices in cultural collective activities. These characteristics and the process followed in group formation distinguish self-initiated groups from older generations of groups, particularly the externally supported groups. The shift towards self-initiated groups was also noted among community groups in Haiti (White & Runge, 1995), in Tanzania (De Weerdt, 2001) and in Ghana (Porter & Lyon, 2006).

This study also attempted a full survey on community groups that exist in Marsabit South District and contacted 300 groups to learn about their activities. Out of them, 153 IGGs were identified and characterised to get an overview of their distribution by size, gender composition and location as well as a detailed list of their activities. More than 80 % of the groups were found to be women's groups. This is particularly notable given that in pastoral areas activities such as livestock trade are reportedly male dominated. Women being the majority and pioneers of IGGs has also been commonly reported in Kenyan agricultural areas in the highlands (Place et al., 2004) and in urban slum areas (Abala, 2013; Fafchamps & La Ferrara, 2012). More than 50 % of the groups were found in manyattas (clan-based "villages") in the rural areas, indicating that pastoralists see IGGs as a livelihood option complementing livestock husbandry, which is their main livelihood activity. Here, groups were valued for their services with regard to providing greater ease to reach livestock and commodity markets, and loans for emergency needs and for supporting individual members' businesses. Often, these three activities were practised in combination. This was attributed to the potential of this particular combination of activities to: (1) boost income generation for the group, (2) stabilise income in the different seasons, (3) make use of synergies between the activities and (4) reduce business risks. Synergistic effects between livestock and non-livestock business activities have also been reported in other studies (Chen et al., 2006; Coppock & Desta, 2013). The diversity and complementarity of group activities can be related to the common practice among the pastoralists of diversifying livestock herds and species (camels, cattle, goats, sheep and donkeys) to obtain diverse benefits, adapt to variability and cope with droughts (Campbell, 1999; Desta & Coppock, 2004; Little et al., 2001; McPeak, 2005; Sato, 1997). These findings on the benefits of a diversity of group activities should inform development efforts and strategies.

Dating back five decades, micro-credit schemes were found to be the earliest group activity in Marsabit South District. This can be understood as an effort to complement the traditional social security system among the Rendille community, called the Maal system, which consists of animal loans and mutual help (Ministry of Livestock Development, 1991; Sato, 1997; Schlee, 1991). The main source of the capital used to support emergency and common felt needs of group members and non-members through micro-credit schemes was the profits from the sheep and goat trade. This indicates that the main objective of collective livestock marketing is to build capital for loans and not necessarily to negotiate for better terms of trade for members. This link between livestock marketing and group loan systems in pastoral areas has to our knowledge not been documented before.

The emergence of the above-mentioned group activities was driven mainly by internally felt problems, such as distance to markets, increasing monetary needs, labour constraints to diversify livelihoods in individual households, and erratic food aid. Especially during dry seasons, when livestock is kept far away from homes and labour is needed critically for livestock management, members profit from the groups as they can cover food and cash needs through loans.

In indicating the factors that influenced the emergence of IGGs, the community members were more outspoken, meaning that they are more concerned and knowledgeable of their own situation than the external agencies. Among the factors, external support was mentioned most often both by community members and external agencies. Although important, external support was equally criticized by most actors interviewed, mainly because of inadequate training, funding and monitoring. Groups were divided over the need for external funding. While external development actors identified a wide range of areas in external support that require improvement, community groups were particularly interested in training that responded to their needs. The criteria used to select groups for support such as the obligation to present written records and submit written proposals were criticized as they do not take into account the high illiteracy rate among the population, a pattern that is common among Kenyan pastoralists (Orodho et al., 2013). In rural Ghana, formal requirements such as the obligation to operate a bank account were seen as a hindrance to enlisting potential beneficiaries for external support (Porter & Lyon, 2006). These criteria as well as vested interests of individual government and NGO staff were seen as the origin of improper selection of beneficiary groups by external agencies.

## 5	Conclusion

External support, particularly from the faith-based organisations, has been instrumental in the development of community groups in the pastoral areas of northern Kenya. In the past three decades, a four-fold increase in the number of groups formed was observed, and this was accompanied by an increase in self-organised groups. The latter trend resulted from weaknesses of externally supported groups, which affected their sustainability. The presence of a problem that requires collective action, but also the supportive internal and external environment, drive the emergence and development of groups. Community groups offer a possibility for income diversification for pastoralists despite labour constraints for domestic and livestock-management tasks. To achieve this, groups prefer a particular combination of IGAs, comprised of sheep and goat trade, microcredit schemes and sale of household items. From the foregoing, community groups are increasingly becoming important in solving felt needs of pastoralists and thus require policy attention. Efforts to promote sustainable community groups should take into consideration lessons learned from the history of group formation as well as insights from the different actors, foremost the views from the community group members.

Acknowledgements

This research was conducted as part of the RELOAD (Reduction of Post-Harvest Losses and Value Addition in East African Food Value Chains) project funded through the initiative for Research on the Global Food Supply (GlobE) by the German Federal Ministry of Education and Research (BMBF) in cooperation with the German Federal Ministry for Economic Cooperation and Development (BMZ) (Grant Number 031A247D). The authors are grateful to the community groups, chiefs, government departments, non-governmental organisations, church organisations and private sector organisations that participated in the study for their invaluable perspectives and time. Raphael Gudere is particularly thanked for his unreserved support in the entire period of field work right from group identification, organising respondents, translation during the interviews and data collection. Daniel Sunyuro is thanked for assistance in field data collection and transcription of audios. Augustine Kilio and Elizabeth Kidenye are thanked for assistance in field data collection, and Gideon Jalle and chief Antonella Mbositan Lekhuyan are thanked for audio transcriptions.

References

Abala, E. (2013). *Challenges facing income generating women groups in informal settlements: a study of selected women groups in Kibera, Kenya.* Master's thesis, Kenyatta University, Nairobi, Kenya.

Adong, A., Mwaura, F. & Okoboi, G. (2013). What factors determine membership to farmer groups in Uganda? Evidence from the Uganda Census of Agriculture 2008/9. *Journal of Sustainable Development*, 6 (4), 37–55.

Aklilu, Y. (2004). Pastoral livestock marketing groups in southern Ethiopia: some preliminary findings. Access to Markets Workshop, Adama Mekonen Hotel, Nazreth 2–3rd November 2004, CORDAID, Institutional and Policy Support Team, AU-IBAR. (pp. 2–3)

Aklilu, Y. & Catley, A. (2009). *Livestock exports from the Horn of Africa: an analysis of benefits by pastoralist wealth group and policy implications.* Feinstein International Center, Tufts University, Medford (MA), USA.

Asante, B. O., Sefa, V. A. & Sarpong, D. B. (2011). Determinants of small scale farmers' decision to join farmer based organizations in Ghana. *African Journal of Agricultural Research*, 6 (10), 2273–2279.

Badejo, A. F., Majekodunmi, A. O., Kingsley, P., Smith, J. & Welburn, S. C. (2017). The impact of self-help groups on pastoral women's empowerment and agency: a study in Nigeria. *Pastoralism*, 7, 28. doi:10.1186/s13570-017-0101-5.

Bauer, M. (1996). *The narrative interview: comments on a technique for qualitative data collection.* London School of Economics and Political Science, UK.

Bender, E. I. (1986). The self-help movement seen in the context of social development. *Nonprofit and Voluntary Sector Quarterly*, 15 (2), 77–84.

Bernard, T. & Spielman, D. J. (2009). Reaching the rural poor through rural producer organizations? A study of agricultural marketing cooperatives in Ethiopia. *Food Policy*, 34 (1), 60–69.

Campbell, D. J. (1999). Response to drought among farmers and herders in Southern Kajiado District, Kenya: a comparison of 1972–1976 and 1994–1995. *Human Ecology*, 27 (3), 377–416.

Chen, M., Jhabvala, R., Kanbur, R., Kanbur, T. H. & Richards, C. (2006). *Membership based organizations of the poor.* Routledge, London, UK.

Chitere, P. O. (1988). The women's self-help movement in Kenya: a historical perspective, 1940–80.

Coppock, D. L. & Desta, S. (2013). Collective action, innovation, and wealth generation among settled pastoral women in Northern Kenya. *Rangeland Ecology & Management*, 66(1), 95–105.

Coppock, D. L., Desta, S., Tezera, S. & Gebru, G. (2011). Capacity building helps pastoral women transform impoverished communities in Ethiopia. *Science*, 334(6061), 1394–1398. doi:10.1126/science.1211232.

Coulter, J. (2007). *Farmer groups enterprises and the marketing of staple food commodities in Africa*. International Food Policy Research Institute.

County Government of Marsabit (2014). *First County Integrated Development Plan 2013–2017*. County Government of Marsabit, Kenya.

De Weerdt, J. (2001). Community organisations in rural Tanzania: a case study of the community of Nyakatoke, Bukoba Rural District. The Nyakatoke Series, Report no. 3, IDS, University of Dar es Salaam and CES, University of Leuven, Belgium.

Desta, S. & Coppock, D. L. (2004). Pastoralism under pressure: tracking system change in Southern Ethiopia.

Dorward, A., Kydd, J., Morrison, J. & Poulton, C. (2005). Institutions, markets and economic co-ordination: linking development policy to theory and praxis. *Development and Change*, 36(1), 1–25.

Fafchamps, M. & La Ferrara, E. (2012). Self-help groups and mutual assistance: evidence from urban Kenya. *Economic Development and Cultural Change*, 60(4), 707–733.

Gugerty, M. K. (2007). You can't save alone: commitment in rotating savings and credit associations in Kenya. *Economic Development and Cultural Change*, 55(2), 251–282.

Holmquist, F. (1984). Self-help: the state and peasant leverage in Kenya. *Africa*, 54(3), 72–91.

Hussi, P., Murphy, J., Lindberg, O. & Brenneman, L. (1993). The development of cooperatives and other rural organizations: the role of the World Bank. Technical Paper No. 199. The World Bank, Washington DC.

Jayne, T. S., Govereh, J., Mwanaumo, A., Nyoro, J. K. & Chapoto, A. (2002). False promise or false premise? The experience of food and input market reform in Eastern and Southern Africa. *World Development*, 30(11), 1967–1985.

Jivetti, B. A. & Edwards, M. C. (2009). Selected factors affecting the performance of women's self-help groups in Western Kenya. *In:* Jayaratne, K. S. U. (ed.), *Proceedings of the 25th Annual Meeting, InterContinental San Juan Resort, Puerto Rico, May 24 to 27, 2009*. pp. 273–281, Association for International Agricultural and Extension Education (AIAEE).

Kabugua, E. N. (2014). *Factors influencing women economic empowerment in Kirima Sub-Location, Ndungiri Location, Nakuru North District, Kenya*. Master's thesis, University of Nairobi, Kenya.

Kariuki, G. & Place, F. (2005). Initiatives for rural development through collective action: the case of household participation in group activities in the highlands of central Kenya. CAPRi Working Paper # 43, CGIAR Systemwide Program on Collective Action and Property Rights, Washington, DC, USA.

Koech, B. K. (2014). *Contribution of women groups in the economic empowerment of rural women: a case of women groups in Bureti Constituency, Kericho County, Kenya*. Master's thesis, University of Nairobi, Kenya.

Krätli, S., Huelsebusch, C., Brooks, S. & Kaufmann, B. (2013). Pastoralism: a critical asset for food security under global climate change. *Animal Frontiers*, 3(1), 42–50.

Lele, U. (1981). Cooperatives and the poor: a comparative perspective. *World Development*, 9, 55–72.

Lele, U. J. & Christiansen, R. E. (1989). *Markets, marketing boards, and cooperatives in Africa: issues in adjustment policy*. World Bank, Washington D.C., USA.

Leyland, T., Lotira, R., Abebe, D., Bekele, G. & Catley, A. (2014). *Community-based animal health workers in the Horn of Africa: an evaluation for the Office of Foreign Disaster Assistance*. Feinstein International Centre. Tufts University, Africa Regional Office, Addis Ababa and Vetwork, UK, Great Holland.

Little, P. D., Smith, K., Cellarius, B. A., Coppock, D. L. & Barrett, C. B. (2001). Avoiding disaster: diversification and risk management among East African herders. *Development and Change*, 32, 401–433.

Markelova, H. & Mwangi, E. (2010). Collective action for smallholder market access: evidence and implications for Africa. *Review of Policy Research*, 27(5), 621–640.

Mayring, P. (2014). *Qualitative content analysis: theoretical foundation, basic procedures and software solution*. Klagenfurt, Austria. Available at: https://nbn-resolving.org/urn:nbn:de:0168-ssoar-395173

McPeak, J. (2005). Individual and collective rationality in pastoral production: evidence from Northern Kenya. *Human Ecology*, 33 (2), 171–197.

Meier zu Selhausen, F. (2015). What determines women's participation in collective action? Evidence from a Western Ugandan coffee cooperative. *Feminist Economics*, 22 (1), 130–157.

Meinzen-Dick, R., Raju, K. V. & Gulati, A. (2002). What affects organization and collective action for managing resources? Evidence from canal irrigation systems in India. *World Development*, 30 (4), 649–666.

Ministry of Livestock Development (1991). Range Management Handbook of Kenya Volume II, 1: Marsabit District. H. Schwartz J., S. Shaabani & D. Walther (eds.), Republic of Kenya.

Molesworth, K., Sécula, F., Eager, R., Murodova, Z., Yarbaeva, S. & Matthys, B. (2017). Impact of group formation on women's empowerment and economic resilience in rural Tajikistan. *The Journal of Rural and Community Development*, 12 (1), 1–22.

Mutinda, M. N. (2017). Contribution of collective action groups on socioeconomic well-being of agro-pastoralists in Makindu Sub-county, Kenya. *Journal of Poverty*, 22 (2), 108–125. doi:10.1080/10875549.2017.1348433.

Nganga, I. N. & Robinson, L. W. (2016). *Factors influencing natural resource management in pastoral systems: case of Tana River County, Kenya*. International Livestock Research Institute, Nairobi, Kenya.

Ngau, P. M. (1987). Tensions in empowerment: the experience of the 'Harambee' (self-help) movement in Kenya. *Economic Development and Cultural Change*, 35 (3), 523–538.

Ngutu, M., Lelo, F., Kosgey, I. & Kaufmann, B. (2011). Potential of pastoralist groups to manage small-stock marketing projects: A case study of groups in Farakoren and Malabot, Marsabit County, Kenya. *Livestock Research for Rural Development*, 23, # 17.

Orodho, J. A., Waweru, P. N., Getange, K. N. & Miriti, J. M. (2013). Progress towards attainment of Education for All (EFA) among nomadic pastoralists: do home-based variables make a difference in Kenya? *Research on Humanities and Social Sciences*, 3 (21), 54–67.

Ouma, E. & Abdulai, A. (2009). Contributions of social capital theory in predicting collective action behavior among livestock keeping communities in Kenya. *In:* International Association of Agricultural Economists

Conference, August 16–22, 2009, Beijing, China.

Place, F., Kariuki, G., Wangila, J., Kristjanson, P., Makauki, A. & Ndubi, J. (2004). Assessing the factors underlying differences in achievements of farmer groups: methodological issues and empirical findings from the highlands of Central Kenya. *Agricultural Systems*, 82 (3), 257–272.

Porter, G. & Lyon, F. (2006). Groups as a means or an end? Social capital and the promotion of cooperation in Ghana. *Environment and Planning D: Society and Space*, 24 (2), 249–262.

Sato, S. (1997). How the East African pastoral nomads, especially the Rendille, respond to the encroaching market economy. *African Study Monographs*, 18 (3,4), 121–135.

Schlee, G. (1991). Traditional pastoralists: land use strategies. *In:* Schwartz, H. J., Shaabani, S. & Walthe, D. (eds.), *Range Management Handbook of Kenya, Marsabit District*. Vol. II, pp. 130–164, Ministry of Livestock Development (MoLD), Nairobi, Kenya.

Schmidt, S. M. (2006). Pastoral community organization, livelihoods and biodiversity conservation in Mongolia's Southern Gobi Region. *In:* Bedunah, D. J., McArthur, E. D. & Fernandez-Gimenez, M. (eds.), *Rangelands of Central Asia: Proceedings of the Conference on Transformations, Issues, and Future Challenges. 2004 January 27; Salt Lake City, UT. Proceeding RMRS-P-39.* pp. 18–29, U.S. Department of Agriculture, Forest Service, Rocky Mountain Research Station, Fort Collins, CO.

Shiferaw, B., Obare, G. & Muricho, G. (2006). Rural Institutions and Producer Organizations in Imperfect Markets: Experiences from Producer Marketing Groups in Semi-Arid Eastern Kenya. Collective Action and Property Rights (CAPRi) Working Paper 60, ICRISAT, Cali, Colombia.

Thomas, B. P. (1988). State formation, development, and the politics of self-help in Kenya. *Studies in Comparative International Development*, 23 (3), 3–27.

Undargaa, S. (2017). Re-imagining collective action institutions: pastoralism in Mongolia. *Human Ecology*, 45 (2), 221–234. doi:10.1007/s10745-017-9898-1.

White, T. A. & Runge, C. F. (1995). The emergence and evolution of collective action: lessons from watershed management in Haiti. *World Development*, 23 (10), 1683–1698.

Winans, E. & Haugerud, A. (1977). Rural self-help in Kenya: the Harambee movement. *Human Organization*, 36 (4), 334–351.

PERMISSIONS

LIST OF CONTRIBUTORS

Taofik Adam Ibrahim, Salisu Bakura Abdu, Muhammad Rabiu Hassan, Suleiman Makama Yashim and Hanwa Yusuf Adamua
Department of Animal Science, Ahmadu Bello University, Zaria, Nigeria

Owolabi S. Lamidi
National Animal Production Research Institute, Shika, Zaria, Nigeria

Jelili Olaide Saka, Opeyemi Adeola Agbeleye, Olukemi Titilola Ayoola, Bosede Olukemi Lawal, Johnson Adedayo Adetumbi and Qudrah Olaitan Oloyede-Kamiyo
Institute of Agricultural Research and Training (IAR&T), Obafemi Awolowo University, Moor Plantation, P.M.B. 5029, Ibadan, Nigeria

Batiseba Tembo and Julia Sibiya
African Centre for Crop Improvement, University of KwaZulu-Natal. College of Agriculture, Engineering and Science, School of Agricultural, Earth and Environmental Sciences, Private Bag X01, Scottsville 3209, Pietermaritzburg, South Africa

Batiseba Tembo
Zambia Agricultural Research Institute (ZARI), Mt. Makulu Research Station, Chilanga, Zambia

Pangirayi Tongoonac
West African Centre for Crop Improvement, University of Ghana, PMB 30 Legon, Ghana

Dirk Hauke Landmann and Jean-Joseph Cadilhona
Policy, Trade and Value Chains Program, International Livestock Research Institute, Nairobi, Kenya

Dirk Hauke Landmann
Department of Agricultural Economics and Rural Development, Georg-August-Universität Göttingen, Göttingen, Germany

Thomas Raphulu and Christine Jansen van Rensburg
Department of Animal and Wildlife Sciences, University of Pretoria, Pretoria 0002, South Africa

Thomas Raphulu
Limpopo Department of Agriculture and Rural Development, Mara Research Station, Makhado, 0920, South Africa

Sudradjat, Oky Dwi Purwanto, Ega Faustina and Supijatno
Department of Agronomy and Horticulture, Faculty of Agriculture, Bogor Agricultural University, Bogor, Indonesia

Feni Shintarika
State Polytechnic of Lampung, Lampung, Indonesia

Cristian Vasco, Carolina Sánchez and Karina Limaico
Facultad de Ciencias Agrícolas, Universidad Central del Ecuador, Quito, Ecuador

Víctor Hugo Abril
Universidad de las Fuerzas Armadas-ESPE, Sangolquí, Ecuador

Médétissi Adoma, David D. Wilson and Ken O. Fening
African Regional Postgraduate Programme in Insect Science (ARPPIS), College of Basic and Applied Sciences, University of Ghana, Accra, Ghana

David D. Wilson
The Department of Animal biology and Conservation Science, College of Basic and Applied Sciences, University of Ghana, Accra, Ghana

Ken O. Fening
Soil and Irrigation Research Centre, School of Agriculture, College of Basic and Applied Sciences, University of Ghana, Accra, Ghana

Anani Y. Bruce
International Wheat and Maize Improvement Centre (CIMMYT), Nairobi, Kenya

Kwadwo Adofoe
Council for Scientific and Industrial Research-Crops Research Institute (CSIR-CRI), Kumasi, Ghana

Natalie Ann Carter and Catherine Elizabeth Dewey
Population Medicine, University of Guelph, Guelph, Ontario, Canada

Delia Grace
Food Safety and Zoonoses, International Livestock Research Institute, Nairobi, Kenya

Ben Lukuyu
Integrated Sciences, International Livestock Research Institute, Kampala, Uganda

Eliza Smith
KYEEMA Foundation, Brisbane, Australia

Cornelis de Lange
Animal Biosciences, University of Guelph, Guelph, Ontario, Canada

Nothando Dunjana, Rebecca Zengeni and Pardon Muchaonyerwa
School of Agricultural, Earth and Environmental Sciences, University of KwaZulu Natal, Private Bag X01, Scottsville 3209, South Africa

Nothando Dunjana
Marondera University of Agricultural Sciences and Technology, Marondera, Zimbabwe

Menas Wuta
Department of Soil Science and Agricultural Engineering, University of Zimbabwe, Mt Pleasant, Harare, Zimbabwe

Renaud Hecklé, Pete Smith and Ewan Campbell
Institute of Biological & Environmental Sciences, University of Aberdeen, Aberdeen, UK

Jennie I. Macdiarmid
The Rowett Institute, University of Aberdeen, Aberdeen, UK

Pamela Abbott
School of Education, University of Aberdeen, Aberdeen, UK

Gracinda A. Mataveia
Department of Clinics, Faculty of Veterinary, University of Eduardo Mondlane, Maputo, Mozambique

Carmen M. L. P. Garrine and Alberto Pondja
Department of Animal Production, Faculty of Veterinary at University of Eduardo Mondlane, Maputo, Mozambique

Gracinda A. Mataveia, Abubeker Hassen and Carina Visser
Department of Animal and Wildlife Sciences, Faculty of Natural and Agricultural Sciences, University of Pretoria, South Africa

Yali Wang and Jiafu Hua
Collaborative Innovation Center of Green Pesticide, Zhejiang A & F University, Lin'an, Zhejiang Province, 311300, P.R. China

Hongshi Yu, Liqun Bai and Jiafu Hu
Zhejiang Provincial Key Laboratory of Chemical Utilization of Forestry Biomass, Zhejiang A & F University, Lin'an, Zhejiang Province, 311300, P. R. China

Wei Gao
State Forest Protection Station, Shengyang, 110034, P. R. China

Elgailani Adam Abdalla, Abdelrahman K. Osman and Aldaw M. Idris
ElObeid Research Station, ElObeid, Sudan

Jens B. Aune
Department of International Environment and Development Studies, Norwegian University of Life Sciences, Aas, Norway

Thomas Raphulu and Christine Jansen van Rensburga
Department of Animal and Wildlife Sciences, University of Pretoria, Pretoria 0002, South Africa

Thomas Raphulu
Limpopo Department of Agriculture and Rural Development, Mara Research Station, Makhado, 0920, South Africa

Josiane C. N. Waltrick
Department of Education of Paraná State, Av. Água Verde, 2140, Vila Izabel, CEP 80240-900, Curitiba, Paraná, Brazil

Gabriel D. Goularte, Nerilde Favaretto, Luiz C. P. Souza, Jeferson Dieckow, Volnei Pauletti and Fabiane M. Vezzani
Department of Soil Science and Agricultural Engineering, Federal University of Paraná, Rua dos Funcionários 1540, CEP 80035-050, Curitiba, Paraná, Brazil

Luciano Almeida
Department of Agricultural Economics and Extension, Federal University of Paraná, Rua dos Funcionários 1540, CEP 80035-050, Curitiba, Paraná, Brazil

Jean P. G. Minella
Department of Soil Science, Federal University of Santa Maria, Av. Camobi, 1000, CEP 97105-900, Santa Maria, Rio Grande do Sul, Brazil

Charles I. Nwankwo, Jan Mühlena, Konni Biegert, Diana Butzer and Ludger Herrmanna
Institute of Soil Science and Land Evaluation, University of Hohenheim, Emil-Wolff Str. 27, 70599 Stuttgart, Germany

Günter Neumannb
Nutritional Crop Physiology, Institute of Crop Science, University of Hohenheim, Fruwirth Str. 20, 70599 Stuttgart Germany

Ousmane Sy
Institut Sénégalais de Recherches Agricoles (ISRA), Bambey, Senegal

Raphael Lotira Arasio and Oliver Vivian Wasonga
Department of Land Resource Management and Agricultural Technology, College of Agriculture and Veterinary Sciences, University of Nairobi, Loresho Ridge, Nairobi, Kenya

Brigitte Kaufmann
German Institute for Tropical and Subtropical Agriculture (DITSL), Steinstrasse 19, 37213 Witzenhausen, and Institute of Agricultural Sciences in the Tropics (Hans-Ruthenberg-Institute 490g), University of Hohenheim, 70599 Stuttgart, Germany

David Jakinda Otieno
Department of Agricultural Economics, College of Agriculture and Veterinary Sciences, University of Nairobi, Loresho Ridge, Nairobi, Kenya

Index